In the Watches of the Night

HISTORICAL STUDIES OF URBAN AMERICA
Edited by Timothy J. Gilfoyle, James R. Grossman, and Becky M. Nicolaides

Also in the series:

The Transatlantic Collapse of Urban Renewal: Postwar Urbanism from New York to Berlin
by Christopher Klemek

I've Got to Make My Livin': Black Women's Sex Work in Turn-of-the-Century Chicago
by Cynthia M. Blair

Puerto Rican Citizen: History and Political Identity in Twentieth-Century New York City
by Lorrin Thomas

Staying Italian: Urban Change and Ethnic Life in Postwar Toronto and Philadelphia
by Jordan Stanger-Ross

New York Undercover: Private Surveillance in the Progressive Era
by Jennifer Fronc

African American Urban History since World War II
edited by Kenneth L. Kusmer and Joe W. Trotter

Blueprint for Disaster: The Unraveling of Public Housing in Chicago
by D. Bradford Hunt

Alien Neighbors, Foreign Friends: Asian Americans, Housing, and the Transformation of Urban California
by Charlotte Brooks

The Problem of Jobs: Liberalism, Race, and Deindustrialization in Philadelphia
by Guian A. McKee

Chicago Made: Factory Networks in the Industrial Metropolis
by Robert Lewis

The Flash Press: Sporting Male Weeklies in 1840s New York
by Patricia Cline Cohen, Timothy J. Gilfoyle, and Helen Lefkowitz Horowitz
in association with the American Antiquarian Society

Slumming: Sexual and Racial Encounters in American Nightlife, 1885–1940
by Chad Heap

For a complete list of series titles, please see the end of the book.

In the Watches of the Night

LIFE IN THE NOCTURNAL CITY, 1820-1930

Peter C. Baldwin

The University of Chicago Press   CHICAGO & LONDON

PETER C. BALDWIN is associate professor of history at the University of Connecticut. He is the author of *Domesticating the Street: The Reform of Public Space in Hartford, 1850–1930.*

The University of Chicago Press, Chicago 60637
The University of Chicago Press, Ltd., London
© 2012 by The University of Chicago
All rights reserved. Published 2012.
Printed in the United States of America

21 20 19 18 17 16 15 14 13 12   1 2 3 4 5

ISBN-13: 978-0-226-03602-1 (cloth)
ISBN-10: 0-226-03602-2 (cloth)

Library of Congress Cataloging-in-Publication Data

Baldwin, Peter C., 1962–
In the watches of the night: life in the nocturnal city, 1820–1930 / Peter C. Baldwin.
p. cm. — (Historical studies of urban America)
Includes bibliographical references and index.
ISBN-13: 978-0-226-03602-1 (cloth: alk. paper)
ISBN-10: 0-226-03602-2 (cloth: alk. paper)
1. Nightlife—United States—History. 2. Night work—United States—History. 3. Municipal lighting—United States—History. 4. Cities and towns—United States—History—19th century. I. Title. II. Series: Historical studies of urban America.
HT123.B245 2012
306.760973'09034—dc23
2011027441

♾ This paper meets the requirements of ANSI/NISO Z39.48-1992 (Permanence of Paper).

Portions of chapter 2 originally appeared in Peter C. Baldwin, "In the Heart of Darkness: Blackouts and the Social Geography of Lighting in the Gaslight Era," *Journal of Urban History* 30, no. 5 (July 2004): 749–68.

Portions of chapter 10 originally appeared in Peter C. Baldwin, "Nocturnal Habits and Dark Wisdom: The American Response to Children in the Streets at Night, 1870–1920," *Journal of Social History* 35, no. 3 (Spring 2002): 593–611.

CONTENTS

1 · Making Night Hideous 1
2 · Lighting the Heart of Darkness 14
3 · Quitting Time 34
4 · Recreations and Dissipations 54
5 · After Midnight 75
6 · Nightmen 104
7 · Incessance 119
8 · Mashers, Owl Cars, and Night Hawks 138
9 · Night Life in the Electric City 155
10 · Regulated Night 179

Acknowledgments 205
List of Abbreviations 209
Notes 211
Index 275

CHAPTER 1

Making Night Hideous

To step into an unlit city street in early America was to enter a world shockingly different from our own. "Scarcely a sound is heard; hardly a voice or a wheel breaks the stillness," wrote the Englishwoman Fanny Trollope of her visit to Philadelphia in the late 1820s. "The streets are entirely dark, except where a stray lamp marks a hotel or the like; no shops are open, but those of the apothecary, and here and there a cook's shop; scarcely a step is heard, and for the note of music, or the sound of mirth, I listened in vain. . . . This darkness, this stillness, is so great, that I almost felt it awful."[1]

Trollope's walk came in the final years of preindustrial night in America, just before miraculous new gas lamps promised to turn night into day. Looking back on the early years of the city, we are struck by how different the experience must have been—imagine darkness so thick that you could hardly see a hand in front of your face! We might assume that everyone stayed inside after dusk. But though it may be hard to recognize it at first glance, the nighttime city was even then a place of human activity—playful, laborious, furtive, even criminal activity, but always something. These pursuits did not scuttle off like cockroaches when the lights came on in the nineteenth and early twentieth centuries; they persisted in altered forms, complicating what might seem a simple progress from the premodern to the modern. To survey the rich experience of night in the city is the purpose of this book. In the chapters that follow we will scrutinize nighttime work, crime, transportation, and leisure in the century when the city became fully visible—the period from the introduction of gas streetlamps about 1820 to full electrification in the 1920s.

American cities about 1820 were still small and primitive by European standards, as foreign visitors liked to point out.² But they expanded at an astonishing pace through the nineteenth and early twentieth centuries, thanks in large part to new technologies. Steam-powered machinery and new manufacturing techniques enabled the explosive growth of urban industry in the cities of the Northeast and Midwest, which this book will focus on. Railroads and their accompanying telegraph lines made cities the hubs for rapid transportation and communication. Inexpensive steamship travel encouraged mass immigration. Machine-cut lumber and nails allowed carpenters to hammer houses together with unbelievable speed. Networks of water, sewer, and drainage lines suppressed epidemic diseases that had once kept populations in check. Street railways whisked people throughout the urban landscape and permitted the city to sprawl miles beyond the industrial and shopping districts. By the end of the nineteenth century, American cities had grown fully as large as those in Europe and had surpassed them in their embrace of modern technology. The traditional skyline of church steeples was becoming obscured by a new sierra of skyscrapers, made possible by steel-framed construction, elevators, telephones, and systems of heating and ventilation. Americans had grown used to viewing cities as sites for limitless growth—upward and outward.³

Advances in lighting technology allowed urban growth to shrug off another restraint. By facilitating nighttime work and transportation, improved lighting let the city expand in time as well as in space. Manufacturers added late shifts at factories that had once closed at dusk. Traffic flowed more smoothly during the night on streets, streetcars, and rail lines that were hopelessly congested during daylight hours.

Anyone could see that railroads, skyscrapers, and electric lights were expanding and transforming urban space. It was trickier to discern how technological change would affect daily life. One of the bolder attempts at prediction appeared in *Caesar's Column: A Story of the Twentieth Century*, an 1890 novel by Ignatius Donnelly, which tried to imagine what New York would be like in 1988. Donnelly portrayed urban life as an odd mishmash of the familiar and the new. His gigantic New York of the future was still reliant on horse-drawn carriages and racked by industrial-era class conflict, yet it enjoyed air conditioning, touch-sensitive computer screens, and transatlantic air travel. Donnelly predicted that the city would be brilliantly illuminated around the clock. "Night and day are all one," he wrote, "for the magnetic light increases automatically as the day-light wanes; and the business parts of the city swarm as much at midnight as at high noon." Artificial lighting held an important place in this popular novel; a cataclysmic uprising began with its interruption. Nonetheless, Donnelly had trouble imagining human behavior in a future city where the sun no longer mattered. He contradicted his description of incessant activity in a later passage, in which he mentioned that the people in 1988 New York journeyed to work in the early morning and returned home to sleep at night, leaving the streets "still and deserted."[4]

Donnelly's confusion was understandable. If artificial lighting allowed people to be active without regard to sunrise and sunset, a change in the pattern of working by day and resting by night could reasonably be expected. Yet, as Donnelly perhaps sensed, it is impossible to predict the effects of technological change simply by noting the physical properties of the technology—say, the superior candlepower of a new streetlamp. Despite popular beliefs that inventions such as the printing press, the automobile, and the personal computer "cause" events to happen, historians of technology caution that such change is rarely so simple. Technologies have to be applied within human societies far more complex than even the systems of retorts, tanks, pipes, valves, jets, shades, and reflectors that combined to produce gaslight. Uses intended by inventors are superseded by those determined by users. Some people have greater access to the technology than others, and some prove more powerful in conflicts over how the technology should be used. Thus inequalities of knowledge, wealth, and power shape the application of any new invention. Technology cannot altogether free us from human society and its historic habits. People don't stay up all night just because there's enough light.[5]

Whether dazzled by the prospect of infinite progress or envisioning the end of civilization, nineteenth-century Americans had difficulty seeing life in the future. Yet it is no easier for twenty-first-century Americans, our vision still clouded by assumptions about technological power, to peer into the past. Before we begin to examine the changing experience of night in the nineteenth century, we'd better let our eyes adjust to preindustrial America. Let's follow one ordinary man on one forgotten night in one unimportant town, through streets as dark as in any city since antiquity. Let's follow a young ship's doctor on a drinking spree in a revolutionary-era New England seaport.

PROVIDENCE, RHODE ISLAND, JANUARY 25–26, 1780

Locked in by the frozen Narragansett Bay, the little Rhode Island port of Providence waited quietly for the weather to break. Frigid air stung the cheeks and gnawed the noses of those who ventured outside during the short hours of daylight; after dark, sensible men and women huddled by their fireplaces. The officers of the *Argo*, though, were determined to go out and celebrate.

The *Argo* had just concluded a glorious career in the Continental navy.

The previous May, in 1779, had begun with Rhode Island's ports still bottled up by the British and with the enemy preying on American shipping along New England's southern coast. But then the men of the *Argo* sailed forth in May to chase off the commerce raiders and attack British merchantmen. They returned to Rhode Island that autumn after capturing a dozen vessels, made all the more joyful by news that the British had evacuated Newport, the state capital. Now, on this frigid January day in 1780, they learned that the Continental Congress had agreed to return the *Argo* to the merchant who owned it, freeing the sloop and its men for the lucrative business of patriotic piracy—politely called privateering. As Dr. Zuriel Waterman wrote in his journal, "Officers concluded to have a Bandge to Night."[6]

Waterman had his own reasons for joining the binge. First, the thirty-four-year-old doctor from Pawtuxet was never one to pass up a lively night of drinking. He chronicled his sprees in raucous detail in the pages of his diary, and later he inscribed this motto in his memorandum book: "Since all is Vanity let us partake of the Dissipation and make it as pleasing as we can." Waterman must also have been delighted to be part of such an illustrious and high-spirited group of officers. While the *Argo* was covering itself in glory the previous fall, he had endured a miserable voyage to the Newfoundland banks as surgeon of the privateer *Providence*. The *Providence* blundered about the stormy North Atlantic for two months without taking a single prize, while Waterman lay seasick in his hammock. He and his shipmates rejoiced when the sloop finally returned home for repairs. Back in port, he signed on as the surgeon's mate for the *Argo*'s upcoming cruise to Antigua, and he was restocking that ship's medicines when he heard the news that it would be returned to private ownership. Waterman and the rest of the *Argo*'s officers now stood to make a lot more money privateering, whether on the *Argo* or on some other vessel (except perhaps the unlucky *Providence*). Why not go out and raise hell?[7]

The officers spent the afternoon drinking grog and cider in Bradford's Tavern, then returned to the *Argo* after dark. They were just getting started. Aboard the *Argo*, wrote Waterman, they "began our frolic with Several stout Bowls of grog & Toddy—raw Drams Slings &c. singing roaring &c. Till wee got too big for the Cabbin to hold us and then sallied out in the street but did not forget to carry a Bottle of Rum with us it being exceeding cold & about 8 o Clock—all in good Spirits and good Spirits in us." The winter was unusually cold, one of the worst of the century. Thick ice covered the harbor and spread down Narragansett Bay to Newport, locking in ships more securely than the British navy ever did. Lately, while traveling between Pawtuxet and Providence, Waterman had been walking on the ice for miles; he would soon freeze his left ear while taking this shortcut on a windy day. The *Argo*'s officers fortunately had a supply of their own personal antifreeze, some of it probably taken from a prize carrying 330 hogsheads of West Indian rum.[8]

"Thus accouttred," Waterman wrote, "we went along [the] street shouting & singing & now & then to cheer our hearts stop & take a drink—the word was Argo!" The street was probably Towne Street, now called Main, a waterfront row of wood-shingled houses and businesses serving what in times of peace would have been a vibrant seafaring economy. The shops of coopers, blacksmiths, and distillers were there, along with warehouses for New England's dried fish, beef, salt pork, and lumber or for West Indian molasses, sugar, and cotton. There too were some of the town's roughly three dozen taverns.[9]

The shouts of "*Argo!*" drew several shipmates out of houses along the way. The growing party "got a Negro fiddler & proceeding up town went in to a house to have a dance." In keeping with time-honored traditions of maritime debauchery, they were an all-male group. No decent lady would be seen carousing in the streets at night, least of all with a mob of drunken sailors. But there were other kinds of women in Providence, and the sailors knew where to find them. A woman was sick in the first house they visited, so they staggered on to another "to make our frolic there but they had got the start of us and had a frolic of their own." The men felt too drunk to join in, Waterman admitted modestly: "One of the women was tumbling to pieces—this business being above our capacity in our present condition we thought fit to pack off." They paused at a house where William Russell, a prominent local merchant, was known to keep his mistress. Prostitutes and concubines were sometimes the targets of harassment by men on drinking sprees, particularly if the women were outside the drinkers' price range. The men of the *Argo*, however, did not break windows or make discordant music. Russell was, after all, a staunch patriot. The party "made no tarry there except for one kiss."[10]

The dozen men moved on to Nathaniel Jenckes's tavern, where those still able danced to fiddle music, apparently without any women. Two of the most profoundly inebriated were carried to bed; a third nodded by the fire in a drunken stupor for a few hours until he made his way home. The others grew hungry for a supper, "but the Landlord affronting us we calld for the Bill." They were not quite drunk enough to be fooled by Jenckes's claim of $120—a huge amount even in the inflated paper currency of wartime Rhode Island. While the party argued with the tavern keeper about the bill, Waterman and a shipmate ducked out and stumbled back to the *Argo* over snowy cobblestones. It was 4:00 a.m. as they came aboard. Unable to remove his boots, Waterman dozed fitfully for an hour until three more of his companions returned. They had been amusing themselves by lighting fires in the street. The three men refueled, grabbed a coffeepot of strong sling for the road, and set out again for more uproarious fun. Before sunrise they succeeded in taking a gun from a sentry, a variation on the hilarious custom of playing pranks on watchmen. It had been a classic spree.[11]

NIGHT AS A TIME OF FEAR

Much of what took place on that cold Providence night could just as easily have happened a century or two later. The raucous pack of drunks would be a familiar nocturnal feature along New York's Bowery in the 1840s, near Pittsburgh's steel mills any payday in the early twentieth century, or on college campuses after a big basketball game today. The night of debauchery became somewhat tamer over time, as police became more effective at suppressing violent disturbances and arresting men for being "drunk and disorderly." Nonetheless, it persisted as a ritual of rebellion against conventional self-restraint. Men demonstrated their devilishness in carefully scripted ways, in established settings and with a standard cast of characters. They whooped and hooted but usually did not cause serious damage.

Though the young men's drinking spree remained just one of many activities that took place in the streets at night, it exerted a disproportionate influence on what can be called the nocturnal culture of the city—the codes of behavior that prevailed there, the underlying values that shaped that behavior, and the demographic profile of the people who were thought to belong. Even as people ventured out after dark in growing numbers and increasing diversity through the nineteenth and early twentieth centuries, the streets were still dominated by young men. Women, when present, remained far more likely than in the daylight to be treated as sexual targets. The rule of law remained shaky, gleefully mocked by otherwise ordinary citizens. Urban night should not be romanticized as offering a haven from oppression. While some relatively powerless young men did seize the opportunity to cut loose from the constraints of their daytime lives, many of those who had the least skills and the lowest standing in society worked by necessity at menial jobs during hours when the more fortunate enjoyed themselves. Young working-class men were disproportionately represented in the new night jobs enabled by improved lighting and other technologies. Young white men who did have free time at night often spent it victimizing people more vulnerable than themselves: women, children, and racial minorities.

As the officers of the *Argo* roared through the streets that frigid night in

1780, most of the people of Providence slept under thick coverlids. Those who heard the noise would have been unsettled. The drunken officers—all either local men or under their authority—were less menacing than the British sailors from the recent occupation of Newport, but still unpredictable. The fires they lit in the streets would have alarmed any townspeople who saw the glow through frosted windows. Fire was a terrible threat to wooden cities, especially on winter nights when water sources froze hard, and Providence's stockpiles of gunpowder added to the danger.[12]

The men were out late at night, and that alone was enough to merit suspicion. Night was known by long tradition in Western society to be a time of crime, immorality, and sickness. The night air itself was thought to carry disease. Supernatural forces were believed to gain strength under cover of darkness, while decent, God-fearing folk took refuge inside the home. Both literally and metaphorically, the contrast between light and darkness was thought to represent the division of good from evil, life from death.[13] In *Contemplations on the Night*, an essay widely published from the 1740s through the 1770s, the English clergyman James Hervey used the fall of darkness as an occasion for meditating on the eternal truths of death and the afterlife. "What a general Cessation of Affairs, has this dusky hour introduced!" Hervey exclaimed. "In every Place Toil reclines her Head, and Application folds her Arms. All Interests seem to be forgot; all Pursuits are suspended; all Employment is sunk away. . . . 'Tis like the Sabbath of universal Nature; or as though the Pulse of Life stood still." Not all of God's creatures followed the dictates of beneficent nature, Hervey warned. Fierce beasts emerged from their dens to prey on travelers. Worse,

> There are Savages in human Shape who, muffled in Shades, infest the Abodes of civilized Life. The Sons of Violence make Choice of this Season, to perpetrate the most outrageous Acts of Wrong and Robbery. The Adulterer waiteth for the Twilight; and, baser than the Villain on the Highway, betrays the Honour of his Bosom-friend. . . . Now Crimes, which hide their odious Heads in the Day, haunt the seats of Society, and stalk thro' the Gloom with audacious Front. Now, the Vermin of the Stews crawl from their lurking Holes, to wallow in Sin, and spread Contagion thro' the Night.[14]

An unfriendly observer would have said many of these things about the *Argo* drinking party. Ship officers blew off steam in this instance, but the dark streets of early America also echoed with the more alarming revelry of common seamen and other workers. Having a lesser stake in society, such men were

FIGURE 1. Eating oysters at night in Philadelphia. This early nineteenth-century watercolor shows a woman and her African American assistant serving oysters by candlelight outside a theater. The only other illumination appears to be moonlight. The young man at right is putting his finger beside his nose as a signal that he is joking with the young woman. The presence of the woman in the street at night, and her eye contact with the young man, suggests that she may be sexually available. Oysters were considered aphrodisiacs. John Lewis Krimmel, *Nightlife in Philadelphia—an Oyster Barrow in front of the Chestnut Street Theater* (ca. 1811–13). Image copyright © Metropolitan Museum of Art/Art Resource, NY.

believed to be more inclined to criminality. In his memoirs, William Otter later reminisced about his exploits as a young apprentice in New York during the winter of 1806–7, when he used the pretense of attending night school in order to roam the city with his drunken friends. Otter and his buddies were returning from a dance one Christmas Eve when they fell in with a mob of sailors who were fighting Irish immigrants in a brawl that roiled through the streets all night. That brawl provided a good excuse for heavy drinking, as Otter and his friends enjoyed ransacking an Irish-owned tavern. Other nights, heavy drinking provided a good excuse for a barroom brawl with sailors, and for the ferocious trashing of another tavern. Otter bragged of his success in beating up his adversaries and tricking the watchmen who tried to restrain him. The violent, hard-drinking milieu of these "jolly fellows" attracted dwindling numbers of men in the later nineteenth century, as middle-class values of temperance and self-restraint took firmer hold, but elements of the tradition have persisted to the present.[15]

Nighttime debauchery was promoted and to some extent contained by an economic subculture of tavern keepers, musicians, cooks, and prostitutes. Taverns and small tippling shops could be found open at almost any hour in early nineteenth-century New York. In Boston in the 1820s, at least a dozen brothels operated openly in one western neighborhood, holding public dances almost every night and filling the streets with light blazing from their windows. Boston mayor Josiah Quincy was determined to suppress violent crime, disorder, and above all the "audacious obtrusiveness of vice," but he cautioned that prostitution itself was a necessary evil. "In great cities the existence of vice is inevitable," he declared in his 1824 inaugural address. "Its course should be secret, like other filth, in drains and in darkness."[16]

More respectable forms of work also took place in cities after dark and before dawn. Bakers labored in the dead of night to knead the dough and bake the loaves for morning customers. "Night soil" collectors and scavengers hauled away the foul materials that the sleeping townspeople did not wish to see or smell. Doctors and midwives attended patients. Servants fetched drinking water and carried messages.[17] Some political and social control endeavors also took place at night. Boston's Sons of Liberty prowled the streets on moonless nights, with blackened faces, to torment British imperial officials in the 1760s and 1770s. Thinly disguised as Indians, they and other Bostonians took advantage of one dark night in 1773 to disrupt British tea imports by throwing the cargo into the harbor. Similarly, on a nighttime raid in 1772, Providence men burned the British schooner *Gaspee* on a shoal south of Pawtuxet. Residents of Boston and other cities enforced community moral standards by organizing mob attacks against brothels at night.[18]

Streetlamps were few and weak before the advent of gas. The whale-oil lamps that New York installed by 1761 were so poorly maintained that they were little brighter than candles; furthermore, New York's lamps in the late 1700s were supposed to be placed 114 feet apart, where they appeared as faint yellow specks engulfed by blackness. City officials in New York, Boston, and Philadelphia attempted to expand street lighting in the late 1700s, but the amazement of rustic newcomers upon entering the "new world of light and splendor" reveals more about rural darkness than about urban lighting. The cities' primitive oil or candle lamps did little to dispel the dark and were often extinguished after midnight or on moonlit evenings. Such lamps were intended not to illuminate the street but to serve as navigation beacons, guiding travelers through the city much as today's runway lights guide airplane pilots. Shop windows illuminated a few patches of sidewalk in emerging commercial districts of the nineteenth century, but travelers who really wanted to see carried their own lights. Streets in poorer areas and at the edge of town were completely unlit, while frequent lamp smashing by vandals left many others in darkness intermittently. Unlit streets could be pitch black—far darker than any unlit space would be under the glowing sky of a modern metropolis. No wonder city people remained familiar with the lunar cycle and with the experience of rounding a corner "from the shade of the buildings into the full light of the moon."[19]

Streets in early American cities were downright perilous on cloudy, moonless nights. Until the mid-nineteenth century, only the most important streets were paved, and then usually with bumpy cobblestones instead of smooth stone blocks. Sidewalks, if they existed at all, were equally bad; a pedestrian might walk a short distance along dry flagstones and then plunge into ankle-deep mud or trip over a wooden step. Travel was obstructed along the sidewalks and street edges by an obstacle course of encroachments: cellar doors, stoops, stacks of cordwood, rubbish heaps, posts for awnings, and piles of construction material. These nuisances posed mortal dangers for the drunk or unwary. People broke their legs or crashed their carriages into unprotected holes left by construction projects. Others drowned by falling off footbridges or stumbling into large puddles. In 1830 a New York watchman running down a dark street toward the sound of a disturbance was killed when he collided blindly with a post. Snowdrifts, frozen ruts, and patches of slick ice added to the difficulty of wintertime travel, although at least the light reflected from the snow made it easier to see.[20]

Criminals posed other serious dangers at night. Thieves crept into homes, looted warehouses, or stole livestock. Gangs of muggers lurked in the shadows, waiting for likely victims. Rapists assaulted women servants on late errands. Ruffians attacked lone pedestrians for amusement, sometimes beating them savagely. "It seems to be now become dangerous for the good People of this city to be out late at Nights, without being sufficiently strong or well armed," observed a New York newspaper in 1749. Even constables and watchmen were afraid to enter certain neighborhoods.[21]

Police protection was limited. American cities began setting night watches in the 1600s, almost as soon as they formed local governments, to guard against fire and to keep order. The nightly setting of the watch was announced by ringing a bell, beating a drum, or firing a cannon. This signal was usually given at 9:00 p.m., though sometimes earlier in the short days of winter or later in the summer.[22] The watchmen's main responsibility was to keep the peace and protect property, not to catch criminals. They simply replaced the ordinary citizens whose presence maintained order during the day. Watchmen typically were assigned to walk the streets suppressing disturbances and questioning suspicious people. They carried lanterns to light their way, rattles to summon help, and such weapons as staves, clubs, or pikes.[23]

The laws were vague about whether people could legally walk abroad at night, but the custom was clear that law-abiding people were to stay indoors. Those found in the streets at night were strongly suspected of being up to no good; they were subject to close questioning and could be detained overnight if their answers did not satisfy the watchman. Women could be arrested for "night walking" if they were suspected of prostitution. Nonwhites were also vulnerable. In 1703 Boston officials ordered all blacks and Indians off the streets after 9:00 p.m. because of their alleged "disorders, insolences, and burglaries." A similar prohibition was enacted that year to control blacks and Indians in Newport. Cities both North and South passed ordinances aimed at controlling slaves' access to public space at night. New York in 1713 forbade slaves over fourteen to be in the streets without a lantern during the hours of the watch, though enforcement appears to have been ineffective. Military guards in colonial Charleston and New Orleans, which were proportionately much larger than the watches in northern cities, also tried with limited success to keep blacks from walking out after dark without a pass.[24] The attempts at curfews and patrolling kept the streets empty without making them safe. As late as the 1820s, wrote Henry Castellanos in a description of New Orleans, streets after sundown "became the property of footpads and garroters [muggers]. Incendiary fires were matters of almost nightly occurrence, as well as burglaries. People ventured out of their houses after dark only at their peril and with great apprehension, and never without a lantern."[25]

Beset by dangers and inadequately protected, the preindustrial American street was relatively quiet at night. Philadelphia, which Fanny Trollope perceived as deathly silent, was actually one of the livelier United States cities at the time, according to Anne Royall, another English visitor in the 1820s. Philadelphia compared favorably with Boston, where

> they have a custom amongst them, as old as the city, singular enough; that is, shutting up their shops at dark, winter and summer, which gives the city a gloomy appearance, and must be doubly so during the long winter nights. I should be at a loss to conjecture how their clerks and young men dispose of themselves, during their long winters. New-York and Philadelphia do as much business after dark as they do in the day, and perhaps more; for the young people then take time to amuse themselves, and the lights which illuminate the shops and stores, give life and activity to the whole city.[26]

SEEING THE ILLUMINATED CITY

At the time these two women made their observations, major East Coast cities were just beginning what would prove to be a century-long project of adopting new lighting technology. Gas lamps and then electric lights would help city people dispel the awful stillness Fanny Trollope felt, without ever quite achieving the universal liveliness imagined by Anne Royall.

At the end of this century of stunning technological change, the American city looked nothing like the places experienced by Trollope and Royall, or by Dr. Zuriel Waterman and his fellow officers. Cities in America's urban industrial core—the Northeast and Midwest—were fully electrified by the 1920s; some 85 percent of urban dwelling units had electric lighting. Better lighting was only one of the differences. A cluster of wooden buildings that looked like early Providence would not have been considered urban at all in the early twentieth century and would have been an unlikely place to find much amusement. Cities now sprawled over more square miles than anyone could stagger across with a bottle of rum in hand. No longer concentrated along a waterfront, urban life had become spatially segregated thanks to economic changes, to the development of streetcar systems, and to deliberate efforts to control the chaotic use of public space. Retailing and finance concentrated into downtown commercial blocks, factory districts stretched for miles along railroad lines, residential neighborhoods were sorted out into those for the rich, the middle class, and the workers. Nightlife had its own spatial order, at least in the larger cities; raucous barrooms and brothels would not be found near the elite restaurants, social clubs, and opera houses, and they certainly were far from the darkened bedrooms of decent citizens.[27]

These changes in urban geography made it impossible for even the heartiest partyers to be as disruptive as Waterman's group had been. Yet for all the changes that accompanied technological and economic progress, certain social traditions proved too strong to overcome, including such lowly traditions as the young men's drinking spree. It was clear by the 1920s that artificial light had failed to "turn night into day." The persistence of morally controversially aspects of nightlife, together with the danger of street crime and the skewed gender profile of night work, hindered any such change. Streets were still controlled at night to a far greater extent than during the day by those who could wield physical violence: police and groups of young men, whose free-spirited pleasures often came at the expense of others. Women remained vulnerable after dark to the danger of street crime and harassment. A woman unaccompanied by a man was often presumed to be available for commercial sex, receptive to seduction, or an easy target for rape. This presumption, shared even by police, was kept plausible by the abundance of male-oriented amusements employing women for their sexual appeal: What legitimate business could a lone woman have in the public streets after dark? Respectable employment for women after dark was limited by the concern for appearances and physical safety. Children too were feared to be vulnerable to evil influences, crime, and sexual exploitation. Reformers struggled to draw them off the streets to uplifting institutions such as boys' clubs. After 1890, state and municipal legislation amplified these patterns of unequal access, deliberately attempting to force women and children inside the home through labor legislation and curfew laws.

Familiar social conflicts, social customs, and power inequalities persisted even as technology offered possibilities for change. As it emerged out of this tangle of influences, modern urban night proved to be something quite different from a simple extension of day. It became a complicated new "space" with its own schedule, its own rules of access, and its own codes of behavior. Amid the wonderful lights that made the nocturnal city shine like day, aspects of preindustrial night survived.

CHAPTER 2

Lighting the Heart of Darkness

A lone gas lamp stood at the corner of Cross and Little Water Streets in lower Manhattan in 1850, illuminating what many Americans considered the heart of urban darkness: the notorious neighborhood called the Five Points. A large and unusually bright lamp by the standards of the time, it seemed to belong in the glittering Broadway shopping district, or amid the Fifth Avenue townhouses of the rich, instead of kitty-corner from the Old Brewery, a tenement crammed with the poorest of the poor. The lamp lit up the filthy intersection not as a convenience to neighbors, or even as a help to night travelers. The flickering patch of brightness was intended as an outpost of order in dangerous territory. Its symbolic value was so great that the city stationed a policeman there to protect it from hoodlums.[1]

Practically in the backyard of New York's City Hall, a stone's throw from Broadway, the Five Points was known as the most vicious slum in nineteenth-century America. Journalists, missionaries, and tourists were fascinated by its desperate poverty, its brazen prostitution, its contagious diseases, its rampant drunkenness, its violent gangs—and by its mix of Irish immigrants and African Americans. Writers portrayed the neighborhood as the epitome of all that was deviant and disordered in urban life. If the ideal American citizen was native-born, white, affluent, healthy, clean, sober, chaste, and law-abiding, inhabitants of the Five Points were the opposite. If decent people kept the early hours God intended, the creatures infesting the Five Points raised hell all night. Shining a bright lamp into such benighted savagery was viewed as a step toward reform.[2]

At a time when gas lighting seemed to be transforming the urban landscape, such optimism was both widely expressed and easily satirized. Several years before the writer George Foster described the lamp at the Five Points, the comic novelist Cornelius Mathews had a pompous, rich character named Halsey Fishblatt urge an alderman to improve street lighting:

> Keep the city well lighted and you keep it virtuous, sir. You should have a lamp in front of every tenement, and where the streets are so narrow that the houses might catch from the wick, you should have men moving up and down with great lanterns, and keep all the thoroughfares and alleys in a glow. You wouldn't have a murder once in a century, and as for burglaries and larcenies, they'd be forgotten crimes, like the Phoenix, sir, and the Megalosaurus![3]

Real-life observers of city life were not as fatuous as Fishblatt. They knew streetlamps alone could not tame the night, yet they hoped a combination of lighting and policing might do the trick. The creation of modern police forces in American cities, from the 1840s through the 1870s, typically began shortly after the introduction of gas lighting. As cops and gas lamps spread nearly simultaneously through the commercial districts, and more haltingly through residential neighborhoods, they were understood to be mutually reinforcing ways to conquer territory from urban barbarians. Gas lamps could be the best policemen—policemen that might allow the nocturnal city to be seen, understood, and safely navigated. Property and public safety were thought to be secure in their presence and at grave risk in their absence.

GAS LIGHTING

Urban slums were proverbially dark from the colonial era through the early twentieth century. Indoor illumination was very expensive in the era before gaslight. Poor people in preindustrial America saved money by burning as little light as possible: a sputtering tallow candle usually sufficed. Wealthier people burned beeswax and spermaceti candles or whale-oil lamps, displaying their wealth to the world by the glow from their windows. In *Moby-Dick*, Herman Melville's Ishmael knew that the brightly lit inns of New Bedford would be too expensive, so he went in search of cheap lodging near the waterfront: "Such dreary streets! blocks of blackness, not houses, on either hand, and here and there a candle, like a candle moving about in a tomb."[4]

The social geography of light and darkness became sharper with the introduction of gas service, which initially focused on the most profitable commercial streets and affluent residential neighborhoods. Public gas lamps first appeared in the United States in 1817, when the Gas Light Company of Baltimore lit a burner at an intersection in the center of the city. The company obtained a city contract to lay pipe through the streets and install more lamps, but the company's limited number of potential private customers—mainly theaters, large businesses, and a few wealthy individuals—gave it little incentive to install an extensive network of gas lines. The company ignored repeated city requests to expand the street lighting system, because such an investment could not be repaid from municipal lighting contracts alone. By 1836 the system could boast only about two miles of pipe, all of it in a compact area west of Jones Falls: from Saratoga Street to the waterfront, and west along Baltimore and West Pratt Streets to Eutaw Street. The vast majority of the city was not lit by gas.[5]

Similar inequality of street lighting developed in other major cities as they built their own gas systems. Systems were in place by 1840 in Boston, Brooklyn, Louisville, New York, and Pittsburgh and were installed in almost every significant city in the Northeast and Midwest over the next twenty years. The gas was manufactured at waterfront industrial complexes, whose stench turned the surroundings to slums, and piped to homes, businesses, and streets that served the wealthy. The southernmost section of New York's Broadway was first lit with gas streetlamps in 1827, enhancing its reputation as a marvel of nocturnal activity. As in other cities, the gas company in New York installed mains at its own expense, and the city government paid to connect them to municipal streetlamps. After lower Broadway, streetlights were next installed on Wall, Pearl, and other commercial streets. Oil-lit blocks, which continued to outnumber gaslit ones until midcentury, now appeared dark in contrast. The novelist George Lippard wrote in 1854 of side streets "dark as grave vaults" in lower Manhattan, while Broadway was "defined by two lines of light, which, in the far distance melt into one vague mass of brightness."[6]

Early gas jets were said to be between six and sixteen times as bright as candle flames and at least three times as bright as the best oil lamps, which at first struck people as very bright indeed. Once the novelty had worn off, critics complained that gas lamps were too dim. The brightness of a lamp varied with the quality of the gas, the pressure in the lines, the design and adjustment of the fixture, and the frequency of maintenance. On the filthy streets of nineteenth-century America, glass in streetlamps quickly became covered with grime nobody bothered to wipe away. Descriptions of the light ranged from brilliant white to sickly yellow to a dull red. Lamps along heavily traveled Chatham Street near New York's theaters and newspaper offices were reportedly as faint as "a row of invalid glow-worms," griped the *New York Times* in 1866; it claimed that the darkness thirty feet from a lamp was thick enough to hide an elephant. Chronic vandalism also interfered with lighting; Boston's lamplighters and police in 1875 reported 8,268 incidents of broken lamps—nearly as many as the total number of lamps in the city![7]

Distances between streetlamps were inconsistent. Some streets were lined with lamps only on one side, and others had irregularly alternating lamps. On an 1870 map of Washington, a five-hundred-foot block of New Jersey Avenue is shown to have only three lamps, while a slightly longer section of East Capitol Street was lit by five lamps on each side of the street. New York City by the 1870s generally had two lamps at each intersection along the avenues, though sometimes none in the middle of the block. Enterprising businessmen managed to get extra lights installed near their businesses, and churches often successfully petitioned the city for a pair of lamps at their front doors even if other lamps already burned close by. Chicago's city council scattered individual streetlights here and there around the city in response to the preferences of the adjoining property owners who would have to foot the bill.[8] As a result, even areas of the city with gas service had varying levels of lighting on the street. Public gas lamps, typically set near the curb on eight-foot iron posts, cast a pool of light strong enough at times to let someone read a newspaper up to five feet away. Pedestrians paused there if they needed to check their watches or count their change. Only in the immediate vicinity of a lamp could faces be recognized. The level of light dropped sharply as one walked away from a corner lamp toward the dark gap in the middle of a block, where the house numbers could not be read from the sidewalk.[9]

The streetlights in some neighborhoods were outshone by light spilling from theaters, saloons, shops, restaurants, and private homes. Public lamps were vastly outnumbered in almost every city by private gas customers, most of whom had multiple gas jets. Only 165 gas streetlights burned in Philadelphia in 1836, the first year of gas service, compared with 2,952 private lights. By 1850, Philadelphia's streets, squares, and market houses were lit by 1,576 gas lamps, while private customers had installed 115,004 gas jets. Illuminated signs, exterior lamps, and show windows shone onto the sidewalks of major commercial streets. "The monster hotels pour out of flood of radiance from their myriads of lamps," observed James McCabe in New York in 1872. "Here and there a brilliant reflector at the door of some theatre, sends its dazzling white rays streaming along the street for several blocks." Gas lanterns shone beside front doors of luxurious homes, and interior lamps blazed from the parlor windows.

Light from businesses and homes helped fill some of the dark spaces between streetlamps.[10]

Gas service spread throughout the urban landscape in the second half of the century as the price dropped to within reach of people with modest incomes. The number of private gas customers in Philadelphia more than quadrupled during the 1850s alone, reaching 41,200 by 1860; by then the number of private gas jets was approaching half a million. Gas streetlamps in Philadelphia increased from 5,604 in 1860 to 26,043 in 1890. The number of public gas lamps in Boston rose from 2,265 in 1859 to 9,525 in 1876; new gas lamps were installed both in Boston's semi-suburban fringe and in the congested streets and alleys of the North End, where they were intended to deter crime. The skies over American cities, once black and starry, began to take on a faint reddish glow in the mid-nineteenth century, marking the city as ever more distinct from the countryside.[11]

Still, darkness lingered in the slums. Gas companies chose to provide service to new neighborhoods only as it seemed profitable. City governments were often unable or unwilling to force the companies to extend gas mains to poor areas, and they saw little need to provide expensive lighting for people who paid little in taxes unless nearby business interests were affected. Municipalities' efforts to recoup their lighting costs by levying special assessments on abutting property owners undoubtedly discouraged poor people from demanding gas lamps on their streets. Oil streetlamps remained common in many neighborhoods long after gas was available. As late as 1876, Boston had yet to install gas streetlamps in many backstreets and alleys where poor people lived. Most working-class families in the 1840s and 1850s continued to light their homes with candles or, at best, lamps burning "camphene"—a mix of turpentine and alcohol. Starting in the 1860s, frugal homeowners upgraded to kerosene, which remained cheaper than gas. Even as light was "democratized" to some extent, the gap between rich and poor persisted.[12]

Maps of Washington, DC, vividly show the division of the city into zones of light and darkness. An 1873 map shows a tightly clustered system of gas mains in the commercial center and the affluent northwest quadrant of the city. Though the system had expanded from 69 miles of mains to 108 miles in the previous two years, the poorer sections to the east and south remained largely neglected. Washington's gas system expanded to serve most of the city by 1891, when electric lights were replacing gas on major streets. A total of 5,607 gas lamps illuminated the city streets and parks in that year, with lines reaching east to Lincoln Square and south to the marshy edge of the Potomac River. Meanwhile, 434 oil lamps still lit the poorer streets—most notably in the African American neighborhood southwest of the Capitol. Residents of the southern and eastern neighborhoods had long complained of unequal treatment, but they continued to suffer an inferior water system and inferior street paving as well as inferior street lighting. Though increasing numbers of Washington's poor lived along alleys throughout the city, the alleys remained almost entirely unlit in 1891.[13]

The unequal advances in gas lighting called attention to the growing division of American cities along functional and class lines. Urban space in the nineteenth century was increasingly differentiated for specific economic uses. Commercial activity concentrated more than ever in central business districts, encouraged by streetcar systems that allowed easy access to the center from throughout the metropolitan area. Downtowns were further divided into retailing, financial, and wholesaling areas. Factory districts grew up along the rail lines radiating from the urban core. Areas of boardinghouses and hotels for single people developed on the fringes of the downtown, as did clusters of saloons, theaters, and brothels. Working-class ethnic enclaves sprouted up a little farther from the downtown or near the factory districts. Thanks in part to changes in the nature of work and the development of a distinct middle-class culture, prosperous families increasingly sought to distance themselves physically from areas of production and from the working class. First in New York, then in other cities, neighborhoods grew more divided along class lines. The superior lighting of affluent areas advertised and exaggerated divisions that were still somewhat hazy.[14]

Midcentury writers seized on differences in lighting as a major theme for their descriptions of American cities, particularly in works promising to expose the "mysteries and miseries" of urban life. Announcing their intentions with such titles as *Sunshine and Shadow in New York* and *New York by Gas-Light*, they suggested that contrasts in lighting allowed perceptive observers to map a new urban geography of wealth and poverty, virtue and vice, safety and danger. The visitor to New York, wrote James McCabe in *Lights and Shadows of New York Life*, is "plunged into the midst of so much beauty, magnificence, gayety, mystery, and thousand other wonders, that he is fairly bewildered. It is hoped that the reader of these pages will be by their perusal better prepared to enjoy the attractions, and to shun the dangers of New York." Physical darkness could be read as a sign of moral darkness, McCabe and many other writers suggested, taking the Five Points as their prime example. The intersection of Park and Worth Streets stood within sight of Broadway—"almost within pistol shot, but what a wide gulf lies between the two thoroughfares, a gulf that the wretched, shabby, dirty creatures who go slouching by you never cross. There everything is bright and cheerful. Here every surrounding is dark and wretched. The streets are narrow and dirty, the dwellings are foul and gloomy, and the very air seems heavy with misery and crime." Navigating by streetlamp, the "mysteries and miseries" authors guided readers through a complex but ultimately understandable landscape.[15]

City residents adjusted their habits in accordance with the belief that greater light meant greater safety. The gas lighting of Broadway in 1827 encouraged New Yorkers to promenade along the sidewalks every evening. George Templeton Strong, who preferred Broadway on his evening strolls in the 1840s, complained that it had become too crowded. By the 1850s, Broadway was busy as late as midnight. Prosperous gentlemen now found it safe to venture throughout much of the city at night, as long as they followed well-lit routes. Joel Ross reported in 1851 that when he was out late, "We are pretty careful to keep out of the grease-lighted avenues when we can; and feel safer to walk a street (if it is not too long) which has but one good gas-light at each end of it."[16] Confident in the power of gas to discourage crime, mid-nineteenth-century observers urged extending the hours of lighting from dusk to dawn, even on moonlit nights. New York's streetlamps were lit almost all through the night by 1860, but this practice was not universal in American cities until the twentieth century. There was also a constant demand for installing gas lamps in more neighborhoods, including some backstreets and alleys.[17]

As improved lighting spread through the city and through the hours of the night, optimistic observers continued to hope it could produce a safer, more orderly society. One advocate of gas lighting asserted in 1871 that

> in our large cities, it has rendered life and property more secure; and those numerous localities, which at night were the haunts of violence and crime, are now comparatively safe; and the lonely traveller (whose purse and person were so frequently in danger from an attack of the highway robber, or footpad) can now pursue his path in comfort and safety. . . . Some may say that this change is due to the advanced civilization of society, and undoubtedly it is so; but gas has contributed largely to that change. The greatly improved illumination of our streets has tended to prevent crime; it has afforded great facilities for persons to assemble at nightly scientific, literary, musical, and other meetings; it has thus facilitated the reunion of society of every grade, and therefore it may be truthfully said that gas has certainly assisted largely in attaining this advancement in civilization.[18]

POLICE

Nighttime police protection was once as weak as an old tallow candle. The night watchmen who patrolled the streets in the colonial and early national eras lacked the full authority of later police officers. The men on the day shift of law enforcement—constables and sheriffs—were the ones with the real power to make arrests. In New York, a night watchman who attempted to seize a suspected criminal was in the same legal position as any other citizen—he could easily find himself on trial for exceeding his authority. Unless he actually witnessed the crime, he was better off simply reporting the matter to a constable in the morning. There were too few night watchmen to make much difference anyway. In 1800, New York had only seventy-two policemen to protect a city of over sixty thousand. Cities would hire more in response to a spate of burglaries or arsons, then allow the rolls to dwindle once the threat subsided. Further, the poorly paid work attracted those who also held day jobs and those with few other employment options—such as old men. Sleepy watchmen liked to spend most of their time in the comfort of the small sheds called watch houses, particularly in bad weather. Watch regulations were filled with rules aimed at punishing watchmen who drank, slept, gambled, or disturbed the peace while on duty. Boston's watchmen were so inattentive that thieves in 1802 successfully broke into a public market only forty feet from a watch house.[19]

City officials were anxious to keep the watchmen awake and walking around. New York, Philadelphia, and other early American cities required them to ring bells and call the hours of the night while making their rounds. This reassured citizens that "all is well" and that their homes were kept safe by vigilant officials. Obviously it also woke up light sleepers and gave criminals ample warning that a watchman was approaching. Boston's city fathers vacillated between two unsatisfactory approaches to policing the dark streets: they intermittently tried having watchmen sneak through the streets to catch miscreants red-handed, only to find the watchmen napping or absent. As late as 1830, Philadelphia had seventy-seven men on its "loud watch" (including forty-eight lamplighters) and sixteen "silent watchmen."[20]

Watchmen were held in contempt by much of the public. Young pranksters enjoyed breaking a watchman's lantern and running away undetected, or knocking over a watch box as the watchman slept inside. Relatively powerless, outnumbered, and disrespected, the watchmen were ineffective as an enforcement arm of the state. The main purpose they served was implied by their name: watching. They were the eyes of the sleeping public. If they saw a fire or a crime in progress, they could raise a "hue and cry" to summon help from the surrounding neighborhood. Otherwise they minded their own business and dozed in front of the watch house stove.[21]

The growth of American seaports into major cities put strains on this rudimentary system of nocturnal law enforcement. Urban crime seemed to increase in the early to middle nineteenth century, though the lack of reliable statistics makes this uncertain. Perhaps people were simply more concerned because they now had greater reason to be outside after dark; perhaps their expectations for safety were rising. In any case, there were numerous complaints about urban crime and disorder. John Pintard, a New York banker, wrote in 1831 that the nighttime streets had become so infested with muggers that his wife feared for his safety whenever he went out on late errands. In 1838 a Baltimore newspaper complained that "our city is infested, disgraced, by a gang of ruffians, who, in defiance of every sense of shame, promenade the streets in company with the most abandoned of the other sex, and at night prowl about, insulting decent females, and, like assassins, waylaying and beating peaceable citizens."[22]

Such complaints, as well as a ferocious outburst of anti-Catholic rioting in the 1840s, helped persuade city officials to follow the example of London in establishing full-time police forces. The new police departments did away with most formal distinctions between day and night policing, providing full arrest powers and better pay to all the officers. In contrast to the uniformed bobbies of London, the new American police patrolled in civilian clothing, bearing only stars for insignia. Many working-class men at this time viewed uniforms as symbols of autocratic authority intolerable to self-respecting freeborn Americans. Until uniforms were finally accepted—in New York in 1853 and most of the rest of urban America by 1880—police continued to blend into the crowd. Their diminished visibility may have made it easier for them to catch criminals, but it limited their effect on citizens' sense of security at night. Criminals could pose as police, and patrolling officers could easily be mistaken in the dark for skulking criminals, as Salomon de Rothschild found during a visit to Baltimore in 1860. "One day I dined out and I returned when it was so dark that people were afraid to set foot in the street," Rothschild wrote:

> To take a short cut I had to go through a little street still darker and narrower than the others, if that is possible. I saw or rather sensed that I was being followed by a man of rather imposing size. I hurried, my shadow imitated me; I stopped, but my companion crossed the street, came up behind me, and was going to put his hand on my shoulder, when I turned, grabbed him around

the waist, making it impossible for him to use his hands, and shouted "Police" with all my might. The man struggled for a moment before saying: "But what are you hollering like that, I am an officer?" At that point, half a dozen policemen who had been running toward the noise came up and asked what was happening. The man, rubbing his arms, said: "I had seen that man very near a store and thought that he had bad intentions."[23]

Rothschild's stalking policeman was acting properly by the standards of the time. An aggressive, proactive approach to crime fighting was exactly what the commanders of the new police forces wanted.[24] Urban crime rates seem to have fallen in the late nineteenth century from a peak in the 1870s, but it is uncertain whether police were effective enough to deserve much credit.[25] Part of the problem was that there were still too few police for cities that were growing rapidly in both population and area. In the period around 1870, Philadelphia had one police officer for every one thousand residents, Boston had one for every six hundred, and New York had one for every five hundred. (By contrast, these cities today have one officer for about every two hundred to three hundred residents.)[26] Police superintendents assigned most patrolmen to night duty, when crime was thought to be most prevalent. New York's Metropolitan Police in 1858 had 358 night patrolmen and 178 day patrolmen. New Haven in 1866 had 17 night patrolmen and 7 day patrolmen; twenty years later, it had 54 night patrolmen and 22 day patrolmen. Brooklyn in 1880 had some 200 patrolmen on night duty and about 100 working the day shift.[27]

Each man on night duty was responsible for a smaller area than his day counterpart, and he could count on having fellow officers within earshot if he had to call for backup by rapping his nightstick on the pavement. Still, there were too few policemen to adequately patrol the urban night. Anxious to protect the entire city, Philadelphia in the 1850s created enormous beats in which the officer passed each point at long intervals. Other police departments conserved manpower by neglecting certain areas. New York, which had only 769 patrolmen on the beat each night in 1876, chose to leave certain streets entirely unpatrolled. Minneapolis patrolled only a minority of its streets.[28]

Officers devoted most of their time at night to regulating working-class leisure, which seemed to observers to be a growing problem. Throughout the middle to late nineteenth century, employers gradually imposed stricter work discipline, forced drinking out of the workplace, and divided workers' lives more sharply into hours of work and hours of relaxation. After the workday ended, workers could retreat from employers' supervision into their own neighborhoods, where they came under the scrutiny of police officers on the lookout for noisy drunks and troublemakers.[29] Between half and three-quarters of all urban arrests from 1845 to 1885 were for drunkenness or disorderly behavior. In Chicago in 1872, for instance, police made nearly fourteen thousand such arrests. Each arrest meant that the officer had to leave his beat to walk the suspect back to the precinct station, sometimes enduring taunts on the way from neighbors or even fighting off attempts to free the prisoner. Not until the 1880s did a system of call boxes and patrol wagons relieve patrolmen of this onerous duty.[30]

The concentration of different activities in different parts of the city produced a new geography of nocturnal crime in the middle to late nineteenth century. This was most obviously true for the prostitution and gambling rackets, which found it profitable to locate within a short walk of the downtowns and the furnished-room districts where single men lived. The increasing liveliness of "bright lights districts"—centers of legitimate consumption and commercial recreation—also presented new opportunities for criminals. Busy saloons and theaters were frequented by pickpockets.[31] Well-lit sidewalks drew crowds of pedestrians and the criminals who preyed on them. New Yorkers from all walks of life chose Broadway for their evening promenade, and pickpockets and prostitutes mingled with the throng despite the abundance of gas lamps and policemen. George Templeton Strong resented the "noctivagous strumpetocracy" along Broadway, but he still objected when Mayor Fernando Wood attempted to crack down on the streetwalkers in 1855. Wood's policy, he wrote, was "dangerous and bad. It enables any scoundrel of a policeman to lay hands on any woman whom he finds unattended in the street after dark, against whose husband or brother he may have a grudge, who may be hurrying home from church or from a day's work, or may have been separated by some accident from her escort, and to consign her for a night to the station house." Broadway streetwalkers weathered such spasms of reform by becoming more discreet. Fearing arrest if they conversed with men on Broadway, prostitutes signaled their intentions mainly by their appearance and manner before scurrying down side streets with their customers. Prostitutes on less prestigious commercial streets could work the crowds more openly. Some in the wretched Five Points stood in doorways displaying their breasts and quoting prices to passersby.[32]

Burglary focused on commercial areas away from the bright-lights districts. Here again, the separation of life from work created a problem for law enforcement. As the demand for commercial space elevated rents, and as middle-class families sought more secluded neighborhoods, the residential population dwindled in what came to be called the "downtown." Congested business streets turned empty and quiet once the bankers, clerks, salesmen, shoppers, and other downtown visitors left in the evening. Much of the city's wealth lay inside warehouses, banks, and department stores, virtually unwatched, a tempting target for thieves.[33]

The new professional police spent much of their time at night patrolling empty downtown streets. Ernest Ingersoll observed in 1878 that

> lower Broadway, dim and gloomy at midnight, is full of police, furiously shaking at the handles of the doors to be sure that all are securely locked, peering through the little peep-holes in the iron shutters, to see that no burglars are at work in the stores where lights are left burning all night, or that an incipient fire is not working insidious destruction; lurking out of sight in shady doorways while they watch suspicious loungers; or standing in groups of two or three on the corners where two posts intersect, and a roundsman has happened to join them.[34]

Policemen could often be found at street corners, which served as strategic vantage points for control of the dark city. Certain corners were informally known as places where neighbors could seek the help of an officer at night, just as certain corners were formally designated as places where officers were stationed during the day. Intersections allowed policemen to survey the streets in multiple directions, and the superior lighting there allowed them to scrutinize passersby and recognize wanted criminals. (Idle young men and boys who liked street corners for similar reasons would spend evenings loitering there if no officer was present to shoo them away. These "corner loungers" would stare into the faces of passing women and make rude or suggestive comments.) Street corners played a crucial role in the navigation of urban night; cities installed street signs on many corner lampposts in the middle decades of the century, or painted the name of the street directly on the post. Philadelphia in the early 1850s installed transparent signs on corner lamps so that the street name would be illuminated when the lamp was lit. The editors of the *North American* newspaper hoped "that all parts of the metropolis might be rendered intelligible at night" in this way but regretted that the city had not installed the illuminated signs in outlying neighborhoods and had failed even to replace broken or missing signs in the center city. Without illuminated street signs, the editors warned, visitors would find the city a "bewildering labyrinth" at night and might be easy prey for tricksters and criminals seeking to misdirect them. Thus Americans hoped street corners that were properly illuminated, policed, and labeled could bring order to the dark city. In these patches of light, law-abiding citizens could check watches and street signs to orient themselves in time and space; criminals could be recognized and observed.[35]

Safety in most parts of the American city depended on community participation. The major streets in working-class districts were relatively safe through the early evening—there were enough people out in these densely settled neighborhoods to deter muggers. Side streets and alleys could be dicey, particularly for strangers who did not know which spots to avoid. Much of the trouble came from street gangs and rowdies who would attack unpredictably, sometimes to rob, sometimes for fun, sometimes to assert their control of the turf. Patrolmen often challenged the right of young men to loiter at street corners, but there were too few officers to make much difference unless supported by the community. Furthermore, certain slum areas were so notoriously dominated by violent gangs that police did not feel safe entering them except in groups.[36] Affluent neighborhoods were well lit but poorly patrolled. If a resident came to a station house to report a crime or ask for assistance, police were ready to respond, but few officers walked beats in these areas. Patrols were further reduced in the late nineteenth century when the new telegraphic call boxes provided an easier way to summon help.[37]

Sparsely settled areas on the fringe of the city remained dangerous through most of the century. As in the early days when only parts of the cities fell within the "lamp and watch district," there was always an unpoliced fringe where drunks and petty criminals felt secure. Working-class districts just outside the boundaries of Philadelphia, such as Moyamensing, were the battlegrounds of rival gangs in the mid-nineteenth century, while Providence's criminals and Sunday-night drinkers found a secure haven over the city line in the Dogtown section of Cranston. Annexations by Philadelphia and Providence—in 1854 and 1868—eventually brought somewhat better police protection to these outlying areas.[38]

Waterfronts and harbors represented another dangerous urban frontier. The dangers of New York's waterfront came to public attention as a result of a sensational murder in 1852, when three teenage robbers killed a night watchman on a boat moored in the East River. Newspaper readers were fascinated to learn that organized gangs of river thieves prowled the unlit regions that were so essential to the city's commerce, stealing cargo and equipment from boats and piers and killing those who got in their way. At that time, wrote George Walling in 1887, the main part of the island [of Manhattan] was bounded by piers and slips, which in turn were fringed on one side by grog shops, rum holes, and all kinds of iniquitous dens, breeding crimes as rapidly as mosquitoes are bred in a swamp. Along the piers ran the swiftly flowing river, a constant source of escape by day or night. Down the North River docks and up the East River docks criminals formed the greater part of the population. They went in gangs. Each gang had its leaders and its rough rules of discipline. Its members lived in the vilest dens, carousing or scheming all day, prowling, marauding and thieving all night.

Municipal authorities quarreled with the gas companies over their persistent refusal to light unprofitable waterfront areas, and they finally secured their compliance. New York also formed a harbor police in 1858 to prevent attacks on ships. The harbor police cleaned up the Hudson River waterfront, then struggled with partial success to limit crime along the East River. Other harbor police units were organized in the 1850s to control crime in and around Boston Harbor, and along the Delaware and Schuylkill Rivers in Philadelphia. Unlike working-class wards, the waterfronts were too commercially important to be left unlit and unguarded.[39]

Together, improvements in lighting and police protection created a greater sense of nighttime security. If the streets were not as safe as citizens would like, they did not conclude that it was foolish to go out after dark or rely entirely on the heavy canes and revolvers that some still carried for protection. Now they also demanded better lighting and policing. If crime prevention depended in part on lighting, they believed, police should see that the lamps stayed lit. "Gas is an adjunct of the police administration, and plays an important part in the prevention and discovery of crime," observed the *Journal of Commerce* in New York in 1873.[40]

EARLY BLACKOUTS

Rumors of all sorts of mayhem spread through the cities on occasions when street lighting was interrupted by accidents, strikes, and disasters that shut down the gasworks. Nowadays, after the New York blackout riot of 1977, such fears may seem rational. But when the gaslights went out in the nineteenth century, people expected massive rioting and looting even though it had never happened before. Their exaggerated fears were the flip side of the exaggerated hopes that light would suppress crime; the panicked rhetoric suggested that far more was at stake than a possible return to earlier levels of street crime. Middle-class writers—newspaper editors, diarists, and public officials—worried that the loss of light would destroy whatever precarious order was evolving in the urban landscape.

The burden of sustaining the social order was a heavy one to place on gas systems that seem pitifully small by today's standards. The mid-nineteenth-century gas company had little in common with the vast continental systems of electricity and natural gas that developed in the twentieth century. Its distribution network of cast-iron pipes served at most a single city. The gas was typically manufactured nearby at a gasworks along the waterfront, by teams of sweating laborers on twelve-hour shifts. Unlike later failures of electric generation, interruptions of gas production did not immediately darken an entire city. If striking laborers let the fires go out in the retorts where gas was produced, the pressurized supply in the massive "gasometer" tanks could light the city through the first evening. Even if fire destroyed the gasometer or cut it off from the distribution mains, not every home or street would lose its light. Enough gas remained in the lines to supply some customers with a weak flame for a few hours; this was particularly true in neighborhoods at higher elevations (usually the wealthier ones), to which the gas would rise. Once the gas was exhausted, though, the city could be completely dark for several nights.[41]

One of the first major "blackouts" in American history (to use the modern term) occurred in 1848 as a result of a spectacular fire that shut down the New-York Gas Company's works in lower Manhattan.[42] As streetlights flickered out and interior lighting failed, the usual activity of the city came to a halt. Stores closed, theaters canceled performances, and printers set type by candlelight. Though the brief interruption returned New York to levels of darkness that most readers would have remembered from their childhood, the *New York Herald* cast the blackout as a catastrophe and a threat to public safety.

> A dismal gloom pervaded every quarter. . . . The calamity will doubtless be hailed with joy by those who prowl about the city to rob their fellow men; for never before has such an opportunity to practice their villainy been afforded them. It would be well for the Chief of Police to prepare for these midnight marauders, and place a sufficient number of men on duty to frustrate their designs.

A day later, however, the *Herald* editorialist was forced to admit that the blackout had proved only a minor inconvenience. The company was producing gas again, and the midnight marauders had evidently behaved themselves. Writers at the less sensational *New York Tribune* spoke of disorder more obliquely, praising the gas company for offering to cover the expense of additional night watchmen to ensure the safety of the darkened district. Their main emphasis in covering the blackout was on the gloomy appearance of the city as it struggled to function with primitive lighting.[43]

A similarly mixed response can be seen in the case of an 1868 blackout in Philadelphia. On Thursday, July 16, 1868, stokers and helpers at the Philadelphia Gas Works struck for higher wages. The gas reserve was soon exhausted, even though the streetlights were not lit that night. On Friday night the center city was completely dark; many large retailers that normally did a brisk evening business were forced to close early, while others struggled to make do with oil lamps. Hotel keepers rushed to buy lamps for their rooms. Grocers tripled their prices on candles.[44]

Philadelphians seem to have experienced the blackout as both an adventure and a threat. "Lawlessness always seeks the opportunity of darkness, but generally, good order was preserved throughout the city, and it was generally agreed, that for one night only, the change was quite romantic," reported the *Evening Star*. More than thirty years had passed since gas lighting was first introduced in Philadelphia, and what began as a novelty had long since faded into the urban background. Indeed, Philadelphians seem to have marveled at the blackout in much the same way as their counterparts in smaller towns were greeting the introduction of gas lighting, and much as urban Americans would greet the introduction of electricity around 1880. "The streets to a late hour were thronged with people curious to observe the appearance of the city under candle-light," reported the *Philadelphia Press*. The *Philadelphia Inquirer* concurred but added that the dark streets became spookily empty as the evening wore on and warned that Philadelphians were at the mercy of robbers and burglars. Fortunately, no crime wave materialized before the conflict was resolved in favor of the workers and gas began flowing again on July 19.[45]

Fears of the darkened city rose to a peak during the 1870s, as seen in the responses to two major blackouts, one in Boston and one in New York. Urban Americans by then had grown even more dependent on gas lighting. Not only were gas systems more extensive than in previous decades, but gaslight was used more intensively, as Americans had grown to prefer brighter homes and streets.[46] More important, though, Americans had grown more fearful of urban disorder as a result of the Paris Commune and the New York Draft Riot. During the Paris Commune of 1871, sensational reports in American newspapers gave readers the impression of a world capital in the hands of a murderous rabble. On a more limited scale in 1863, working-class New Yorkers had resisted the Civil War draft by taking over sections of the city in a five-day explosion of arson, looting, lynching, and gun battles with security forces. Perhaps acting in the European revolutionary tradition of "lantern smashing" during moments of urban upheaval, the draft rioters had destroyed a small gas manufactory at the Hudson River end of Forty-Second Street, thus darkening the surrounding neighborhood. Troops and armed citizens had then garrisoned the remaining gasworks in New York. "Fortunately the riotors [*sic*] did not carry out their threat of burning the gas-works," wrote the editors of the *American Gas-Light Journal*. "Had they done so, the effect would have been calamitous indeed. The streets would have been in perfect darkness, and wild bands of marauders could have roved about, and the reign of pillage and assassination would have been complete."[47] Editors of the *New York Times* warned readers repeatedly in the 1870s that a new uprising could erupt at any time; beneath the surface of city life lurked "a powerful 'dangerous class,' . . . who burrow at the roots of society, and only come forth in the darkness and in times of disturbance, to plunder and prey upon the good things which surround them."[48]

In Boston in 1872, fear of a blackout contributed to worsening the fire that destroyed much of the downtown, and fear of a crime wave kept Bostonians on edge after the gas system finally shut down. The fire broke out on Summer Street on Saturday evening, November 9, 1872. Escaping gas fed the flames, particularly as collapsing stone warehouses burst the gas mains. Gas leaked into the sewers and set off explosions that blew manhole covers high in the air. The flames kept gas workers from reaching shutoff valves in downtown Boston, and company officials were reluctant to shut down the whole system until the police finally ordered them to do so Sunday evening. The fire raged out of control until Sunday afternoon; fourteen people were killed and 776 buildings were destroyed, many of them in the heart of the business district. As the press eagerly emphasized, most of the losses had been borne by prosperous merchants. The fire was presented as leveling Boston figuratively as well as literally; in the language of the headlines, "rich men [were] beggared in a day." *Boston Globe* writers personified the fire as "the red terror" and "the fire fiend," bent on destroying Boston's wealth.[49]

Tales of disorder far surpassed reality. Some of the rumors appear to have originated from a misunderstanding in which police made over two hundred arrests of people carrying shoes and other merchandise from the burning district; they had been given permission by the owners of doomed shops and warehouses, but they had to spend the night in jail anyway. The police chief later acknowledged that there had been very little real looting.[50]

Lighting the Heart of Darkness 31

BOSTON—SCENE IN CHAUNCEY STREET—MERCHANTS DEFENDING THEIR GOODS AGAINST THIEVES AND ROUGHS.—[SEE PAGE 934.]

FIGURE 2. A night of chaos. This engraving shows an almost entirely imaginary scene: the looting of Boston businesses during the fire of 1872. The press and public of large cities in the 1870s feared that urban slums harbored a dangerous underclass that might "come forth in the darkness and in times of disturbance, to plunder and prey upon the good things which surround them," as the *New York Times* warned. This particular scene never occurred, but the possibility of such events influenced Boston's response to the fire and to the subsequent loss of gas lighting. "Boston—Scene in Chauncey Street—Merchants Defending Their Goods against Thieves and Roughs," *Harper's Weekly*, November 30, 1872.

Enterprising reporters and book publishers, though, produced vivid tales of mayhem in the burning district. The *Globe* reported that packs of subhuman looters lurked in the darkness, awaiting their chance:

> The "birds of prey," who fatten on the calamities of their fellows, hovered about, eager to pilfer upon any opportunity that offered. They skulked in the shadows of the houses, watching with greedy eyes for a chance to gratify their rapacity. . . . They crouched in dark niches, ready to dart forth and pounce upon any neglected trifle. They leered horribly in the faces of timid women, and uttered coarse jests in their hearing. Their brutal faces, low brows, small, snaky eyes, narrow lips and heavy jaws showed them to be the botchings of nature's handiwork, and these distinctive and repulsive features were seen in

distressing numbers in every direction. They grew so bold at last that the military were called in to keep them in check, and as the steady tramp of the soldiers was heard through the din, and their swords and bayonets reflected back the fire in cold and fitful flashes, this scum crept away to the dark places from which they emerged.[51]

Rumors spread through a gas-deprived Boston Sunday night that a train full of criminals was on its way from New York to plunder the darkened Hub. The police and the militia turned out in force to meet the ruffians at the station, but they never arrived. Still, Bostonians told each other that criminals had flocked to the city, and versions of the rumor were reported as fact in Boston newspapers and respected outside publications. Fears of lawlessness were heightened by delays in restoring gas service. Workers for the Boston Gas Light Company shut off the mains leading into the still-smoldering burned district so that gas could safely flow to the rest of Boston. On Monday, November 11, company officials turned the gas on again. Explosions were soon heard throughout the city, gas leaks started new fires downtown, and service was shut down for another two nights. "For two days we have been without gas— it was a fearful time for old Boston," wrote the merchant William Brooks in his diary on November 13. Perhaps embarrassed by its earlier sensationalism, the *Globe* became the voice of reason on the final day of the blackout, reassuring readers that police had found no significant increase in crime.[52]

Some of the most extreme expressions of blackout panic came during an 1873 strike—ultimately unsuccessful—that disrupted production at the gasworks supplying lower Manhattan. The circumstances were disturbing: overt class conflict produced a blackout in the heart of American commerce, and the strikers attempted to spread the shutdown to gas companies serving other parts of the city. Unlike in Philadelphia in 1868, the novelty of darkness now drew few admirers. Streets were nearly deserted despite active patrolling by police and firemen. Newspapers reported fears of mugging, murder, looting, and organized raids on the banks of Wall Street.[53]

Harper's Weekly warned in a front-page editorial that a prolonged blackout could plunge New York into anarchy.

> The imagination could hardly picture the possibilities of such a reign of terror. The police force, barely adequate to control the dangerous classes under the most favorable conditions, would be powerless to prevent such an outbreak of ruffianism as the city has never yet witnessed. Pillage, robbery, and murder would take possession of the streets, and the light of burning houses would

shed a terrible illumination over the city. No honest man would dare venture abroad after night-fall, and there would be little security at home. Every dwelling-house and business establishment would have to be converted into a fortress.

Most of this editorial was a harsh diatribe against strikes and the immigrants who supposedly caused them, yet the text was accompanied by a half page of cartoons that suggested alternative reactions to the blackout. Only a few cartoons echoed the message of the editorial, most notably one showing a stereotypical Irishman, "the cause of the trouble (as usual)." The rest suggested themes that had appeared and would appear again in response to blackouts from the mid-nineteenth century to the late twentieth century. One of the most common themes in coverage of blackouts was that the failure of light temporarily erased the blessings of technological progress and threw the affluent back on their own individual resources. In one cartoon in this issue of *Harper's Weekly*, for instance, an old lady in antiquated dress states disdainfully that "we didn't have this trouble in the good old time." In another, a candlestick comes to the aid of an ineffective gas lamp.

While trying to put a lighthearted spin on a traumatic experience, some of the cartoons had a subtly disturbing edge: they presented the blackout as affording a moral holiday from everyday life. Two suggested that the blackout was an occasion for heavy drinking, while two others indicated that darkness offered erotic opportunities. One cartoon of a kissing couple is captioned as follows: "In the darkness Mr. B. very naturally mistook their pretty chambermaid for Mrs. B."[54]

Far less consequential than class upheaval, these peccadilloes still called attention to the fragility of urban order, for they represented moral failings among respectable citizens. New Yorkers did not have to be reminded that drunkenness and sexual immorality tempted men and women at night. More and more people were venturing into the streets after dark, reassured by the bright lights and uniformed police and given more leisure time by the new schedules of work emerging in the nineteenth century. After stopping work for the day, city people had the time and money to seek amusement in the city's public spaces. It remained uncertain whether gas lamps and police only made it more appealing for them to go wrong.

CHAPTER 3

Quitting Time

Evening spread slowly across the open countryside. In the scantly wooded hills of mid-nineteenth-century New England, men in the fields sensed shadows reaching from stone fences, colors deepening before dusk. Chores eased away as gradually as daylight. While day's afterglow shrank to a stripe on the horizon, there were still cows to milk and scythes to sharpen. After dark, peeling apples blended with social visits with neighbors. At the harvest moon, or at the birth of calves, special tasks might keep country folk from their beds long into the night.[1]

Such memories followed the thousands of young people who left their farms for the cities of the industrializing Northeast, where along with other shocks they encountered a radically different experience of time. Nothing seemed gradual in cities after 1830, particularly the structure of the workday—"minced into hours and fretted by the ticking of a clock," in the words of Henry David Thoreau. For twelve hours or more, young men and women waited on customers, tended looms, or copied legal documents by hand. The confinement was difficult to bear. In furtive defiance of their employers, they relieved the tedium with whispered conversations and occasional pranks. They stole glances out the window at the river flowing past the mill, at hills and trees, at clouds, at passersby in sunny streets, or at the blank walls of buildings. As they daydreamed, the shuttles moved back and forth, back and forth. The hand skittered right, lifted to dip the pen, then dropped back to begin another line.

Then they stopped abruptly. The workday ended with the clang of factory bells or the bolting of store shutters, and workers were suddenly, thrillingly, freer than ever before. With ears ringing or eyes aching they stepped outside. It was night. The air, spiced by horse manure but carrying a fresh breath of leaves, felt cool after the dead funk of the mill or shop. Whatever sunset colored the horizon was blacked out by rooflines. Lamplight flowed from shop windows. Streetlamps shone on paving stones. At this moment, far from family guidance, coins weighing pleasantly in their pockets, young men and women faced the possibilities of city life. "The evening," wrote one former mill worker, "is all our own."[2]

This was the moment so alluring to young people and so worrisome for their elders. Urban work regimens threatened to create a double life, a day of confined drudgery and an evening of limitless freedom. Turned out onto the street at night, would these new city dwellers improve the time before sleep by worshipping and learning, or would they cast aside the moral habits of their upbringing? Preachers, travelers, and journalists who wrote about urban life kept coming back to the problem of "quitting time." Writers paid particular attention to two groups of young people: the girls and women who worked in the cotton mills of Lowell, Massachusetts, and the boys and men who worked as clerks in New York City. Both of these cases raised issues about young people's behavior away from family supervision. In both cases, writers observed that the dangers of city life were heightened when work ended after dark.

Employers, reformers, and other concerned citizens managed the problem of evening freedom differently in the two cities. In Lowell, boardinghouses served as surrogate homes to police the morality of mill girls and to limit their leisure to two or three hours before a 10:00 p.m. curfew. The early curfew discouraged attendance at dances, nudging women toward early evening activities such as shopping or attending morally instructive lectures, prayer meetings, and evening classes. New York clerks, on the other hand, were left to their own devices. Those who lived in boardinghouses there were usually free from troublesome curfews. Furthermore, store clerks were regularly kept on the job until 9:00 or 10:00 p.m., pushing their leisure into later hours when amusements were fewer and less wholesome. Clerks and employers fought intermittent battles over closing hours in which each side tried to score points by invoking the danger of the saloon and the brothel. By stretching the regular workday past sundown, employers took a significant step toward what in the later nineteenth century became a radical disjuncture between work schedules and natural rhythms. By squeezing leisure out of the day and into the night, they also turned urban nightlife into a more salient site of conflict.

COTTON MILL GIRLS

Night work in industry grew out of an older artisanal practice of completing work by candlelight during the short days of fall and winter. In spring and summer, craft workers began work as soon as they could see and ended when they no longer could. They worked exceptionally long days around the time of the summer solstice, when the sun rose in Boston before 4:30 a.m. and set after 7:30 p.m. As autumn days shortened to less than twelve hours, employers would "light up" the workplaces, establishing a winter quitting time defined by the clock. Even the earliest American textile mills followed the practice of lighting up in the evenings. Spinning mill account books from the pioneering Rhode Island firm of Almy and Brown in the mid-1790s show numerous entries for the purchase of candles, tallow, and whale oil.[3]

Though stretching the day into early evening was nothing new, it was practiced on a greater scale in the second quarter of the century as water-powered textile mills proliferated and grew larger. Investors built mills for spinning cotton or wool fibers and weaving cloth along quick-moving sections of many New England rivers, as well as along creeks and rivers in Pennsylvania, New York, and New Jersey. Boston investors founded the city of Lowell on the Merrimack River in the early 1820s, developing it into the nation's leading textile center by 1850. By that time, in a city of thirty-three thousand people, nearly twelve thousand worked for the major manufacturers, mainly in tall red-brick factories along canals that fed the millraces, or along the Merrimack River itself. Most workers in the early decades were young women from rural New England who worked for a few years before returning home; from the mid-1840s they were joined and increasingly replaced by Irish immigrants. Their employers spent fortunes to keep the workrooms lit and busy after sunset. The enormous Lowell mills consumed tens of thousands of gallons of illuminating oil annually in the 1830s and 1840s—spermaceti from the whaling fleets of Nantucket and New Bedford, olive oil from Mediterranean groves, lard oil from the humble swine of American barnyards. In 1850, the Lowell mills began supplementing oil with manufactured illuminating gas.[4]

Still, midcentury factory owners only tentatively separated work schedules from the hours of sunrise and sunset. To minimize their use of expensive lamp oil, employers relied mainly on natural light from the windows, and thus on the varying length of the solar day. They typically set hours by the sun in the morning and the clock in the evening.[5] A Pennsylvania operative testified in 1837 that "there are no fixed hours—no set time of working at our mill, but we work from the time we can see, in the morning, until the light is insufficient to see in the evening. In the early part of the winter season we begin work at daylight, and continue until eight and a-half o'clock at night; when the days get shorter, we are called up before day, to breakfast, and go to work at daylight; we are called to work by the ringing of a bell." Even with extensive use of artificial light, mill days in Lowell in 1845 ranged from a low of eleven hours, twenty-four minutes in December to a high of thirteen hours, thirteen minutes in April.[6]

Winter evenings in the countryside were times of rest and conviviality; winter evenings in mill towns were shortened by what workers called "long and dreary hours of lamp-lit and exhausting toil." Whether in the early Slater mills of Rhode Island or in the huge spinning and weaving mills of the midcentury Merrimack Valley, the season of night work began and ended each year at the equinoxes: the first day of "lighting up" took place on or about the twentieth of September, and "blowing out" took place about the twentieth of March. By the 1840s these days had become significant annual events, celebrated in Lowell with parades and dances after work, particularly at "blowing out," when workers could begin to enjoy the spring twilight after quitting time. Mill owners appear to have used lighting up as an opportunity to seize greater control over the length of the workday. The time of day had not yet been standardized, so clocks in each town were expected to conform more or less to sun time, with noon representing the moment of the sun's zenith. Taking advantage of the usual imprecision in clock time and their workers' lack of pocket watches, manufacturers often set their own clocks twenty or twenty-five minutes slower than "true Solar time" would indicate, thus squeezing extra work out of the mill hands each night.[7]

Lit only by flames, mills grew hot and stuffy as the evening progressed. Workrooms were sealed in for the winter behind double panes of glass; the dozens or hundreds of lamps in each room pushed temperatures as high as ninety degrees by quitting time. Furthermore, complained a labor newspaper in 1846, workers suffered "weakness of eyes, from the performance of difficult work by lamp-light, from watching constantly the fine moving threads, in the operation of machinery, looking for defects in texture, 'drawing in,' &c." The poor quality and high cost of lighting discouraged factory owners from attempting to run their mills through the night with shift workers. Workers were willing to tolerate some night work, but they sought to limit its duration by agitating (unsuccessfully) for a ten-hour workday.[8]

Mill hands relished their free time in the evenings. Having been isolated amid their noisy work all day, they gathered eagerly with their friends at quitting time, and walked down the street in pairs and small groups. Although the long hours of labor left workers little time for rest and recreation, observed a former mill girl, "the time we do have is our own. . . . [W]hen finished we feel perfectly free, till it is time to commence it again."[9]

By today's standards, this free time may seem remarkably unfree. Lowell mill girls were permitted just a few hours each evening between quitting time and the curfew enforced by their boardinghouses. Anxious to ensure an uninterrupted flow of new workers from the countryside, mill owners in the 1830s and 1840s boasted to visitors about their success in preserving the morality, sobriety, and chastity of female mill hands, who were typically what we would now call teenagers and young college-age women. The mill owners strongly encouraged employees to live in company housing near the mills, and most workers complied. Boardinghouse keepers were required to enforce a strict 10:00 p.m. curfew, to prevent improper behavior at any hour within the boardinghouse, and to report any boarder guilty of improper conduct or of failing to attend church.[10]

Boardinghouses were large, crowded, and completely lacking in privacy. They were typically brick row houses, three stories tall, with extra living space in the attic. Each house accommodated from twenty-four to fifty young women; boarders slept two to a bed in communal bedrooms. In some boardinghouses the only common room was a dining room packed with tables and chairs; other houses had a sitting room as well. Many young women spent their evenings there or in their bedrooms, reading novels, writing letters, mending, or chatting with fellow boarders. Peddlers often stopped by to sell books, shoes, and candy to women who lacked the money or the inclination to shop in the stores. The cramped rooms—lit by oil lamps and scented by the mill hands' sweaty bodies—were less appealing to many boarders than "going upon the street," as they called any trip into the city's public spaces.[11]

In the brief interval before curfew, mill girls had to balance their desire for amusement with the demands of basic personal maintenance. Boardinghouse suppers in the 1840s did not begin until after 7:30 in the fall and winter; women with plans for the evening undoubtedly skipped the meal or bolted their food as quickly as they had their breakfasts and lunches. Tasks such as mending or washing clothes stole time, and physical exhaustion imposed further limits. One short story in the *Lowell Offering* described a new mill girl's first evening: "O, how glad was Susan to be released! She felt weary and wretched, and retired to rest without taking a mouthful of refreshment. There was a dull pain in her head, and a sharp pain in her ankles; every bone was aching, and there was in her ears a strange noise, as of crickets, frogs, and jews-harps, all mingling together; and she felt gloomy and sick at heart." Most young women mustered the energy to go out at least occasionally, though it could be a struggle. "I well remember the chagrin I often felt when attending lectures, to find myself unable to keep awake," recalled another woman who had worked as a Lowell mill hand in the late 1830s. "I am sure few possessed a more ardent desire for knowledge than I did, but such was the effect of the long hour system, that my chief delight was, after the evening meal, to place my aching feet in any easy position, and read a novel."[12]

Mill girls enjoyed strolling after supper along the canals and especially along Merrimack Street, the main commercial street. "Extensive ranges of well-furnished shops, brilliantly lighted up, and continuing so until nine o'clock, or a little later, indicated the prosperity of the place," observed a British visitor, "whilst a very large number of the young women of the factories, were seen enjoying the evening air in the streets, and not a few availing themselves of the facilities offered them for shopping." By 1849, there were some ninety stores in the heart of the downtown. Besides the shops, another common evening destination was the post office, where mill hands jostled their way through a throng to retrieve letters from home. Crowds in the room reserved for ladies were such that it could take up to an hour to claim a letter, and some women reported having their pockets picked.[13]

Young women felt safe on the streets of Lowell because of their large numbers and the relative scarcity of men. Women in their twenties made up fully 25 percent of Lowell's population in 1840, outnumbering men that age more than two to one. Teenage girls outnumbered teenage boys more than three to one. The gender imbalance was particularly striking in the central section of Lowell where women spent most of their evenings: the few blocks between the Lawrence and Merrimack mills to the north and the Appleton and Hamilton mills in the south. Corporation boardinghouses clustered in this area, and the main commercial district stood in the middle. Young women seemed to own the downtown shopping streets once the mills let out, wrote one local observer. "Let any man stand at a given point on Central or Merrimack streets, for half an hour, of a pleasant evening, and probably two to three thousand people would pass him, keeping the wide walks densely crowded all the time. Of this dense moving mass three fourths are female." Walking in a city was a novelty for many Lowell pedestrians—for so many, in fact, that sidewalk traffic was snarled by greenhorns who had not yet learned to navigate a crowd.[14]

Women pedestrians were occasionally harassed by rude young men. These "third rate soap lock rowdies," who aped the rough working-class style of New York's "Bowery b'hoys," would gather in the evenings to "insult females" with unwelcome advances and coarse jokes. Lowell women learned to avoid passing certain street corners or saloons on their walks. Other hangouts for the loafers were unavoidably central, such as on occasion the intersection of Merrimack and Central Streets. The loafers were sometimes reported to be unemployed ruffians, sometimes idle rich boys, and sometimes shop clerks. When the streets grew nearly deserted after the 10:00 p.m. curfew in the boardinghouses, young men drifted to backstreet taverns or to brothels on the fringe of town.[15]

Once settled in to life in Lowell, mill hands sampled a range of entertainments unheard of in their rural villages. Evening schools offered women a chance to improve a general education that had been inadequate in their earlier years. Other classes allowed them to pursue aesthetic interests or gain practical training. In 1842 Fisk's Academy offered classes on penmanship, drawing, painting, arithmetic, bookkeeping, and music. Prayer meetings and singing classes drew other women, while smaller numbers would polish their creative writing in collective "improvement circles."[16]

Mill owners and local boosters were proud to point to the popularity of public lectures as evening entertainment. Lowell's emergence as an industrial city in the 1830s and 1840s coincided with the rise of the American lyceum movement, aimed at promoting local self-education and providing wholesome leisure activities. Lyceum associations formed in most of the major towns and cities of the northern states, as well as in many New England villages. By the 1840s, the lyceum had evolved into a series of entertaining lectures on a wide range of topics, often by speakers from outside the community. Lowell's lyceum in the 1840s was the Lowell Institute, which in 1845 sold twelve hundred subscriptions for its popular lecture series at City Hall at seventy-five cents each. A typical evening's schedule at the Lowell Institute included musical performances in addition to a lecture on science, history, or travel. "The City Hall is commonly crowded full," wrote the Reverend Henry Miles, who had spoken there in 1843. "Many of the female operatives attend, and the opportunity is justly prized by them of deriving more entertainment and instruction than most of them could receive at home." Mill supervisors encouraged attendance at the evening lectures by sometimes permitting mill hands to leave work a bit early. Additional lectures were held at city churches, and concertgoers could hear orchestral, string, and vocal music at Mechanics Hall or City Hall. Blackface minstrel shows were particularly well attended.[17]

Provincial New England frowned on the theater. In 1841 a writer for the *Lowell Offering* boasted that public disapproval of the theater's "demoralizing tendency" had caused the institution's failure in Lowell, a sure sign of the town's advanced morality. Too refined to attend anything that called itself a theater, Lowellians in the 1840s and 1850s instead enjoyed their theatrical performances at the Lowell Museum. The museum in its earliest years had advertised "a large variety of specimens of Natural History, Paintings, Engravings, Statuary, Wax-work and curiosities," in its building at Central and Merrimack. "Ladies and Families are informed that the strictest order is maintained and that they can with perfect propriety visit the Museum without the company of a gentleman." The museum closed each evening at 9:00 p.m. But even the thrills of seeing a celebrated dwarf or the Fejee Mermaid in a respectable setting weren't enough to draw an adequate crowd. The museum added performances of singing, dancing, ventriloquism, and plays in its thousand-seat "lecture room." The terms of its city license in the late 1840s allowed the museum to stay open until 10:30 but to offer only moral entertainment. This latter rule was loosely enforced. Plays ranged from Shakespeare all the way down to *Rosina Meadows*, the lurid tale of a New Hampshire girl who ends up killing herself in a Boston brothel.[18]

Susan Brown's diary from 1843 reveals some of the evening amusements of a mill girl. Brown attended at least thirteen lectures, performances, and concerts on weekday nights, including talks on geology, antislavery, and magnetism, and a presentation of the *Conflagration of Moscow* at the Lowell Museum. On Sundays, she reported going to church services and to afternoon lectures on temperance and other moral topics (though apparently these lectures failed to dissuade her from buying gin one rainy evening in June).[19] Except perhaps for the gin, these all seem to be perfectly proper ways to spend one's free time. Yet even the Lowell Institute lectures at City Hall could be less than orderly, disrupted by young men and women talking, flirting, and engaging in "unbecoming conduct" in the gallery. "While the music was playing last Wednesday evening, these ill-bred rowdies of both sexes . . . actually began to dance in their seats," according to an 1842 complaint in the local newspaper.[20]

Dancing was a touchy subject in 1840s Lowell, where six or seven public halls provided facilities for dances and parties. Most other entertainments claimed to have some element of moral improvement and ended in time for mill girls to comply with the boardinghouse curfew. Not so with dances. A writer to a labor newspaper warned in 1846:

> The manner in which the public dances are conducted in our city . . . must be injurious to the physical, mental, moral and social happiness of all connected with them. . . . They are often crowded with the worst characters that infest our city; and are not unfrequently turned into scenes of drunkenness and open disgrace; the influence of which must be of the worst kind, to the

moral feelings of the young. They are generally continued during the night; and the consequences following such constant exercise, must be most injurious to health after a day of toil.[21]

The real issue was not exhaustion but chastity. Although premarital sexuality was common in rural New England, as elsewhere, commentators assumed that the state of female virginity in Lowell spoke volumes about the nature of American industrialism. They wrote of evening entertainments as if a single misstep would start an innocent maiden on a downward path to the brothel. One such misstep, argued a local moralist in 1849, was when a woman entered a dance hall unattended. She would find herself dancing with perfect strangers and relying on them for refreshments and for escort home. Mill girls sometimes stayed out until the dance broke up between 11:00 p.m. and 3:00 a.m. Either the curfew was not enforced in some boardinghouses or the mill girls were sleeping elsewhere. "A strange darkness rests upon the where, but in regard to the how, suspicion cannot shoot wide of the mark," warned the writer. The newspaper editors of the sensational local *Vox Populi* expanded on this theme in a short novel, *Norton*, published in 1849. The novel claimed that "the brothel is filled with many a victim from here [Lowell]; for where could a lecherous man find a better field for his hellish lust than here."[22]

News of improper sexuality in Lowell spread to the Boston press and appeared even in the chaste pages of the *Lowell Offering*, a literary magazine written by mill hands. Harriet Farley warned in an 1845 editorial that innocent country girls were preyed on by lustful young men and shopkeepers. "And when it is remembered that these girls are afar from fathers and brothers, that they are ignorant of the gallantries . . . of city gentlemen, that they are young, guileless and confiding, it may be imagined that much unhappiness—to use the gentlest term—is the result."[23] Some mill hands undoubtedly concealed pregnancies by leaving town before their condition became evident. Others took drastic measures. Abortions were performed in Lowell at least as early as 1837; their availability was such an open secret that women came from distant rural areas specifically for that purpose. By 1847, "French Periodical Pills" were advertised in the newspaper with the promise that they would "regulate the system and produce the monthly turns of females without hazarding life." A Mrs. F. Morrill advertised in 1850 that in addition to selling "monthly medicines" she provided medical treatment for women at her downtown office until 9:00 p.m. or at her home afterward. "A sure Cure Warranted in all Suppressions, Irregularities, &c.," she declared.[24]

Mill managers, when faced with demands for shorter working hours in the 1830s and 1840s, warned that an increase in evening leisure time would encourage immorality among the workers. The very idea of a ten-hour working day, reported a clergyman, "was held in the utmost scorn; for the time that would be saved to the operative, would be spent in vanity and wickedness—in loafing around town, and cutting up all manner of shines!"[25]

THE CLERKS OF THE GREAT METROPOLIS

Men were the more frequent targets of such objections. During the ten-hour movements of the 1830s and 1840s, when workers in the northeastern states organized to demand shorter hours, arguments over quitting times often focused on the problem of leisure. Did working people need free evenings to maintain their health, care for their families, worship God, improve their minds, and perform their duties as citizens of the republic? Or, as employers claimed, did they need to be kept busy to save them from their own sinful impulses? Labor activists, some of them linked with the Washingtonian temperance movement, insisted that drinking would decrease if workers were not so exhausted at night. White-collar workers stood apart from the labor struggles of the era, but their own fights for shorter hours were framed in the same terms.[26]

If in the nineteenth-century imagination the Lowell mill was the archetypal urban destination of the adventurous young woman, the New York shop or office was the destination of the ambitious young man. Young clerks figured prominently in antebellum advice literature and fiction, working hard by day and struggling by night to maintain their moral bearings amid the city's temptations. One sentimental tract published by the American Sunday School Union serves as an example. Reuben Kent, an innocent New England boy, takes a clerkship in New York. A dissipated young man at his boardinghouse introduces him to the pleasures of the nocturnal city, including gambling. Fortunately, Reuben sees the error of his ways and stops short of the fatal step into the theater. Readers at the time were undoubtedly familiar with the widely publicized true story of Richard P. Robinson, a Connecticut youth who became a store clerk in New York in the mid-1830s. Though working for an eminently respectable merchant, Robinson fell into vice in his leisure hours and finally murdered a prostitute with a hatchet.[27]

The outpouring of warnings about the nighttime city came amid some of the fastest urban growth in American history, when the population of American cities grew from a mere 500,000 in 1830 to 3.8 million in 1860. New York City quadrupled in population from 202,589 to 813,669 in these years. As the dominant port for European imports and a center for trade with the North American interior after the completion of the Erie Canal in 1825, New York eclipsed its rival ports of Boston, Philadelphia, and Baltimore. It surpassed Philadelphia by the 1830s as the nation's financial center. By midcentury, the city's developed area had expanded northward to what is now Midtown from the original center at the foot of Manhattan Island. Further development was limited by the slowness of land transport, but ferries allowed the rise of satellite cities flanking lower Manhattan on either side. Ferries shuttled constantly across the East River to Brooklyn and over the Hudson to Jersey City.[28]

Brooklyn and Jersey City initially developed as bedroom communities for affluent New Yorkers, but they soon gained economic importance as well. Brooklyn's East River waterfront in the 1850s and 1860s was lined by warehouses, grain elevators, sugar refineries, oil refineries, iron works, white lead works, distilleries, and other industries that had trouble finding space in Manhattan. Jersey City's growth was powered by industry and transportation. The Colgate Soap Company moved there from New York in 1847, joined over the next twenty-five years by small industries producing iron, oil, tobacco, bricks, wagons, and pottery, as well as by stockyards and an abattoir. Jersey City in the 1840s and 1850s was the terminus of the Morris Canal and of the Cunard transatlantic lines; it became a railroad hub in the 1850s and 1860s as railroads built their New York depots there on the western shore of the Hudson. By 1870, Jersey City was America's seventeenth largest city, with a population of 82,546; Brooklyn was the third largest, with 396,099 people.[29]

These early New York suburbs had a homegrown white-collar workforce as well as an affluent class of commuters. "Clerks"—a vague term that could be applied to almost any white-collar employees—found work in the railroads and industrial firms, where they managed company operations, copied documents, and kept the accounts. Clerks also worked the counters and kept the accounts of local retail stores. Brooklyn's Fulton Street was lined as early as 1834 with shops catering to the commuters who rode the Fulton ferry to New York; by the mid-1850s, this retail district had been rebuilt with large brick dry goods stores, banks, and other businesses. The Fulton Street district in the 1860s stretched far from the ferry, past the City Hall; retailers also located on Myrtle Avenue, Atlantic Avenue, and Columbia Street. Smaller, poorer Jersey City had a substantial retail trade, much of it on Montgomery Street. More than four hundred men calling themselves shopkeepers were listed in the 1860 census for Jersey City, and more than two hundred teenage boys were listed as clerks.[30]

Manhattan, of course, was the great center of all buying, selling, money

Scene in a Fashionable Shoe Store, on Broadway, New York.

FIGURE 3. Men serving women. Retail clerks complained of their long working hours in the mid-nineteenth century. They and their supporters in the "early closing" movement blamed frivolous women for shopping in the evenings. Here men are shown deferentially attending to women's needs. "Scene in a Fashionable Shoe Store, Broadway, New York," *Sporting Times*, January 15, 1870. Courtesy of American Antiquarian Society.

handling, and paper shuffling—occupying the days (and perhaps crushing the souls) of legions of real-life Bartlebys. The 1840 US Census counted 3,620 dry goods stores, groceries, and other retail shops in New York, along with 417 commercial houses involved in foreign trade and 918 commission houses that wholesaled the imports. By 1855, nearly 14,000 clerks of all kinds lived in New York, outnumbered as an occupational category only by servants and laborers.[31] Their major areas of employment extended no more than a mile from City Hall Park, the triangle at the intersection of Broadway and Chatham Streets (now Park Row). The retail blocks of Broadway stretched north to Houston Street from A. T. Stewart's "Marble Palace" dry goods emporium at the northwest corner of the park. South of the park was the booming financial district of Wall and Pearl Streets. A printing house district of newspapers and publishers abutted the park on the east and extended a few blocks toward the East River, merging into the shipping and wholesaling district along the waterfront. Retailing spread north in the 1860s along Broadway and Sixth

Avenue, while less prestigious merchants catered to a working-class clientele in the Bowery or on Chatham Street.[32]

In the mid-nineteenth century clerking was still regarded as an apprenticeship for future merchants, but it was undergoing a major transition. Like apprentices, clerks had once lived with the families of their employers and had been subject to their supervision, but in the antebellum decades merchants and clerks lived separate lives. As employers enjoyed domestic privacy in homes safely distanced from the commercial districts, clerks and other employees were left to find their own housing. For young men who had come to the city on their own, this typically meant a boardinghouse.[33]

Boardinghouses at their worst could be cold, shabby, cheerless places. Shared bedrooms were crowded, common rooms small and poorly lit, meals a silent assembly of strangers forking down cabbage and leathery meat. Everyone agreed the boardinghouse was a sorry surrogate for a real home. The atmosphere was said to be so unwelcoming that it practically forced young men to spend their evenings elsewhere. Contemporary descriptions emphasize the loneliness that overtook a young boarder newly arrived from the countryside, the lack of family supervision that left him open to temptation after quitting time. Recalling his early days of clerking in Brooklyn in 1851, Stephen Griswold used language that could easily have been taken from antebellum advice literature:

> For a time all went well; the hurry and bustle of the city, all so strange and fascinating to me; the new occupation, calling into play an entirely different line of thought; the new surroundings, all combined to ward off any feeling of loneliness or homesickness. A few weeks of this, however, sufficed to wear away the novelty, and a full sense of my solitary conditions rushed over me. . . . I began to look around for companions, or at least for some place where I could spend my evenings, when the time dragged most heavily.[34]

A walk through the nighttime city relieved boredom only at great psychic and moral cost, claimed the authors of midcentury advice literature. Alone in streets filled with strangers, the country youth grew aware that he was free from public scrutiny, free to do almost anything he wanted without consequences. Moralists warned that the youth was bombarded by new stimuli and enticed by new opportunities for amusement. The Reverend John Todd wrote in 1841 that the spirit of urban night was wholly at odds with the spirit of rural night. In the countryside the voice of God is heard in the silent darkness. "But when you come into the great city, every thing tends to shut God out of the human mind. . . . If you walk the streets, you see nothing, hear nothing, think of nothing, but what man is doing. . . . The glitter and blaze and thunder of night make the power of man still more perceptible."[35]

Stephen Griswold recalled being saved from such temptations as he strolled through Brooklyn's streets by a notice posted for the Reverend Henry Ward Beecher's Plymouth Church. He began attending services there and soon met an old friend from his native Connecticut who drew him deeper into evening church activities. Griswold joined a Bible class and eventually became a church usher. "From this time on I had no reason to complain of any lack of social life." Many other young men similarly found evening activities and social contacts within a Christian community. The diary of Michael Floy Jr., a pious young New Yorker who lived and worked with his family in the 1830s, records numerous evenings spent at prayer meetings or in charitable activities. Floy took food to the homes of poor families, and he taught evening classes at the African Adult School. On other evenings he would stay home studying theology or mathematics or writing in his diary. He believed these activities improved the time that might otherwise be wasted on frivolity or sin. "'What is the use of continually poring over dry mathematics?' father and James have often said. I get no money by it and would starve at it, but I have been kept while a careless youth from bad company, from frequenting Taverns, Theatres, or houses of ill fame." Diary keeping itself encouraged a reflection on daily challenges that gave a young man a sense of control over his life. Writing letters home helped the individual feel connected to his family and its values.[36]

Merchants who worried about their diminishing influence on employees' evening hours promoted religious and educational activities for young men. The Young Men's Christian Association (YMCA), upon opening at the New York City Lyceum in 1852, was heralded as an effort "to ralley around the young stranger and save him from the snares of this wicked city." Mercantile libraries drew young men in Manhattan, Brooklyn, and many other American cities, while lyceums offered lecture courses. In the middle-class setting of the library, aspiring clerks enjoyed comfortable evenings surrounded by their peers, improving their taste and developing useful knowledge. Young men's moral and social reform organizations proliferated, as did various singing, debating, and athletic societies, often supported by wealthy donors. Even activities that were purely recreational, rather than aimed at self-improvement, were increasingly valued for keeping young men out of trouble. Reading, asserted the *New York Times* editors in 1853, was an alternative to "a dangerous study of the book of city life, whose illuminated pages are the theatre, drinking-saloon, and gaming parlor; and whose obscene ones are stolen from . . . the 'Devil's Bible.'" Such were the hopes for saving the lonely young men of the boardinghouses.[37]

Boardinghouses were not always the grim impersonal spaces that critics made them out to be. The young men who lived there formed quick friendships with each other, as many young women did in Lowell. Like college students today who complain about dorm life while enjoying its social opportunities, male boarders passed many an evening chatting and smoking out on the front steps. Tiring of this or seeking shelter from the weather, small groups would head for a nearby tavern. Of course, this kind of social life disturbed moralists just as much as the dangers of isolation. An unprotected youth might be tempted to enter a saloon alone, but he would find himself just as surely on the road to ruin if he went there with his fellow boarders. Thus introduced to the comforts and conviviality of the barroom, warned an 1849 advice manual, the "unsuspecting youth repeats his visits, finds out other similar resorts, and finally is in the habit of being abroad every night and is found at his boarding-house only for his meals and late lodgings. He visits all the distinguished saloons, refectories, bowling alleys, theatres, gambling-hells, and other abodes of affiliated infamy." Some boardinghouse keepers cramped the style of would-be sporting men by locking up early, but these were regarded as cranks. Unlike in Lowell, in New York a young boarder did not have to worry about a curfew. "I could enter at what hour I pleased—subservient to no control after the business of the day was over," recalled the notorious Richard Robinson.[38]

New York clerks could go out for a night on the town confident that their employers would never know. Affluent families' move to remote neighborhoods left young single men to their own devices in the older parts of the city south of Canal Street. Observers were struck by the youthfulness of the daytime crowds on Manhattan's streets; when the family men went home at night, the imbalance downtown was even more extreme. The 1855 New York census found that young adults between twenty and thirty-five made up 48 percent of the population of the Second Ward and 44 percent of the Third Ward—the two wards immediately north of the Wall Street financial district. Many clerks lived in the boardinghouse district west of lower Broadway, much of which fell within the Third Ward. Though a slight majority of all New Yorkers were female, the majority of downtown residents were male.[39]

In the young man's world of lower Manhattan, workplaces, boardinghouses, and nightspots were often just a few blocks apart. During the day this allowed a convenient return to the boardinghouse for a midday meal. At night it meant that a short stroll could take a man across the moral divide from virtue to vice. Men who might shudder to enter the rough bawdy houses east of Broadway in the 1830s and 1840s could find more appealing bordellos close at hand on Church, Chapel, or Thomas Street, just north of the Third Ward. A wide range of businesses clustered in the blocks immediately surrounding City Hall Park, including the famous Park Theater, the Astor House hotel, Barnum's Museum, and various saloons and oyster houses. In the late 1840s and 1850s, A. T. Stewart operated his "Marble Palace" dry goods store across the street from the park's northwest corner, employing some two hundred clerks practically next door to New York's largest concentrations of brothels.[40]

It was easy in these circumstances for a young man to live a double life—by day politely folding silk for lady shoppers, by night staggering from barroom to gambling hall. A broad subculture of young clerks embraced the life of the sporting man, proudly advertising their sympathies with the brothel murderer Richard Robinson in 1836 by harassing his accusers, dressing like him, and turning out by the thousands to catch a glimpse of his trial. Polite society turned a blind eye while vice was practiced more and more openly. Walt Whitman claimed in the 1850s that most young men in Brooklyn and New York were regular brothel customers who "feel perfectly at home in the most infamous places."[41]

Thus the daily end of work marked a radical shift of power. Clerks in the New York area were turned loose to spend their evenings however they wanted while merchants washed their hands of direct responsibility. "The sorts of amusements to which they are most attached, are not properly scrutinized," complained an editor for the *Brooklyn Eagle* in 1842. "True, numbers may be found in libraries, reading-rooms, and a few other places of a like character, set apart for their use—but what of the mass? Are not their enjoyments, to say the least, of a more questionable character?"[42] Closing time was now freighted with greater importance than before work and leisure became so clearly separated. What would happen if businesses closed earlier? Could clerks be trusted with more free time? These issues came to a head during efforts of dry goods clerks to promote "early closing" agreements among merchants.

Though clerks in many business houses, banks, and law offices ended work in the late afternoon, retail clerks worked unusually long hours in antebellum New York. With the coming of gaslight in the 1820s, shops stayed open later in lower Manhattan than in the sleepy, dark streets of Philadelphia or Boston. A writer for the *Working Man's Advocate* observed in 1830: "A large proportion of the stores in this city [New York] are kept open from seven in the morning till ten at night; fifteen hours; and some one must be behind the counter to attend to business all that time. This appears to us very unnecessary, and very oppressive on this class of citizens." Clerks resented the encroachment on their precious free time. "Monday our hour of closing . . . changed to 9, a very unpleasant arrangement, as it deprives me of a valuable leisure hour every evening," wrote Henry Patterson, a New York dry goods clerk, in the early 1840s. According to his diary, at least, Patterson preferred to spend his evenings attending lectures, participating in his debate club, and reading at the New York Mercantile Library.[43]

In contrast to the powerful mill owners of Lowell, shopkeepers in New York had limited control over their own hours. A merchant who decided to close early would lose business to his competitors. Shopkeepers therefore pushed themselves and their employees to the brink of exhaustion to draw customers as late as 10:00 or even 11:00 p.m. Any agreements to close the shops earlier could be undermined by one or two holdouts, as reportedly occurred among the storekeepers in New York's Chatham Street in the 1820s. Philadelphia's retail merchants attempted to set standard closing hours in 1835 with the stated goal of "granting our young men an opportunity of improving their minds by reading, or of taking such recreation as they may stand in need of in common with others of their age and circumstances." The merchants themselves would also benefit from such an effort to set limits on competition, at least in theory. Clerks who joined the Philadelphia movement requested a twelve-hour day ending at 7:00 p.m., but merchants proved unable even to stick to a 9:00 p.m. closing. The movement fell apart as some grocers persisted in staying open late. An early closing effort in Brooklyn in 1842 collapsed for similar reasons. Nonetheless, early closing movements perennially reappeared in American cities for the rest of the nineteenth century. Opening hours were oddly uncontroversial; rather than shortening the workday by shifting it later, out of phase with the usual daily work cycle, merchants and clerks focused almost exclusively on the proper quitting time.[44]

The early closing movement in New York, Brooklyn, Jersey City, and more distant urban centers drew support from a broad coalition. The initiative in the 1840s and 1850s often came from merchants, cheered on by politicians, clergymen, journalists, and grateful clerks. At an 1863 mass meeting in Brooklyn, the Reverend T. L. Cuyler blamed late closings for immorality among the clerks. By the time the shops closed, "Where are they to go? Their evening is gone. The religious meeting is closed; the lecture and the concert are nearly over—all proper means of recreation are closed for the evening. But the lights are shining in the grog-shop, in the place where the dice box is rattling and in the billiard saloon, to tempt the wearied clerk who finds no other means of recreation open to him." Fervently denying the existence of labor-management tension, early closing advocates sought a united front against a common enemy: the customer. They suggested that the real issue was whether late shoppers would exploit the men behind the counters. Frivolous ladies were said to be shopping at night when tired clerks and merchants longed to be off improving their minds. In speeches, resolutions, and editorials, early closing advocates urged ladies to change their habits in support of the movement. A procession of four thousand clerks from New York, Brooklyn, and Jersey City marched through the Williamsburg neighborhood one night in May 1863, carrying torches and illuminated signs with mottoes such as "Ladies, do your shopping by daylight."[45]

Criticizing recreational shopping was an inspired tactic. It allowed the movement both to shift blame to a female scapegoat and to distract attention from the troublesome fact that early closing allowed clerks more free time. Americans in the mid-nineteenth century were struggling to sort out a tangle of shifting beliefs about the value of leisure. Brooklyn's own Reverend Beecher was among the most prominent and most confused American intellectuals to engage the issue. In the early 1840s he had preached a traditional Protestant message about the redemptive value of work and the dangers of idleness, disseminating this view by publishing *Seven Lectures to Young Men* in 1844. Beecher continued to espouse this ascetic code when speaking to working-class audiences, but in the 1850s and 1860s he also celebrated the restorative benefits of leisure. He warned that young men stunted their personal growth by working too hard.[46]

What was a young clerk to do? As an aspiring member of the middle class, should he have his evenings free to cultivate his intellect? Or would idleness devolve into sloth, failure, and sin? The clerks and their sympathizers took the first position, of course. "All other classes in the community are furnished, both with the means and TIME for mental culture," complained W. B. Jones, an activist Brooklyn clerk, in 1850. "We stand alone, as though our class had forfeited all claims upon libraries and lectures."[47] The clerks often spoke of their leisure as if it would be devoted solely to the hard work of self-improvement. The editors of the *Brooklyn Eagle* argued more generously that recreation and socializing made life worth living, and that a grim regimen of fifteen- or sixteen-hour days denied clerks that right. The editors acknowledged that some clerks might spend their evenings in immoral pursuits, but they denied that this was reason enough to oppose the early closing movement. "If the objection be valid, it ought not to stop with clerks, but should be applied to other classes and pursuits, and some method be devised for keeping them at work in their offices and shops lest they should acquire bad habits!" they added sarcastically.[48]

Beecher threw his support to the early closing movement in the 1850s and 1860s. In 1864, in front of fifteen thousand people at an early closing rally in New York's Union Square, he asserted that "every human being should have the opportunity of enjoying those social pleasures which are the exclusive right of no particular class." This opportunity should not be deferred until a young man reached the age of thirty, nor should the decent majority of young clerks be denied free time just because an immoral minority might abuse the privilege. Beecher and *New York Tribune* editor Horace Greeley told the cheering crowd that work and leisure could be balanced painlessly; if shops closed early, the same business would be done more efficiently in a shorter time without wasting gaslight.[49]

Unfortunately for the clerks, the growing acceptance of recreation was accompanied by a reevaluation of evening shopping. In the 1840s and 1850s, the *Brooklyn Eagle* had contended that ladies thoughtlessly postponed their shopping until they could drag their husbands to look at fripperies. Defenders of evening shopping argued that working men and women had no other opportunity to shop (in contrast with the lazy aristocrats who whiled away their days buying things). Early closing, therefore, would privilege white-collar clerks and affluent customers at the expense of the poor.[50] Without abandoning the necessity argument, shopping advocates after 1860 often added that evening shopping was a perfectly acceptable, even desirable, practice in a modern city. The editors of the *Eagle* came to agree. They dismissed the call for Brooklyn to follow the example of those in "a certain rural village in Connecticut called Hartford," where the early closing movement remained successful. "The habits of the people in large cities differ from the primitive customs of the rural population, who rise when the rooster crows, and go to bed with the pullets. We rise later and sit up later; all our leisure time is in the evening, for recreation or shopping." No longer just a feminine frivolity, evening shopping was reimagined as an essential characteristic of urban life to which ladies were entitled. In 1886, a Mrs. Jones of Baltimore wrote to the *Journal of United Labor* to complain that the labor movement had forced stores to close early while saloons stayed open late. "I am opposed to women being deprived their privileges while men are not. Now, many women cannot get out in daytime to do their little shopping, and have to wait till their husbands come home to stay with the children."[51]

Dry goods clerks achieved intermittent success in a number of cities in the late nineteenth century, inspiring early closing efforts among other retail clerks as well as among tailors and barbers. These efforts were encouraged also by the eight-hour movements of the post–Civil War era, but now employers were more likely to resist the cause than to lead it. Frustrated clerks adopted such aggressive tactics as picketing or smashing display windows. Merchants who agreed to close early often reneged when competitors stole their customers by closing late. Hours in the 1870s and 1880s were as long as ever; on Saturdays and during busy seasons, exhausted dry goods clerks completed sixteen-hour stints at 11:00 or 12:00 at night.[52] In the final decades of the century, news items about closing hours noted the presence of women as clerks as well as shoppers. Saleswomen in New York reported being kept on their feet twelve to fourteen hours a day with only half-hour lunch breaks; some fainted or became sick with exhaustion. Sympathizers complained that the long hours were more than the weaker sex could bear. Calls for early closings emphasized physical well-being and personal pleasure instead of the need for manly self-improvement.[53]

Over the course of the nineteenth century, mainstream American opinion grew reconciled to the value of entertainment for young men, and somewhat more accepting of participation by young women.[54] Men felt less obliged to defend their evening activities as forms of intellectual or spiritual labor and more confident of their right to pleasure. Women gained some of the access to urban night once enjoyed by men. Groups like the YMCA and YWCA still tried to provide guidance to otherwise unsupervised young adults, but there was little hope of restoring the discipline once exerted by Lowell's boardinghouses. Not every form of urban entertainment was equally acceptable, of course. Throughout the nineteenth century, discussions of night drew sharp moral distinctions between early and late hours. Activities of the sort once smiled on by Lowell's manufacturers—church services, classes, lectures, and concerts—continued to end early and were almost universally accepted. Those that ended late remained disreputable, even long after the era when boardinghouse keepers enforced curfews. In the emerging schedule of nightly leisure, as in the schedule of daily work, time was minced into hours and fretted by the ticking of a clock.

CHAPTER 4

Recreations and Dissipations

Walt Whitman watched from an imaginary perch along Broadway in 1856 as an endless stream of New Yorkers passed by. Hour by hour, he noted the different classes and activities in the street, from the predawn rumbling of butchers' carts to the midnight journeys of hacks, the equivalent of taxicabs. The morning procession of pedestrians began about 5:00 a.m., in Whitman's telling, with "twos and threes, and soon full platoons, of the 'industrial regiments' . . . uniformed in brick-dusty shirts and overalls, battered hats, and shoes white or burnt with lime, armed with pick, spade, trowel or hod." As these men scattered to construction sites, shopgirls walked down Broadway toward the bookbinderies and tailor shops. "Mingling with them, and flocking closer, for it is now eight or nine in the morning, come the jaunty crew of the downtown clerks," whose fashionable clothes covered physiques weakened by fast living. Merchants and bankers strolled to their offices in midmorning. After 4:00 p.m., Whitman wrote, "the successive waves of the morning tide now begin to roll backward in an inverse order—merchants, brokers, lawyers, first; clerks next; shop-girls and laborers last."[1]

Nineteenth-century newspapers and magazines published many such attempts to understand the daily schedules of urban life, usually distinguishing people by wealth or occupation. A few writers tried to produce nocturnal counterparts to Whitman's essay. Their descriptions of night paid some attention to economic class, but more to morality, manners, and work ethics—personal attributes central to an emerging middle-class identity. One's job, after all, was not always obvious to an observer at night; by scrimping and saving on necessities, people of modest means could go out dressed as if they were much wealthier. How men and women chose to spend their leisure time, though, revealed whether they shared the values by which middle-class people distinguished themselves from the rest of the population: particularly self-discipline, bodily restraint, and recognition of the importance of work.[2] The specter of work the next morning loomed over every evening's pleasure, with the sole exception of Saturday night, "the poor man's holiday."[3] Would workers use the interlude for rest and "recreations" that restored their capacity for effective labor? Or would they squander it on "dissipations" that further depleted their energies? Would workers greet the dawn refreshed or bleary-eyed?

One of the most elaborate descriptions of urban night was an 1872 essay by the Reverend T. DeWitt Talmage, titled "After Midnight." Talmage had both the moral credentials of a Brooklyn clergyman and an interest in the seamy side of New York that eventually made him an easy target of satire. Talmage's essay divided night into four "watches," each with its distinct activities and moral character. The first of these three-hour periods begins at 6:00 p.m., as the city's business establishments gradually shut down. Workingmen trudge home to their families, and young clerks are turned loose to throng the streets. In the second watch, a few hours later, "all the places of amusement, good and bad, are in full tide. Lovers of art, catalogue in hand, stroll through the galleries and discuss the pictures. The ball-room is resplendent with the rich apparel of those who, on either side of the white, glistening boards, await the signal from the orchestra. The footlights of the theatre flash up; the bell rings and the curtain rises." The avenues are filled with thousands of pleasure seekers. The city in the first two watches is "beautiful and overwhelming."[4]

The character of night changes after midnight, Talmage continued. "The thunder of the city has rolled from the air. Slight sounds now cut the night with a distinctness that excites your attention. You hear the tinkling of the bell of the street-car in the far distance; the baying of the dog; the stamp of the horse in the adjoining street; the slamming of a saloon door; the hiccoughing of the inebriate; and the shriek of the steam-whistle five miles away." A few workingmen hurry home from night jobs, doctors and ministers are out caring for the sick and dying, but most of those still awake are up to no good. Robbers, burglars, and arsonists do their work at this time. Seemingly decent people indulge in vicious pleasures. Drunks stagger through the streets or pass out on the saloon floor; men throw away their fortunes on cards, then commit suicide. "Deed of darkness, unfit for sunlight or early evening hour! Let it come forth only when most of the city sleeps, in the third watch of the night!" In the fourth and final watch, as revelry subsides with the approach of dawn, working people emerge from their homes to begin anew the daily cycle of toil.[5]

Talmage was a showman given to flamboyant language and gestures, but his chronology of night contains useful insights. Even as entertainment options expanded through the mid-nineteenth century, commentators confirmed that there was an intricate schedule of evening activity, with moral overtones changing by the hour. Nightlife varied from the energy of New York to the emptiness of New Bedford, from the staid refinement of Philadelphia or New Haven to the crude provinciality of Louisville or Terre Haute.[6] Yet everywhere, it seems, respectable entertainments ended early and people who aspired to respectability went home to sleep.

FIRST WATCH

What Talmage called the first watch, from 6:00 to 9:00 p.m., was the least dissociated from day. As described in the previous chapter, both rural and urban Americans worked through early evening. Quitting time for individual urban workers marked an abrupt transition from work to leisure, but a more complicated transition was visible each evening in the collective life of the city. Staggered quitting times made possible by artificial light meant that some people were out enjoying themselves long before others, such as shop clerks, could make their way home. Anyone might be out walking with utmost propriety, whether to return home, breathe some fresh air, buy necessities, or shop for pleasure. Early evening crowds on a commercial thoroughfare such as Broadway were what one writer called "a heterogeneous mass," representing all types within an urban population that was "promiscuously intermingled." Streets were at their most respectable as long as daylight lingered. After dusk, women found it wise to have a male escort. "In a city, or in any lonely place, a lady must avoid being alone after nightfall, if possible. It exposes her, not only to insult, but often to positive danger," cautioned an 1869 etiquette manual. Still, pairs of ladies could safely stroll the larger gaslit streets, and some modestly cloaked women ventured out alone on errands and social visits.[7]

Church, of course, was a laudable destination for women and men alike. As an unmarried schoolteacher in Washington, DC, in 1871, Jane Briggs Smith attended numerous evening church events, sometimes traveling there alone. The 1894 diary of Mary Bainerd Smith (apparently no relation) recorded her walking to a Philadelphia neighborhood church several evenings a week: along with her mother or sisters, she faithfully attended prayer meetings, evening worship services, missionary meetings, and other events. Church activities were among the few respectable public gatherings in small towns in the first half of the century. Every evening in Cincinnati, claimed Fanny Trollope in 1832, "brings throngs of the young and beautiful to the chapels and meeting-houses, all dressed with care, and sometimes with great pretension; it is there that all display is made, and all fashionable distinction sought. The proportion of gentlemen attending these evening meetings is very small, but often, as might be expected, a sprinkling of smart young clerks make this sedulous display of ribbons and ringlets intelligible and natural." Even in the largest cities, revivals, vestry meetings, choir practices, and the like created a nightly pulse of pedestrian and vehicular traffic through the streets. "It is pleasant to encounter groups making their way homeward through the quiet street from week day religious services," wrote one New Yorker in 1852. "It is a practical recognition that there is an unseen world, and that things eternal are worth thinking of on other days than the Sabbath."[8]

In practical recognition of the world of work, church events tended to end early. The virtues cultivated through religious devotion were seen in Protestant churches as fitting seamlessly into the life of a devoted worker; neither the values nor the hours were in conflict, and membership helped newcomers make advantageous connections. A twenty-four-year-old Bridgeport law clerk, George Watson Cole, observed on joining the choir of the Park Street Congregational Church in 1875 that "it will tend to help me in getting acquainted here and is a capital opportunity." Choir practices ran from 7:30 until 9:00 or 9:30, leaving Cole time for reading, cribbage, or other pursuits before bed. Cole also attended Sunday evening worship services, which ended in time for him to be back in his boardinghouse by 9:00. Carolyn Healey, the teenage daughter of a Boston merchant, infuriated her father one October night in 1839 by returning from a Sunday school teachers' meeting at the late hour of 10:05. "I thought it very hard, and burst into tears for I knew that if I had been to a party or the Theatre or—a concert he would not have complained."[9]

Meetings of voluntary associations also drew people out in the early evening. Numerous benevolent organizations sprouted up in the antebellum era—to free the slaves, shelter widows and orphans, reform prostitutes, dispatch missionaries to the heathen, and resist the scourge of demon rum. These benevolent efforts were often extensions of church work, or at least inspired by evangelical Protestantism. Michael Floy Jr., a devout young Methodist in 1830s New York, interspersed his evening attendance at prayer meetings with charitable visits to poor families and with service as a teacher at an "African adult school in Allen-Street."[10] Christian faith mingled with secular aspirations in the evening schools and other cultural organizations that emerged in the

FIGURE 4. The proper way to spend an evening. This illustration was used in the Reverend T. DeWitt Talmage's 1879 indictment of immoral recreation, *The Masque Torn Off*, to show what people should be doing instead: conversing, reading, playing music, and enjoying other parlor activities that could be shared with family members and friends of both sexes. The tube descending from the light fixture carries illuminating gas to the table lamp. T. DeWitt Talmage, *The Masque Torn Off* (Chicago: J. Fairbanks, 1879).

antebellum era—including literary societies, lyceums, musical societies, Young Men's Christian Associations, and lending libraries. Supporters of these associations claimed that mind and soul would be uplifted together. "The lyceum is, in truth, a week-day church a little humanized and enlarged," asserted *Putnam's Monthly Magazine* in 1857. Lyceum lecture schedules featured a range of scientists, reformers, travelers, authors, and an especially large contingent of Protestant clergymen. Notwithstanding their wide variety, the lectures tended to be rooted in the oratorical style of a Congregational sermon, increasingly inflected through the late 1840s and 1850s by stage performance. Combining moral instruction with entertainment, they covered nonsectarian, apolitical topics suitable for a wide audience. Such was their moral tone that pairs of unescorted ladies could attend evening lectures without concern for their reputation. Lectures did not succeed in luring mass audiences away from the theater, as some promoters had hoped; indeed, they drew many people who would otherwise have gone to evening prayer meetings. Still, the growth of all these cultural organizations provided wholesome recreational options for city people, and an evening spent at a meeting, school, lecture, or library ended in plenty of time for a good night's sleep. As late as 1876, evening library hours in New York ended at 7:00, 9:00, or 10:00, depending on the library.[11]

The regimen of the workday prevented most urban Americans from socializing until after their evening meal, but then they were free to pay calls on friends and relatives. The compact size of midcentury cities made frequent short visits easier for the majority of people who could afford to travel only by foot. An extraordinary illustration of this practice is the 1849 diary of a young Philadelphia bookkeeper, Nathan S. Beekley, describing a year filled with evening visits to the homes of friends and romantic interests.

Beekley's early life could almost be the prototype for a Horatio Alger novel. He had grown up in poverty in rural Chester County, the son of a shoemaker who had to rely on public assistance to pay for his children's schooling. Leaving the family home in East Nantmeal township, Nathan Beekley had moved a few miles east to the growing manufacturing town of Norristown, where his uncle, the Reverend Nathan Stem, was Rector of St. John's Episcopal Church. Beekley found work setting type for a printer, probably either in Norristown or in Pittsburgh, before resolving in his early twenties to try a new career in Philadelphia, just a short train ride from Norristown. On January 4, 1849, he wrote, he "was introduced into the Counting House of Messrs. Reeves, Buck & Co., where by diligence and attention to business I hope to remain." He did indeed remain, hunched over the ledgers as a monotonous succession of days, months, and years dragged by in the North Water Street offices of the iron merchants.[12]

During the year covered by his diary, Beekley lived for his evening amusements. Unlike the friendless youths of Victorian advice literature, he built a large social network of friends and relatives rooted in his Norristown and Chester County connections and to a lesser extent in his own boardinghouse. That boardinghouse was conveniently located at Franklin Square about half a mile west of his employers' offices. He walked to work and to anywhere else he needed to go, like most other Philadelphians at a time when the only alternatives were the costly omnibus or carriage. Beekley enjoyed the friendship of several fellow boarders but preferred not to spend his evenings at home unless it was raining or snowing; on those nights he would read, sit by the fire, play checkers, and bemoan the leaks in his only pair of boots.[13]

Pleasant evenings found him out and about. Eager to find a wife and personally devout, Beekley often escorted young ladies to and from evening church services. He passed many evenings at the home of the "Misses West," five sisters in their teens and twenties who lived in the North Mulberry Ward just west of Franklin Square and who had friends in Norristown. Also in the neighborhood, on the South Mulberry Ward side of Sassafras Street, was the home of Elizabeth and Sophia Snowden, aged twenty-one and nineteen. These were the daughters of Thomas Snowden, a prosperous manufacturer of surgical instruments who presided over Philadelphia's Common Council; they would have been far out of Beekley's league except for their fortunate kinship to some of his friends from Uncle Nathan's church. Beekley that winter was enamored of yet another woman with Norristown connections, cryptically recorded as Miss K. P. R——y: "one of the few, <u>very</u> few young ladies with whom an evening can be spent most agreeably without either music or dancing." After he visited "this model of a lady" on several winter evenings including Valentine's Day, a decisive rebuff forced him to turn his attentions elsewhere. He became the devoted escort of Fanny Clement, a librarian in her early thirties who also lived in South Mulberry Ward. Fanny introduced him to a Miss L. Davids, whom Beekley regarded with the critical eye that he affected toward women in the privacy of his diary. "Like most young ladies, she was very talkative and exceedingly trifling and nonsensical, having but one idea—fun," he wrote. His most frequent visits through the rest of the year were to Elizabeth Snowden and Fanny Clement, but Beekley managed to work the fun-loving Miss Davids and several others into the rotation while keeping an eye open for new prospects. His office, his social circle, his lodgings, and his favorite entertainment spots were clustered tightly enough for him to drift easily from place to place. He began one November evening by spending a little time at the Franklin Institute (five blocks south of Franklin Square), then he called on Fanny Clement, who was out of town; he "then went with friend Perry to see Miss Warren but she was also out—then we called on the Wests and they were out; so we gave up on the ladies as being beyond our reach." Despite all his efforts, he glumly concluded in his December 31 entry that he had finished the year "no nearer being married."[14]

Dropping by someone's house unannounced might seem a breach of Victorian formality, but the practice among affluent city dwellers followed an elaborate etiquette. One of the most important rules was that <u>while gentlemen could call on ladies, ladies should never call on gentlemen</u>. Beekley was startled when some Norristown friends violated this rule one day by stopping by his boardinghouse; the three young ladies "represented themselves as <u>cousins</u>, thus excusing what would otherwise be considered a very improper proceeding." Calls even by good friends were received in the parlor, not in the more private zones of the home associated with work or care of the body. Proper parlors were equipped with a standard arsenal of armchairs, sofas, draperies, carpets, engravings, knick-knacks, bookcases, and a piano. Only a narrow range of activities was permissible in this space—conversation on light topics during the day; conversation, music, and perhaps parlor games in the evening. Deportment was tightly regulated. An 1869 attempt to codify the ritual of calling, a chapter in Frost's *Laws and By-Laws of American Society*, filled eight dense pages with ominous prohibitions and commandments including an important edict observed by Beekley: people without carriages should not make calls in wet weather when they might drip on the carpet.[15]

Some American women set days and times when they were "at home" to callers. If they did not, etiquette manuals advised callers to observe a proper schedule anyway: formal "morning calls" should not be made until after noon and should last no more than ten or twenty minutes. Evening visitors were supposed to arrive about 8:00 and leave no later than 10:00 to avoid interfering with the family's schedule of meals and sleep. As evening visits were said to imply familiarity, they were not supposed to be paid by casual acquaintances—a rule Beekley ignored on nights when his friend Perry tried introduce him to the elusive Kate Warren. This rule was one of many rooted in aristocratic England that were inappropriate to American middle-class life. "In this country, where almost every man has some business to occupy his day, the evening is the best time for paying calls," observed an 1860 etiquette manual. The distinction between the formal morning call and the less formal evening call was impossible to maintain; instead, morning calls were made and received mainly by women, while men paid visits of all sorts in the evening. The bounds of propriety could turn the call into a constrained performance. Many people dreaded the stilted conversation of the parlor. Men and women found it difficult to communicate across the gender divide because of their dissimilar interests and experiences. Yet good humor, warm friendship, and music could breathe life into the cold form of the social visit.[16]

Small parties developed spontaneously when several friends happened to drop by, or they could be planned with less formality than a large evening party. Dinner parties, once held only by the elite, were spreading to the middle class by the 1860s. They usually started about 5:00 p.m., after gentlemen had completed their day's work. Although "dinner" was customarily a hearty midday meal, fashionable hosts and hostesses observed the same time shift that made a 3:00 p.m. visit a "morning call." A few affected aristocratic disdain for early hours by starting their dinners as late as 8:00 p.m. (Even later meals called "suppers" were held during large, formal parties, often after midnight.) Dinner guests were expected to arrive punctually and to leave after two or three hours. Other sorts of small evening gatherings, called soirees, began about 8:00 p.m. and often ran until midnight or later. They featured conversation and music as their main activities. Wealthy hostesses were known to hold private concerts in which guests would listen to ensembles of professional musicians. Hostesses might vary the routine by focusing the soiree on whist, dancing, or parlor theatricals. Amateur theatricals ranged from simple charades to tableaux vivants and costumed dramas complete with makeup, improvised stages, curtains, scenery, and props. They came into vogue in the 1850s and 1860s just as the professional theater was improving its image and expanding its audience.[17]

THEATER, A BRIDGE INTO DEEP NIGHT

On evenings when he made no calls, Beekley could often be found in the theater or concert hall. The theater in the mid-nineteenth century was firmly established as "the principal place of public amusements" in American cities, *Putnam's Monthly Magazine* observed in 1854. Though constantly denounced by clergymen, it drew throngs every night but Sunday. Adopting gaslight early, enthusiastically, and extensively, theaters stood out as among the most visible places in the nocturnal cityscape, with large lamps and reflectors at the entrances and colorful illuminated signs above. One quasi theater in New York, P. T. Barnum's American Museum, was topped by a quicklime beacon that cast "a livid, ghastly glare for a mile up the street."[18]

Public performances in colonial cities had been sporadic, regarded with suspicion for their indecent plots, suggestive dialogue, immoral actors, and disorderly atmosphere. Plays were banned in most of the northern British colonies in the later 1700s in the belief that they undermined morality. To mollify the authorities of Newport, a 1762 performance of *Othello* was advertised not as a drama but as "moral dialogues." The actors cleverly promised to finish "at half past 10, in order that every spectator may go home at a sober hour, and reflect upon what he has seen, before he retires to rest." Similarly, plays in Providence and Boston in the 1790s were billed as "moral lectures" to evade laws against theatrical performances. Low attendance inhibited the growth of theater in pious New England even after such laws were abolished, yet theater spread into cities and towns throughout the United States in the early nineteenth century and thrived in New York and Philadelphia. Rising wages made attendance possible even for the working class.[19]

Evening performances in the early Republic might begin at any time during the first watch of the night, depending on the season, and run at least into the second watch, sometimes the third. Playbills and newspaper advertisements from the late 1700s and early 1800s show that theatrical performances typically began at 6:00 or 6:30 p.m. in autumn and winter when the sun set early. Theater managers or theatrical companies adjusted curtain times as daylight hours changed, and they occasionally printed announcements like this one on a March 1783 playbill in Baltimore: "On Account of the Length of the Days, the Doors, in Future, will be opened at six o'clock and [the performance will] begin precisely at seven." Performances in May, June, July, and August started at 7:00 or 7:30.[20] This seasonal variation persisted well into the nineteenth century. In 1829, curtain times at the theater in Providence ranged from 5:45 in January to 7:45 in August; in Baltimore they ranged from 6:30 in January to 8:00 in June.[21]

In a trend so gradual as to be barely perceptible, curtains rose later and later in the second quarter of the century. Theater managers were disinclined to start past 8:00 at any season, but they gradually extended their summer hours to performances in spring and fall.[22] Curtains in the 1840s rose at 7:00 or 7:30 even in the dark weeks around the winter solstice. On January 17, 1849, the night Beekley went to hear Madame Anna Bishop in *La Sfogato* at the Walnut Street Theatre, the sun had been down for two hours before the curtain rose on the opening comedy. By 1860, the announcement of winter hours at Wallack's Theatre in New York seemed no more than a gesture at tradition: starting on December 17 the curtain would rise at 7:30 instead of 8:00! Even that minuscule adjustment was abolished in major cities by the mid-1870s, when performances of all sorts—lectures, comedies, operas, and "Parisian Can-Can Dancers"—began at 8:00 p.m. year-round.[23]

Winter's early hours disappeared so gradually over so many years that the process cannot be explained by the introduction of gas lighting in theaters. The Chestnut Street Theatre in Philadelphia introduced gas in late 1816, the first American theater to do so. Its playbills and advertisements bragged about the innovation, but its curtains still rose at 6:15 for the next several winters, only fifteen minutes later than in previous years; more than thirty years afterward, the theater was still starting an hour earlier in winter than in summer. There was no reason to expect anything different. Unlike mills, shops, and other interiors, theaters did not rely at all on natural daylight, which barely penetrated the performance hall anyway. Nor were early winter hours any use in helping people travel through unlit streets: a walk to a northern theater in January was just as dark at 6:00 as it was at 8:00. Seasonal adjustments in performance time make better sense as ways to conform to working schedules. Theater managers had to adopt later hours in June and July to avoid conflicting with the common habit of working until sunset. Even a curtain time of 7:30 or 8:00 around the summer solstice meant starting during twilight, which could discourage attendance. Theater impresario Sol Smith recalled what happened when he tried to hold the opening ceremony for a new theater in St. Louis on July 3, 1837:

The Prize Address was delivered ... to an audience of about ten people, who thought it worth their while to go to the theatre in time to see and hear the beginning of the performance. It being summer-time, eight o'clock came only about half an hour after sun-setting; so, as nobody in St. Louis thinks of going to the theatre or any other amusement "before dark," it was all accident that there were a dozen or so of people (all strangers) present at the time advertised for the beginning of the exercises on this occasion—the opening of the first real theatre west of the Mississippi River.[24]

Before curtains began rising in unison in the 1870s, performance times on any given night might vary from one establishment to another depending on schedules of seasonal adjustment, types of venue, and sorts of performance. One May evening in 1852, performances in Baltimore were advertised to begin at 7:30, 8:00, and 8:30; on a December night in 1860, performances in New York were advertised for 7:15, 7:30, 7:45, and 8:00.[25] Performance times tended to be later in venues billing themselves as "gardens," some of which had ornamental grounds where visitors could stroll on pleasant evenings. Instrumental and vocal concerts were more likely than dramas to start late, often beginning at 8:00 even in late autumn or winter when most dramatic performances began at 7:00 or 7:30. Blackface minstrel shows, promoted as respectable "concerts" during their height of popularity in the 1840s and 1850s, also started at 8:00 p.m. Opening times for Italian operas varied. When the Garcia opera company made its celebrated New York debut on November 29, 1825, in *The Barber of Seville*, the Park Theatre management announced that the curtain would rise at 8:00 p.m.—an hour nearly unheard-of for November theatrical performances in the 1820s. Opera historians agree that this was part of an effort to promote Italian opera as an elite cultural form, setting it apart from the more familiar English musicals (called "operas" at this time) and from translated and adapted versions of Italian operas.

Combined with high ticket prices and exhortations in the press for refined behavior, the late hour signaled that foreign-language opera was for a special audience—those with fashionable schedules and lots of money. Ladies and gentlemen of good breeding would not have to mingle with the plebeians who infested the theater on other nights. Such an elitist marketing formula could not succeed for long in mid-nineteenth-century America. New York theater managers William Niblo and Ferdinand Palmo sought in the 1840s to draw somewhat broader audiences and scheduled performances of French and Italian opera accordingly: at 7:00 or 7:30. Differences in starting time faded

FIGURE 5. Going to the opera. Foreign-language opera gained popularity as an elite entertainment in the mid-nineteenth century. In contrast to the theater, opera deliberately excluded rowdy behavior in order to provide a refined environment for affluent women and men. Nonetheless, women attending such events were carefully protected by men from contact with the physical and social dangers of the nocturnal street. "A Wet Night at the Opera," *Sporting Times,* May 9, 1868. Courtesy of American Antiquarian Society.

away in the 1850s, 1860s, and 1870s as performance halls moved toward the standard 8:00 p.m. opening.[26]

For ordinary city people to accept regular performance times of 8:00 required changes both in their daily schedules and in the nature of the performance. As noted in the previous chapter, the change from working by the sun to working by the clock was part of a gradual trend toward a precise, uniform quitting time regardless of the season. City dwellers—especially the growing number of middle-class people who worked indoors—became free to go to the theater at about the same time year-round. Further, the modern schedule of urban work by gaslight was shifted later than the practice of working by daylight that had traditionally prevailed in the city and that persisted in the countryside. As fewer city people rose at the crack of dawn, fewer felt pressured to go to bed early.[27] The experience of theatergoing in the mid-nineteenth century was also being transformed by changes in class culture and heightened social conflict. Theaters, once tenuously controlled meeting places of all sorts of people, were increasingly differentiated by class and cleansed of their morally controversial qualities. Time at the theater, formerly unpredictable and interactive, was remade into a standardized package of consumption.

The moral tone of plays and actors continued to trouble many Americans through the mid-nineteenth century. Theater was "a school of vice," a writer for the *New York Evangelist* claimed in 1844. "Who ever attended a theater without hearing language or sentiments, or allusions, which either do or ought to offend every pure and delicate mind, and tinge the cheek of modesty with the blush of shame?" Scenes of murder and other horrors were said to arouse morbid passions, leaving spectators with a craving for excitement that distracted them from the quieter work of self-improvement. Indeed, New York theaters seemed to be taunting the clergy by producing plays with salacious or diabolical titles such as *The Lady and the Devil, The Married Rake, The Savage and the Maiden, Rake's Progress, The Wife of Seven Husbands, The Devil to Pay, Devil's Ducat,* and *H–ll on Earth, or The Devil's Daughter*. Other critics railed against the actors, depraved hypocrites who mouthed fine sentiments, thus demonstrating how sinful character could be concealed. Diarists expressed similar concerns. "It is not often that I will tolerate a play," wrote Tracy Patch Cheever, a twenty-eight-year-old Boston lawyer, in 1853. "There is so much lack of art . . . together with such a conspicuousness of moral inferiority in many of the performers, that high thoughts and noble deeds coming through them, seem mockeries and leave none of the force of life behind them."[28]

Greater concern focused on the behavior of the audiences—not because the behavior was getting worse but because of rising expectations among the urban middle class. Spectators of all classes had been accustomed since the beginnings of American theater to talking, eating, drinking, and walking around during the performance. The activity on the stage was considered only part of the evening's attraction; socializing and people watching were equally important. Though inattentive to the performance at times, members of the audience could also choose to interact with the performers by hissing, heckling, or cheering while a scene was in progress. Brazen young men might even wander onto the stage and interrupt the scene by talking to the actors. Believing that the audience was sovereign, theatergoers felt entitled to demand encores of songs and favorite scenes and to boo actors who displeased them. Audiences expressed annoyance by yelling and stomping to drown out the performance or by barraging actors with all sorts of debris, from rotten eggs and vegetables to chairs and dead animals. Minor disturbances were to be expected almost every night. Ladies who claimed refinement stayed away on nights when a full-scale theater riot was rumored, but their men did not. In fact, high-spirited young men from good families—called "bloods" or "bucks"— were among those causing the trouble. The divide between different styles of audience behavior ran along the lines of age as much as class. Affluent people continued to patronize the theater in the early nineteenth century, generally accepting the wide range of status and behavior within the audience as a reflection of the normal condition of society. Everyone was there, each in his or her proper place. The "better sort" occupied comfortable seats in the boxes, with a good view of the stage. Men and boys of the "middling sort" sat or stood on the grubby wooden benches in the "pit," the ground floor in front of the stage. Poorer people and people of color sat high above in the cheap bench seats of the gallery. Just below the gallery, the third tier of boxes was becoming known for its prostitutes and for the men who went to arrange trysts with them. Men of any class could drink at the bar, often located behind the third tier.[29]

The theater's atmosphere seemed more troubling in the 1830s, 1840s, and 1850s amid the clash of two growing cultural forces. On the one hand, a self-consciously "respectable" urban middle class was distinguishing itself from its inferiors partly by greater self-control and sensitivity to disorder. On the other hand, a newly assertive working class bristled at any deviation from its notions of American equality. Public behavior of all sorts was judged as a politicized expression of class culture. Affluent men and women, unlike their predecessors in the more securely hierarchical society a generation earlier, now recoiled from behavior that had long been tolerated. Anna Cabot Lowell Quincy found on an evening at a Boston theater in 1833 that "our comfort . . . was somewhat disturbed by our next neighbors, the next box being partly filled by a party of half sort of gentlemen, who had been dining together and who had apparently 'passed the genial bowl' more freely than soberly. They were fortunately too much stupefied to be noisy, but were very disagreeable." Though she enjoyed the performances, "the Theatre, however, struck me this evening as more disagreeable than in any of our late frequent visits to it. . . . It is certainly, even at the best, no fit place for 'an elegant female.'" Drunkenness and rowdiness marked people as beyond the pale of middle-class culture: at best a "half sort of gentlemen."[30]

Already by the 1830s, such behavior was becoming identified with working-class men, and theaters where it was tolerated were shunned by most affluent people. In an era when status-conscious families were moving away from mixed-class neighborhoods in the city center, the idea of mixing with rude and immoral people in the theater seemed repugnant. A lady entering or leaving the theater might brush shoulders with a "bedizened strumpet," or might even be mistaken for one. A young man who left his box seat to use a chamber pot might be tempted before his return to drink at the bar or flirt with the prostitutes. At the very least it was difficult for anyone to avoid hearing uncouth language from the gallery, or smelling liquor on the breath of men in nearby seats, or catching a glimpse of the third tier. Respectable theatergoers in smaller cities like Bangor, Maine, and New London, Connecticut, had to choose between withdrawing from the theater altogether or enduring behavior they found offensive. In larger cities with multiple theaters, respectable men and women attended venues that required passive spectatorship and surrendered the others to working-class audiences.[31]

The Bowery Theatre in New York was the most famous of the working-class theaters, where depending on one's prejudices one could see the vigorous democracy of America or a barbarous mob that mistook liberty for anarchy. George Foster, the sensationalist New York writer, saw the latter. He called attention to

> the loud and threatening noises from the pit, which heaves continually in wild and sullen tumult, like a red-flannel sea agitated by some lurid storm—the shuffling and stamping of innumerable feet in the lobbies—the unrepressed exuberance of talking, the laughing, children-nursing, baby-quieting, orange-sucking, peanut-eating . . . unconventionality of the "dress circle"—the roaring crush and clamor of the tobacco-chewing, great-coat-wearing second tier—the yells and screams, the shuddering oaths and obscene songs, tumbling down from the third—mingling with the compulsive howls and spasmodic bellowings of the actors on the stage.[32]

The proper bourgeois response to this catalog of horrors was revulsion, or at best sociological curiosity. Some such as Nathan Beekley avoided the theaters with the rowdiest behavior. Others who detected incivility in every theater refused to attend any. "I detest the theatre, the crowd of horrid, vulgar people disgusts me," wrote Sidney George Fisher, a young Philadelphia gentleman, in 1837. Midcentury advice literature described the theater as a gateway to hell, filled with prostitutes, drunks, and lechers.[33]

Late hours compounded the problem of the audience. Until the 1850s or 1860s, an evening at the theater really could mean an entire evening, from the opening of the doors at dusk through the final encores somewhere around midnight. In between, the audience could watch a full-length drama, one or two shorter comedies or farces, and various songs and dances between the plays. Special occasions were more extravagant still: "Tragedy! Comedy! Opera! Ballet! Spectacle! And Melo-Drama! All on the Same Evening," screamed the playbill for January 1, 1844, at New York's premier theater, the Park.[34] Contrary to the playbills, curtains did not always rise at "precisely" the advertised time. Demands for encores could stretch out the evening even longer. Members of the audience felt free to leave early, just as they felt free to arrive late. It was common to skip the "afterpiece," a short farce that closed the evening. Of course, many in the audience stayed until the very end, and everyone had the privilege or temptation to do so.[35]

Those who criticized the theater raised several objections to late hours. One was that Saturday evening performances extended after midnight in violation of the Sabbath. Other complaints focused on the harm to theatergoers. By upsetting the natural regularity of sleep, the theater was said to be a menace to health that each year drove hundreds to an early grave. Job performance also suffered. A late night at the theater, the Reverend John Todd warned young men in 1841, "unfits you to meet the duties of the next day. Let a book-keeper sharpen his pen the next morning, and rub his aching head, and see if he can accomplish much; or let the mechanic try and see if he can use his tools, command his thoughts and hands as well the day after going to the theatre, as if he had not gone."[36]

Furthermore, the late ending put the remaining audience on the street at an hour when all decent folk ought to be in bed—and when many of those still awake were far from decent. Here was a problem similar to the danger of mingling with theatergoers of different status and moral character. The range of people and activities inside the building made theater a bridge between respectable recreation and disreputable dissipation; the late hours meant the theater spanned an important temporal divide, taking respectable people into the morally dubious hours around midnight. Stepping into the dark street along with the rakes, harlots, and other stragglers, a young man might be tempted to continue the night's dissipations elsewhere. Almost every theater "is flanked by taverns, recesses, and houses of ill-fame," warned the Reverend H. Daniel in an 1859 magazine article. Men who believed their sins would be concealed by darkness succumbed to these temptations. "Thus late hours, intemperance and licentiousness, dark and melancholy trio! appear naturally and necessarily to associate themselves with the amusements of the theatre."[37]

REFORMED THEATRES AND CONCERT HALLS

New York's elite responded to the declining reputation of theater by attempting in 1847 to establish its own alternative: the Astor Place Opera-House, with aristocratic pretensions and no connection to older traditions of theatergoing. Instead of a pit with benches, it had a sloped "parquette" with upholstered seats. The gallery was small. The Opera House pointedly lacked a third tier of boxes and further discouraged the presence of prostitutes by prohibiting unaccompanied women. Opening night drew the cream of New York society to display their finery in a setting that incidentally contained an opera. The Opera House added dramas and concerts to its schedule when opera alone proved insufficient to pay the bills, but it continued to favor an elite audience. Popular resentment toward this bastion of snobbery culminated in May 1849, when a mob of patriotic working-class men attacked the building during a performance of *Macbeth* by the controversial British actor William Charles Macready. The militia suppressed the riot with a volley that killed twenty-two.[38]

By the time of this debacle, Moses Kimball and P. T. Barnum were developing a successful alternative that excluded the audience traditions of the theater without excluding quite so much of the audience. Their popular museums in the early 1840s featured—along with stuffed birds and a purported mermaid—what were euphemistically termed "lecture rooms" where one could see variety shows, freaks, and other entertainments. First Kimball in Boston and then Barnum in New York expanded these offerings to include comic plays, melodramas such as *The Drunkard*, and even Shakespearean tragedies purged of the naughty bits. Barnum boasted in 1853 that his three-thousand-seat lecture room had proved to be

> a superb and convenient place of FAMILY AMUSEMENT where strangers and citizens may attend, confident of being entertained without having their moral instincts offended, as all improper persons are excluded and no intoxicating beverages are to be obtained on the premises—a place to which all may escort their wives and children with a certainty of listening to the most eloquent dramatic productions of the age, without witnessing a scene or hearing

an expression calculated to bring a blush to the cheek of innocence, or suggest an impure thought to the mind of the unsophisticated.

This formula did draw surprising numbers of women of all classes, many of whom would never have set foot in a place called a theater. Imitators in other cities opened similar "museums" in which exhibits were secondary to the performances or nonexistent. By excluding liquor and presenting preachy reform melodramas, the museum theaters went much further than previous family-oriented venues such as Niblo's Garden. Establishments such as Wallack's that honestly called themselves theaters sought an affluent clientele by selectively adopting innovations of opera houses and museums: excluding prostitutes, replacing the pit with the parquette, and hiring ushers (called "police") to maintain decorum.[39]

Managers of reformed theaters did not deliberately exclude working-class men and women, who attended in significant numbers in the 1860s and 1870s, yet they made no secret of their wish to impose the manners of the middle-class parlor on what had once been a tumultuous public space for all classes. The closing of the third tier and the exclusion of prostitutes were practically universal by the 1870s. The managers also put an end to most of the eating and drinking inside the performance hall. They turned down the house lights during performances to focus attention on the stage. Ushers suppressed noisy outbursts with the support of like-minded members of the audience. Rioting grew rare after the Astor Place fiasco, as did all sorts of interactions between the audience and the performers. Working-class audiences who wanted the lively atmosphere of traditional playhouses could go to their own theaters or join mixed-class crowds at minstrel shows; when they entered a reformed theater they were on middle-class turf.[40]

While losing its function as a meeting ground for different classes, the reformed theater also ceased to serve as a bridge between moral and immoral leisure and from evening to deep night. The curtain now fell at a prudent hour, another innovation that helped improve the image of theaters. (Early hours had previously been touted in defense of the equally dubious circus.) Similarly, although in the 1840s Barnum's American Museum had started its performances at least as late as traditional theaters, the museum closed for the night at the sober hour of 10:00. Reformed theaters in the 1850s and 1860s shortened their own hours by eliminating the afterpieces and in the 1870s were limiting the evening's program to a single drama, as moralists had been suggesting for years. Meanwhile, old traditions lingered at the Bowery Theatre, which offered very long performances into the 1860s.[41]

By shortening the performance and reducing the number of interruptions from encores and audience noise, theater managers in the 1860s and 1870s were better able to predict when the final curtain would fall, to the point where some began announcing the closing time in their advertisements. A person entering a playhouse for an 8:00 p.m. performance could now expect to leave at 10:00, 10:30, or 11:00, after silently watching the stage for 120 to 180 minutes. Times of 10:15, 10:40, and 11:10 were also advertised, giving at least the impression of precision. One unusually time-conscious New York theater in 1875 advertised times for each of eleven performances within a single variety show: "Prof. Davis and his dogs" at 9:25 would be followed by a ballet at 9:40, by Miss Andrews's singing of "Sweet Spirit" at 10:10, and so forth.[42] Wealthy theatergoers were expected to arrange for their carriages to line up in front of the theater at the appointed hour. The rest of the audience could use the information to plan their return by omnibus, horsecar, or commuter railroad. Transit companies and hack drivers could schedule service more efficiently, having vehicles ready at the moment the crowd spilled out of the theater, and improved public transportation diminished the need for theater managers to provide their own omnibus service, as a few had in the 1830s and 1840s.[43] The downside of relying on tightly scheduled transit was that theatergoers began to worry about missing their ride home if a performance ran unexpectedly late. This danger gave many people another excuse for leaving during the final minutes of the performance, a persistent disruption that managers, performers, and some members of the audience struggled to suppress.[44]

Theaters that successfully reformed audience composition, behavior, and scheduling produced an experience more like that of a concert than an early nineteenth-century play. Concerts had traditionally been refined events, with more exclusive audiences and strict codes of behavior even during the eighteenth century. Affluent families considered an appreciation of music a mark of gentility to be cultivated by marriageable young women. Some nineteenth-century vocal concerts, like those given by the antislavery Hutchinson family singers, were suffused with Christian benevolence or reformism. Other, overtly "sacred" concerts were held in churches. Whether sacred or secular, concerts were seen as "an innocent public amusement" even by some of those who denounced the immorality of the theater and ballroom. Anna Quincy Thaxter, a Bostonian in her early twenties who agonized about whether it was proper to enter a theater, had no doubts about the concert hall. She enjoyed many instrumental or vocal concerts in 1846 and 1847, including a performance by the Hutchinsons.[45]

Concerts at this time varied in length from two to four hours. Shorter ones allowed Anna Thaxter to be home by about 10:00 p.m. Longer ones followed the same spirit as the theatrical programs, piling on popular and classical works that lasted till about midnight. As many as thirty pieces of music were played in an evening, counting encores. These "grand concerts" offered something for everyone but left even aficionados exhausted. Audiences would begin trickling out of the auditorium during the latter half of such marathons, discreetly if they could but sometimes in such numbers that they prevented others from hearing. One jam-packed grand concert in New York in 1844 ran so long that audience members tried to escape through the windows. "Monster concerts" became rare in the late nineteenth century as conductors discouraged encores, weeded out popular tunes, and tried to stop people from leaving early. Nineteenth-century writers who objected to late hours rarely mentioned concert attendance; the concert hall lacked the associations that made the theater questionable.[46]

By the 1870s, an evening at the theater or concert hall could be envisioned as a passage from one domesticated interior to another. On the way there, omnibuses and streetcars shielded passengers from the mud, the filth, and the poor people in the street. The city policeman at the theater door protected the arriving audience from beggars and thieves. Inside, the management ensured that the audience was "respectable," its behavior proper, the entertainment inoffensive. Curtains fell at a predictable hour, so the return home could be planned. Even weather was less of a concern. Bitter cold, pouring rain, or sweltering heat had once meant poor box office receipts for theaters and physical discomfort for the few intrepid souls who shivered or sweated through the play. Theaters had often canceled performances because of weather, and they had customarily closed for the summer. All this began to change in the middle years of the century. "No postponement at any time at this Establishment as the Grand Entrance from Broadway ... is protected from the weather," Niblo's Garden promised in 1838. William Niblo assured the public in January 1841 that his theater was warmed to such a comfortable temperature (fifty-five degrees) that they could remove their hats and coats if they wished. Theater owners who built new performance halls in the 1840s and 1850s installed central furnaces in place of the ineffective stoves that once decorated pits and lobbies. They installed ventilation systems to allow performances through the hot summer months. Combined, these reformed conditions helped shield audiences from all sorts of unexpected encounters with the urban environment—in both its social and its physical aspects.[47]

An evening out was still not entirely safe for women. Though unescorted ladies could safely enjoy matinee performances after about 1850, they would be foolish to attend an evening show without a male protector. Even in the most refined venues, men peered at women through opera glasses or gazed down their décolletage from seats in the gallery. Obnoxious men sat next to any attractive woman who appeared to be alone. Etiquette manuals cautioned gentlemen not to leave their lady companions unattended in a theater for even a moment and warned women to comport themselves modestly to avoid drawing attention. After the performance, of course, women should never travel home alone.[48]

Some people did not want to be so insulated from the nocturnal city. They savored the experience of the night streets—the odd mix of characters, the occasional serendipitous encounter, the cool air on one's face, the hint of danger. Others sought out the night for the same reason moralists condemned it: for the concealment it provided. When most of the city was sleeping and streets were too dark for faces to be easily recognized, those still awake felt a sense of freedom. People at night seem to have been more willing to indulge forbidden thoughts and impulses. Richard Robinson, the young New York clerk who frequented brothels at night, mused about this experience in a mid-1830s diary entry. "In the morning the mind is calm, and reason vigorous and imagination asleep," he wrote, "but at night the passions are up. They love the late hours. Sometimes, at the dead hour of midnight, a thought is roused up from the deep caverns of the mind, like a startled maniac, which all the energy of reason can scarcely re-cage!"[49] Night was when many people indulged in pleasures that had to be restrained during the day: drinking, dancing, and sexual expression.

The mere fact of breaking out of the lockstep of the workaday schedule added excitement to amusements that were perfectly innocent and extra spice to those that were not. This feeling seems to have been particularly strong in younger people. Theater audiences in the 1870s were composed disproportionately of people in their late teens and twenties, which helps explain why the theater's effect on youth continued to be of such concern to moralists.[50] For young men with money to spend, an evening that began in the theater often continued after midnight in the saloon, the oyster cellar, the brothel, or the gambling hall. Though all these places were open at earlier hours as well, they were in their glory after the bells chimed twelve.

CHAPTER 5

After Midnight

Hours given to late night pleasures were usually stolen from sleep. Hence, amusements after midnight were by definition "dissipations": they squandered valuable energy when it should be replenished. Anyone who risked exhaustion in this way was supposedly undermining his capacity for economic production or, in the case of women, her capacity for reproduction. Worse, agreed many critics, the preferred amusements of the night owl all led to poverty, sickness, and sin. The ballroom was considered a gateway to vices practiced more flagrantly in taverns, brothels, and gambling halls. Heavy drinking left people hungover the next morning; sex left them depleted and diseased; gambling destroyed their desire to work.

"The villains who work entirely by night, sleep all the day; so do the miserable courtesans," claimed the popular writer Ned Buntline in 1848. "The gamblers and very fashionable young men sleep more than half the day away, because they are up nearly all night. The 'Five Points' are quiet as a grave-yard in the day-time, but noisy as Pandemonium at night. The classes who parade the streets by day, are of entirely different grades from those who fill the thoroughfares at night."[1] Gaslight, it seems, did not always bring the values of daytime into the dark city; it might instead make the dead of night less frightening and more enticing. The people who haunted urban public spaces after midnight were said to be those who scoffed at the decency of the sunlit world: the decadent rich, the irresponsible poor, the criminal underworld, and—most troubling of all—the corrupted sons and daughters of the middle class.

DANCING

Proper middle-class activities reflected the economic virtues of sobriety, industry, chastity, and self-restraint and best thrived in the absence of people who resisted these values. Theater reform targeted those below the middle class on the social hierarchy, but strumpets and "Bowery b'hoys" were not the only ones rejecting middle-class culture. Another challenge came from above. Lacking a genuine aristocracy, American cities nonetheless contained an upper crust of newly rich merchants, bankers, and manufacturers. Some members of this urban establishment, having acquired the luxuries and courtly manners of the European aristocracy, were keenly aware of what separated their American version of "gentility" from its noble origins: they, unlike the gentry they emulated, had worked for their money. This was not always a point of pride. Sidney George Fisher looked back with regret in 1857 on having worked as a Philadelphia lawyer for a railroad company. Disgusted by business and businessmen, he was glad to have retired at age forty-five so that he could enjoy a more truly genteel life of literature, contemplation, and social events.[2]

Though sons of wealthy merchants usually followed in their fathers' footsteps or pursued a profession in medicine or law, some devoted themselves to leisure as a prelude or an alternative to a career. Arrogant young "bloods" and foppish "dandies" became stock characters in urban fiction as well as a real presence in the city. Modeling themselves after their conceptions of the British or Continental elite, they in turn were emulated by men of modest wealth, social-climbing "swells." Journalistic and fictional descriptions of all these types suggested that pretension was their dominant quality. Young men affected aristocratic mannerisms even if their fathers had started life as tradesmen, or even if their families were not wealthy at all. They squandered money on luxuries and fashions. Not having to work, they kept conspicuously late hours. "Took a lecture from my mother, for being out after midnight," writes a dandy in a fictional 1852 diary. "Told her, in return, that two o'clock was a fashionable hour—just about the right time to retire."[3]

The cultural dissent embodied in the dandy was, like him, often considered harmlessly amusing. Monocled, languid, effeminate, the dandy was out of place in the hardworking commercial city. But the desire to emulate aristocracy was stronger and more widespread than middle-class Americans cared to admit. Held in check during the workday, it was unleashed at night in one of the most popular social institutions of midcentury America: the ball. The fashion reformer Abba Goold Woolson wrote in 1873 that "the greatest objection to these entertainments is that there is nothing in them distinctively

American." Balls with late hours, expensive clothing, and baroque etiquette were "servile copies of similar assemblies in decayed monarchies."[4] Late night dances were popular among people who could not in any sense afford aristocratic lifestyles. Through the 1820s, 1830s, and 1840s, urban Americans of all classes flocked to public balls throughout the winter months, ignoring the vehement opposition of the clergy. "If there is anything New Yorkers are more given to than making money it is dancing," wrote expatriate New Yorker Thomas L. Nichols.

> During the season, that is from November to March, there are balls five nights a week, in perhaps twenty public ball-rooms, besides a multitude of private parties, where dancing is the chief amusement. The whole city is made up of clubs and societies, each of which has its balls. . . . There are in New York fifty or sixty companies of volunteer firemen, with from fifty to a hundred members; and each company every winter gives one or more balls, in which their friends are expected to take part. There are twenty or thirty regiments or battalions of military volunteers, and each one has its ball. There are hundreds of societies and lodges of Freemasons, Odd Fellows, Sons of Temperance, Druids, and various Irish, German, trade, and benevolent societies, which must have at least their annual dance in winter. . . . Then there are assembly-clubs of young men who unite for the sole purpose of dancing, subscribe for a dozen cotillion parties. . . . Besides all these, are the great balls, held at the Academy of Music or largest theatres—as the General Firemen's Ball, the ball of the Irish Benevolent Society, &c.[5]

Promoters insisted that charitable balls were a benefit to society: everyone enjoyed a restorative break from labor while raising money for worthy causes. Setting aside their differences, the wealthy danced alongside the working class to benefit charities connected with volunteer fire companies.[6] The mixing of classes served a noble purpose in itself: unlike the theater, at balls the crowd dressed and comported itself in ways congenial to its wealthier members, who were pleased to imagine themselves uplifting fellow citizens by example. This was not an unrealistic fantasy; outside the ballroom, Americans from humble backgrounds read etiquette manuals for just this sort of instruction. One 1846 work reassured readers that in their egalitarian nation, "every man has the right, and should have the ambition to be a gentleman—certainly every woman should have the manners of a lady." An 1837 ball in Philadelphia led Sidney Fisher to reflect complacently on the refinement of American society:

Last night went to the Citizens ball in the aid of the Firemen's fund. . . . [S]everal thousand people were present. It was certainly an interesting sight and I was much pleased with the respectable appearance, & proper quiet behavior of the multitude. There were certainly many ridiculous & awkward figures, but I question very much whether any other country could show, from the same classes of society, so many well-dressed, & well-behaved people. . . . These public balls have a good effect in bringing the different classes together occasionally, & tend to produce a more kindly feeling on both sides. The higher order are impressed with respect by witnessing the multitudes of decent, good-looking people, among those whom they are apt to regard with contempt, & the lower are gratified by being in the same room with persons whom they consider above them, by having an opportunity of seeing and observing their appearance and manner, & by feeling that they do not disdain to mingle with them and partake of the same amusements with themselves.[7]

Fisher had a few misgivings about this vision of amiable hierarchy. "I like a public ball— I mean a select one," he wrote a year later. The problem with the public ball was that it was not very select; almost anyone could attend who could afford a dollar or two for a ticket. Libertines and mistresses might mingle unrecognized into the well-dressed crowd, along with swells, gamblers, and con artists.[8]

Members of the elite grew hesitant about mingling with common folk in the late 1830s and 1840s, following the divisive presidency of Andrew Jackson. Even if ballroom etiquette still looked up to the European aristocracy, ballroom dancing had trickled down through the urban population, thanks to dancing instructors, dance manuals, public balls, and the rising affluence needed to buy some approximation of evening dress. In response, the upper crust held more exclusive "public" balls with limited ticket distribution or retreated to dances in private homes. *New York Herald* editor James Gordon Bennett called attention in 1849 to an extravagant ball to be held (where else?) at the Astor Place Opera-House. Tickets were six dollars, nearly a week's wages for some workingmen. "It is the offspring of an attempt by certain classes of our people to make themselves exclusive, to hold themselves up as superior to the world around them, and to introduce some of the refinement and many of the evils which characterize the corrupt, aristocratic circles in Europe, into society in New York," Bennett charged.[9]

Exclusivity was well advanced by 1859, according to a dance manual published in that year: "Formerly it was not considered improper or derogatory

FIGURE 6. In a dance hall. Dancing was popular throughout all ranks of urban society in the nineteenth century, though clergymen and moral conservatives warned that it aroused dangerous feelings. In this particular case, the warnings may have been appropriate. The engraving appeared on the cover of the *Weekly Whip*, to illustrate an article about commercial dance halls that were fronts for prostitution. "Scene in a House of Fashionable Resort," *Weekly Whip*, February 12, 1855. Courtesy of American Antiquarian Society.

for ladies and gentlemen to attend public balls, and share in their performance; but as the population augmented and the ball-room habitués degenerated into a mixed assemblage, the more refined portion of the community avoided them. Dancing, therefore, among the most cultivated and élite, is confined to parlors and private assembly rooms." Marred by same rough behavior found in taverns, public balls were left to those of modest incomes and modest social aspirations. Prosperous families gave their own parlor dance parties, which could be crowded, sweaty affairs with barely enough room for dancing; rented ballrooms offered more comfortable accommodations. Elite balls became occasions for competitive consumption and social posturing.

This trend culminated in the early 1870s with Ward McAllister's formation of the Patriarchs, a group of twenty-five New Yorkers at the pinnacle of society whose Patriarchs' Balls raised snobbery to a high art.[10]

Whether public or private, large or small, the dancing party presented another problem: the sublimated sexuality of dancing might dangerously arouse the young people who composed most of the crowd. Although twenty-first-century Americans may be puzzled to see this perennial concern applied to waltzes and polkas, nineteenth-century observers considered dancing the most extreme publicly sanctioned expression of eroticism. Where else could an unmarried woman appear with shoulders and upper chest exposed to the public eye, swaying in the arms of men she barely knew? The free-love advocate Thomas L. Nichols described crowded ballrooms as thinly concealed orgies: couples held each other in ardent embrace, legs touching, bosoms throbbing with pleasure. An evening of dancing was said to leave young men with their desires inflamed and their moral senses dulled by wine; far too many found the ballroom to be the antechamber to the brothel.[11]

All this makes the ballroom sound more interesting than it was. Men at the time complained more about boredom than excessive excitement. Conversation was notoriously vapid; it was constrained by the formality of the event, the noise of the band, the fear of committing a faux pas, and the awkwardness between men and women who had just been introduced. Young men filled silences with patently contrived flattery of their partners before returning them with relief to their chaperones. Men preferred clustering in the doorways or withdrawing into smaller rooms to play cards, eat, and drink with their friends. Young ladies sat waiting to be asked to dance. "If going into Society consists in habitually participating in such comfortless, joyless, insipid exhibitions of extravagance without results and folly without amusement, I have shewn more wisdom in staying at home hitherto than I gave myself credit for," the young George Templeton Strong wrote in his diary after attending balls on two successive nights in 1845. "Dissipation it is in the strictest sense of the word. Rational speech there is none, and none is expected; people leave their common sense in the dressing rooms with their cloaks and hats, and one finds himself the next day unfit for business and wholly stupefied and done up without having had anything in the way either of amusement or edification to show for it. So I'll go to no more balls."[12]

Writers of advice literature would have viewed this decision as the only healthful one. Breathing inside the stuffy ballroom was considered dangerous, as dancers inhaled air that had been tainted by other lungs or depleted of oxygen by the voracious gas lamps. Stepping outdoors was worse: cold night air gave the overheated body a shock it might never recover from. Most vulnerable were those fashionable ladies whose dresses exposed tender skin to the elements. "Great care should be taken, in leaving a heated ball-room, to put on sufficient wrappings before going into the outer air," warned Sarah J. Hale. "Many young and beautiful girls have lost their lives by inattention to this point." Fatigue was more dangerous still. Some people managed to take naps in the early evening before a ball or were able to sleep late the next morning, but those with regular working hours had to give up most of a night's sleep. The men and women who danced with such youthful grace were just a short step from the graveyard.[13]

Balls and formal evening parties began in earnest about 10:00 p.m. in the major East Coast cities of the mid-nineteenth century. (They started earlier in smaller towns and inland cities such as Cincinnati and Chicago.) The hostess was supposed to be ready by 9:00 to welcome early arrivals, but most people did not show up for another hour or two. "The gentility of a party is estimated in no small degree by the hour," wrote a disapproving visitor to New York in 1828, a few years after the 10:00 p.m. start became common there but before it had spread to other cities.

> If you want to be tolerably genteel, you must not go until half past nine—if very genteel, at ten—if exceedingly genteel, at eleven;—but if you want to be superlatively genteel, you must not make your appearance till twelve. The crying absurdity of this arrangement, in a society where almost every person at these parties, has business or duties of some kind to attend to by nine o'clock the next day, must be apparent. . . . It is a pitiful aping of people abroad, whose sole pursuit is pleasure, and who can turn day into night and night into day, without paying any other penalty but the loss of health, and the abandonment of all pretensions to usefulness.[14]

Dancers typically paused near midnight to take a late supper, and they returned to the dance floor with inhibitions loosened by wine or champagne. "All that repays one for the weariness of preparation for an evening party are the hours after midnight," a young lady reportedly told a magazine writer in 1846. "All before that time is so formal; the rooms are so crowded then, and the gentlemen too are so hungry till they get their suppers, that they are like unfed bears; but after supper they become amiable. . . . Then they find out that we are well-dressed and look particularly attractive, and are ready to dance to our heart's content. Then it is that we dare to whirl through the waltz, and gaily trip it in the Redowa, the Polka and Mazourka." These hours were, she added, the best for flirtation and conquest. Perhaps George Strong's problem was just that he left too early![15]

A new dance fad, the German cotillion, emerged in the 1840s and grew in popularity from the 1850s through the 1880s. "The German" was half dance and half party game. Typically reserved for after midnight, it involved vigorous and often ludicrous motions that broke the decorum of the ballroom. Its leader could choose from among dozens of established "figures" or make alterations at will. In the figure called "the kangaroo" or "the fan," the lady selected one gentleman as her waltz partner and handed her fan to another. The hapless gentleman with the fan was supposed to hop around the waltzing couple, fanning them. "The figure is intended to create a hearty laugh at the expense of one gentleman at a time," explained an 1878 dance manual. Among the numerous other figures that composed the German were ones whose very names suggest their deviation from formal dancing: "the ladies mocked," "blind-man's bluff," and "the sea during a storm." According to another dance manual, the figure of "the rope" literally involved stretching a rope across the ballroom to trip leaping gentlemen, while "the race" was even livelier: "Whips and reins are the necessary properties for this figure. . . . Partners are chosen and the ladies proceed to drive the gentlemen who race from one end of the room to the other, obedient to the whip and rein." Teenagers and young adults were said to be delighted by the intricacy and ever-changing novelty of the dance, along with the freer physicality that prevailed (especially if the chaperones dozed off). Dancers at a successful ball would continue the German for hours, sometimes until daybreak.[16]

It was not uncommon for people to attend several late night parties each week during the frenzied height of the dancing season before Lent, and some attended more than one in a single night. Nor were the late hours limited to the upper or even the middle classes. An 1843 account of a firemen's ball in New York noted that a fire alarm about midnight forced some of the men to take a break from the festivities. Inevitably, late night partyers did not feel their best at work in the morning. The Reverend T. DeWitt Talmage warned of dire consequences. "What will become of those who work all day and dance all night?" he asked. "A few years will turn them out nervous, exhausted imbeciles. Those who have given up their midnights to spiced wines, and hot suppers, and ride home through winter's cold, unwrapped from the elements, will at last be recorded suicides." Throwing away sleep, health, and savings in a fit of conspicuous consumption, the midnight revelers made themselves incapable of the honest labor American society depended on.[17]

Balls and evening parties fell within the woman's sphere, at least in theory.

Ladies, after all, were thought to have the primary responsibility for upholding etiquette and raising the tone of social life. In the commercial republic of America, middle-class ladies supposedly stood above the grubby practice of work that occupied middle-class men, instead presiding over the domestic realm as surrogate aristocrats. They were also said to be the most graceful, congenial, and beautiful, qualities that shone brightly in the ballroom. On a practical level, they were more likely to have taken dancing lessons, and they certainly had more elaborate evening dress than gentlemen. If the evening party was to take place within a private home—woman's domain, of course—the invitations would come from the hostess. Finally, gentlemen at a ball were expected to treat women with a deference bordering on servility.[18]

Women's power in the ballroom had its limits. There was the physical problem of the clothing, for one thing; motion was limited by a carapace of whalebone, wire, and layered fabric. Fashionable evening attire always included a corset and, depending on the styles of the moment, might involve stiffened petticoats or a crinoline, train, or bustle. Sleeves set below the shoulder inhibited arm movement. Further, though the parlor was a feminized domestic space, the ballroom was not. Women were in public, exposed to the gaze of men, and were expected to comport themselves accordingly. They were escorted to and from the ball by a gentleman or a chaperone. Inside the ballroom, as if still in the dark street, a lady must never be unattended; the gentleman requesting the honor of a dance had to stay with her if she chose to take a turn around the ballroom at the end of the dance or if she wanted refreshments, never leaving her until she was returned to the safety of her protector.[19]

DRINKING

There was no ambiguity about who was in charge outside the ball. At other amusements after midnight, women were either absent or openly used as sex objects. Male power was at its zenith in the dark city through the third watch of the night.

The quintessential place of male leisure was the tavern. The "public house" in an eighteenth-century American city had offered much more than food and drink. It served as a public space shared by a wide range of men, sometimes with song and conviviality, sometimes with barely suppressed tension. Clubs and societies convened around its tables. Larger taverns provided rooms for public meetings and polling places. News could be found there, freshly arrived from the ships, passed by word of mouth, or printed in the newspapers—and so could gambling on cards, dice, billiards, and cockfights. Quarrels quickly

FIGURE 7. Inside a tavern. Taverns and saloons, the most common recreational spaces in the nocturnal city, drew an almost exclusively male clientele. This engraving from the *Weekly Rake* shows a relaxed group drinking, smoking, joking, and reading the *Rake*. "The Pewter Mug, on a Saturday Night," *Weekly Rake*, October 22, 1842. Courtesy of American Antiquarian Society.

turned into fistfights, usually taken out onto the street. Some taverns became notorious for brawls; almost all had customers who staggered home drunk late at night.[20]

Until the early nineteenth century, Americans were in the habit of taking small drinks throughout the day. There was nothing unusual about a lady's starting a cold morning with a glass of brandy or a workingman's joining his fellows for a whiskey break in the afternoon. By 1830, though, most middle-class families had banished liquor from their homes as part of their turn toward respectability and self-restraint. American consumption of alcohol dropped sharply between 1830 and 1850. Temperance literature warned young men that success in business depended on their sobriety. The temperance movement assumed that the natural delicacy of women would make them shun the bottle, and temperance literature portrayed women as the pious victims of male drunkards. Middle-class people did their best to avoid or suppress drinking in public. They moved out of the mixed-class neighborhoods and theaters that they could not control, and they struggled to impose state prohibitions on liquor sales. At least they got their way within the workplace. Churchgoing businessmen who once treated employees to ale during working hours now demanded strict sobriety on the job. Nevertheless, the physical separation of the classes gave workingmen more freedom to drink unobserved after work. They gathered in their own drinking establishments—taverns, kitchen barrooms in private homes, and groceries or dram shops where whiskey was sold by the glass—to pack a day's worth of fun into a few boisterous hours. Drinking culture survived as part of working-class male identity. Despite the success of the Washingtonian movement in the early 1840s, temperance was never as popular among workers as among the middle class.[21]

Many businesses served the needs of the thirsty city man. (Women looking for a public drink had few alternatives to the working-class grocery.) The saloon, destined to be the dominant urban drinking place after 1870, had already appeared by the 1840s. The saloon occupied a single long room dominated by a bar with bottles and "flashy glasses" along one wall. Most men stood at the bar, since there were few seats in the room, an arrangement that encouraged faster turnover and also made it easier for new arrivals to mix into the crowd. Men readily sang together or bought rounds, though they still settled disagreements with their fists. In addition to saloons where only liquor and cigars were sold, an 1865 report identified several other types of establishments within a single New York neighborhood: "lager bier saloons, where much more lager bier than strong liquor is consumed, and with less rapidity. Ale-houses similar to lager bier saloons, though generally more quiet and partaking somewhat of the character of restaurants; billiard rooms, restaurants, concert saloons and corner groceries." An English observer in 1859 identified additional types while remarking on the American propensity for barhopping. "Jonathan can't keep still, but rushes first to the bar-room of the hotel where he had dined, has a drink, thence to the confectioner's saloon, then to a cigar ditto, next to an oyster ditto, and . . . most likely 'smiles' [takes a drink] at each of them."[22]

By 1855, there were over 5,500 liquor vendors in New York, slightly fewer than one for every one hundred people. They were located mostly in the lower wards where businesses and younger men were concentrated. One rough waterfront neighborhood, the Fourth Ward, had a drinking place for every forty-nine people in 1865. Here, as in other districts in lower Manhattan, corner saloons and cellar dram shops occupied the lower floors of tenement houses. Many an urban laborer entered such places at the end of the workday, seeking a bit of pleasure that he could not find in his cramped lodgings. "In a few steps, he can find jolly companions, a lighted and warm room, a newspaper, and, above all, a draught which, for the moment, can change poverty to riches, and drive care and labor and the thought of all his burdens and annoyances far away," explained Charles Loring Brace. "The liquor-shop is his picture-gallery, club, reading-room, and social salon." The Fourth Ward saloons undoubtedly served many customers who lived elsewhere but worked in the district: printers, leather workers, and longshoremen. Nearly unbroken lines of saloons could be found along Water Street near the wharves, and on William Street near the major newspapers and printing houses.[23]

Drinking places drew different clienteles depending on their location, management, prices, and atmosphere. Lager beer saloons in ethnic German neighborhoods ranged from tiny storefronts to grand halls filled with German music and jovial families; men would go to these "beer gardens" with their wives and children and leave "at a seemly hour." Saloons for English-speaking men were distinguished from one another by the occupations and subcultures of the regulars. "There are grades to Porter Houses as well as to Courtezans, Churches, Theatres, or any other 'fancy places,'" according to slang-ridden article in the *Weekly Rake*, a New York "sporting" paper that covered all those amusements except churches.

> The lowest grade is a "three cent ratgut" shop, where tired coal-heavers, fireboys, emancipated cobblers, and others of like caste, resort to play "seven up" and "rounce," till the bell, tolling the "witching hour of night," warns them to seek their cot bedsteads to regenerate their frames for the next day's labor. The staple subjects of conversation are Fires, Prize Fights, Chatham Melodramatic performances, and "gallus" women. Then comes a superior sort of "three cent" shops, which rank a peg higher: These are supported by retail dry good's [*sic*] clerks, mechanics who "steam it some," cartmen, and "young men with small means" but great expectations. The all-absorbing topics of conversation here are politics on a small scale, balls, parties, the rights of working men, &c., &c.[24]

At the other end of the spectrum, wealthy gentlemen who would never be seen in a back-alley dram shop could drink with their peers in restaurants, hotels, and private clubs. Elite restaurants—notably Delmonico's—drew some of New York's leading businessmen, who stood at the bar drinking brandy, or drank wine and champagne at the dinner tables. Luxury hotels such as the Astor House or St. Nicholas also attracted well-heeled drinkers to their dining rooms and barrooms; swells liked to loiter in the lobbies and entrances.

Though women had separate dining rooms, men and women constantly encountered each other in the corridors, which in the 1850s were known as "flirtation galleries."[25]

Wealthy men who drank at the hotels and restaurants often did so in the company of small social clubs. Philip Hone, a semiretired New York merchant in his fifties, described an evening in 1835 with one of these all-male peer groups: "I went to the Book Club . . . at nine o'clock. This is a club which meets every other Thursday evening at Washington Hotel, where they sup, drink champagne and whiskey punch, talk as well as they know how, and run each other good humoredly. I have been admitted a member. . . . This is a very pleasant set of fellows. They sit pretty late, however, for I came away at one o'clock and left the party seated at the supper-table." Over the next few years he enjoyed similar evenings with the Reading Club, the Kent Club, and the Hone Club, named in his honor. These gatherings differed from men's dinner parties mainly because the same men attended regularly. Club dinners were not necessarily well behaved. The Harvard student Benjamin Crowninshield described several that "degenerated with a few exceptions into a general and exceedingly jolly and fiendishly noisy drunk." After a meeting of the Hasty Pudding Club in January 1857, he and the other members piled into a sleigh and rode to the Parker House in Boston; they drank and sang until nearly 2:00 a.m., then capped off the night by smashing windows and gas lamps on the ride home.[26]

The drunken sprees of Harvard bloods were merely upscale versions of the group binges common among young working-class men; even the club, as an institutional framework for bingeing, could be found throughout urban society. "All over the city you can find these little associations of ten or a dozen individuals," observed *New York Herald* editor James Gordon Bennett in 1844. "They are of all degrees of respectability—as the phrase goes—from the choice recherché club of bon vivants, with a Judge at their head, who riot on champagne suppers at Clarke and Browns, to the nest of small store keepers and tradesmen who celebrate their orgies amid the fumes of a Bowery cellar." Young men from a range of class backgrounds used the volunteer firehouse as an informal social club, with especially heavy drinking after a fire.[27]

Larger social clubs for elite men grew in popularity after 1850, despite the temperance movement. Many men belonged to several. Boston Brahmins founded the Somerset, Union, and St. Botolph clubs between 1850 and 1880, while their New York counterparts founded the Union League, Manhattan, University, and Knickerbocker clubs. Elite Philadelphians formed their own Union League and the Rittenhouse Club. Similar elite clubs appeared during this era in Providence, Pittsburgh, Chicago, and other cities. Inside the elegant clubhouses, members could enjoy gourmet dining, billiards, reading rooms, and a host of other amenities. Men used them as after-hours spots. Critics charged that long evenings at the club interfered with men's domestic life and their ability to work.[28]

The leading late night drinking spot for all classes was the oyster saloon. Oysters were abundant and cheap in the mid-nineteenth century. New York's street vendors and downscale oyster cellars offered them on the "Canal Street plan": all you could swallow for twelve and a half cents. Fresh oysters were shipped by canal or rail to the cities of the interior, but the oyster saloon thrived in New York as nowhere else. In addition to the bargain-basement outlets on Canal Street near the Bowery, more impressive ones had been clustering near the theaters since the 1820s. These saloons could be found two or more to a block along parts of Broadway in the 1850s, their red lamps and gaudily lit signs beckoning pedestrians. "The oyster cellars, to which you descend from the sidewalk of Broadway, are twenty-five feet in width, by a hundred or more in length, and many of them are fitted up with great luxury—plate-glass, curtains, gilding, pictures & c.," Thomas Nichols wrote. "The fashionable saloons upon the ground floor . . . are frequented day and night by ladies as well as gentlemen."[29]

By day, oyster saloons provided quick, cheap meals. By night, the ones on Broadway sucked in theatergoers before and after the shows, wrote George Foster, along with "rowdy and half-drunken young men, on their way to the theater, the gambling-house, the bowling-saloon or the brothel—or most likely all in turn." Men crowded up to the marble bar where bartenders feverishly poured drinks and shucked oysters, while busy eaters tucked into their meals in curtained booths. Some underground oyster saloons stayed open all night. No proper lady would descend into these establishments at any hour, despite their glittering decor. Inside, paintings of voluptuous nudes alerted even the naive that these cellars were devoted to male pleasure. Oysters themselves were thought to stimulate lust and enhance sexual performance; sleazier oyster cellars allowed male and female customers to test this theory in private rooms at the back. "The women of course are all of one kind," wrote Foster, "but among the men, you would find, if you looked curiously, reverend judges and juvenile delinquents, pious and devout hypocrites, and undisguised libertines and debauchees."[30]

Underground drinking spots had a bad reputation. Journalists, novelists, reformers, and travel writers often chose cellars to represent sinks of debauchery or the wretched dwellings of the poor. Urban space was divided in two in this trope: the city of decency and prosperity stood in sunlight while a pathological underclass lurked in dark caverns below. The dives where poor people drank were described as places of death and damnation, into which souls descended as if into hell.³¹ More insidious than such repellent groggeries were the elegant oyster cellars that might dazzle an unsuspecting visitor. Descriptions of the cellars emphasized their jumble of incongruous elements. Thanks to profuse gas lamps, the subterranean depths were brilliantly lit through the darkest night. They were public places concealed from public view, tombs full of life, wide awake during hours meant for sleep, and filled (as writers almost always mentioned) with mirrors. Social distinctions were destroyed as upstanding citizens mingled with harlots and wealthy bloods drank alongside criminals. Inside these palatial cellars the high and the low allied against moderation.³²

Such descriptions expressed middle-class nightmares of disorder, and—worse—were essentially truthful observations of an urban subculture that placed leisure instead of work at the center of life. This subculture originated in England among a diverse range of urban men collectively called "the fancy": aristocrats and wealthy young dandies who came together with urban wage laborers to enjoy the traditional manly amusements of drinking and pugilism. The American fancy was similarly heterogeneous and similarly focused on the hedonistic pleasures of the bachelor: drinking, whoring, and gambling. Those who tried to make a living through their connection to these enterprises, or to the stage or prize ring, came to be known as "sporting men." The subculture grew more conspicuous in the middle decades of the century as many other affluent men retreated from the tavern and the fun-loving crowd that gathered there. Ned Buntline attempted to distinguish "fancy men for the upper ten-thousand" from "fancy men for the lower million," but the commonalities were stronger than the differences. A way of life defined by male leisure was in direct opposition to the values of hard work, domesticity, and self-discipline. It posed a serious challenge to middle-class culture.³³

A new kind of urban drinking place—the concert saloon—exploded in popularity during the Civil War era and spread sporting culture more widely than ever before. Based in part on British music halls, concert saloons drew crowds of men and teenage boys every night to enjoy variety shows and to savor what were advertised as "the best of wines, liquors and segars, served by the PRETTIEST FEMALE ATTENDANTS in the city." Men were admitted for little or no charge, were seated at tables in front of a low stage, and were encouraged to drink by waitresses working on commission. The acts included singing, blackface minstrelsy, comedy, gymnastic stunts, and dances touted as displaying the charms of beautiful women. Old theaters and saloons were converted to new use, and new buildings constructed especially for the purpose. Part of the concert saloons' appeal was that they encouraged an informal spectatorship that was otherwise disappearing from stage performances. Men enjoyed the freedom to drink, talk, and walk around during the acts as they could no longer do at most theaters. In December 1860 George Templeton Strong went to one of the Broadway's largest, Canterbury Hall, named after a famous music hall in London. It was, Strong told his diary, "a queer place. No women in the audience, which was made up mostly of raffish men drinking lager at little tables and smoking. The performances (ballet-gymnastics, singing, and so on) were respectable enough. Perhaps the ballet dancers' skirts were half an inch shorter than in *Robert le Diable* at the Academy of Music. There was some very fair comic singing . . . and good music by a small orchestra."[34]

Each element of the concert saloon may have seemed tame, but the combination of alcohol, waitresses, and dancing women struck people as indecent. The "pretty waiter girls," whose obvious purpose was to flirt with and be ogled by the customers, were suspected of prostitution. New York newspapers a year after Strong's visit denounced concert saloons as "obscene dens" and a "truly diabolical form of shameless and avowed Bacchus and Phallus worship." A grand jury declared concert saloons a threat to public morality, and the state legislature passed a law regulating them. New York police promptly raided Canterbury Hall, arresting its proprietor and fifteen waitresses.[35]

The relative tameness of sexual display at these places made them all the more dangerous, in the view of some observers. Like oyster cellars, concert saloons transgressed moral and social boundaries. They were "half-way houses between the first-class and the gutter," introducing sexual display into respectable amusements, drawing together men of all qualities, and letting the curious contemplate vice without risking their reputations. Brilliantly lit facades, like those of theaters, provided false reassurance to anyone who misread light as a marker of decency. The editors of the *New York Times* pointed out that owners of concert saloons, "instead of carrying on their curious calling in the half-privacy courted by legitimate houses of prostitution, . . . storm the citadel of the public senses. With this view in mind they have chosen Broadway, and called to their aid all the resources of gaslight, transparencies, cartoons, and huge vermillion posters." The *Philadelphia Inquirer* charged that crowds of men loitered on the sidewalks outside, creating a nuisance for the surrounding neighborhood. Concert saloons continued to flourish in major cities through the 1860s and 1870s, often in cellars, despite efforts at regulation. In 1866, police counted 223 of these places in New York, serving nearly thirty thousand men daily. In 1872 there were still some 75 concert saloons in New York. There and in other cities, some establishments featured private rooms, waitresses who were unambiguously prostitutes, and dances that "could not possibly be surpassed in vulgarity, grossness, and disgusting lewdness." They attracted thieves who preyed on unwary drunks.[36]

WHORING

Concert saloons were just the tip of an enormous iceberg of prostitution. Opportunities for sex pervaded American cities at night. The faded beauty at the cigar store, the painted woman in the third tier, the pretty waiter girl, the teenager strolling alone under the park lamps—all were likely to be "fast." Young girls who peddled flowers or swept the street crossings often used their work as a cover for prostitution. Seemingly respectable female shoppers were willing to return after closing time to exchange sex for dry goods. Men brought women of questionable virtue to the private rooms of oyster saloons or to volunteer firehouses. Others made the acquaintance of women at museums or ice-cream parlors, then escorted them to houses of assignation. Brothels provided women and girls of all ages for men and boys of all ages.[37]

Walt Whitman, hardly a chaste moralist, was disturbed by the extent of sexual vice among the young men of New York and Brooklyn. "Though of course not acknowledged or talked about, or even alluded to, in 'respectable society,' the plain truth is that nineteen out of twenty of the mass of American young men, who live in or visit the great cities, are more or less familiar with houses of prostitution and are customers to them," Whitman wrote in 1857. Like many others, he blamed the situation on a lack of paternal guidance for the teenagers and young men who came to cities to find work. The Reverend Warren Burton declared in 1848 that brothels in and around Boston, numbering several hundred, had spread licentiousness among young men flocking to the New England metropolis. Boys as young as fifteen or sixteen were taken to houses of prostitution by companions from their workplace or boardinghouse or were lured there by streetwalkers. George Thompson, the sensational novelist, argued that it was impossible to suppress prostitution in cities where transient men had no other social interaction with women. "The passions must be got rid of, or their exercise provided for in some way. There are thousands of men, whose animal natures are so fierce, that nothing can control them. All the laws in the universe will not check such men in the enjoyment of their lusts. You might as well try to dam Niagara." Other commentators argued that any attempt to block the safety valve for bachelor lust would put innocent maidens at risk.[38]

There was enough reticence on the subject that only a few diarists acknowledged visiting prostitutes. In the same year that Reverend Burton warned of the spread of vice in Boston, the Harvard student Randal W. McGavock wrote from personal experience: "Went to a party at Mrs. McGee's and had a glorious time. Such kissing and such girls I never saw before." On another late night in 1848, McGavock recorded with winking emphasis that he had "called on a *woman*, etc."[39] Even this degree of candor was rare. Diarists usually mentioned prostitutes, if at all, as a social problem that did not touch them personally. Individual encounters were either elided from their accounts of their day or recorded for private memory in words too vague for others to understand.

Though nineteenth-century Americans were secretive about their own entanglements with prostitutes, they wrote a lot about prostitution in general. Trying to assess its prevalence, nineteenth-century journalists, clergy, and other observers gave estimates ranging so ridiculously high that 20 percent, 40 percent, or even 70 percent of the young women in New York would have been counted as prostitutes. (A more recent historical study has estimated that only 5 to 10 percent of young New York women ever prostituted themselves.) Contemporaries in other cities also complained of rampant prostitution, which at various times was said to be especially flagrant in Baltimore, Pittsburgh, New Orleans, Chicago, and Philadelphia.[40]

Popular writers often remarked on the streetwalkers of New York's Broadway. "As soon as the sun sets over the Great City, Broadway, and the streets running parallel with it, become infested with numbers of young girls and women, who pass up and down the thoroughfares with a quick, mysterious air, which rarely fails to draw attention to them," wrote James McCabe in 1868. "These are known as street-walkers, and it would seem that their number is steadily increasing. . . . They are chiefly young girls, seventeen being the average age, but you will see children of twelve or thirteen among them. . . . The neighborhoods of the hotels and places of amusement are the most frequented." These stretches of sidewalk were heavily traveled by evening pleasure seekers and—like the nearby shop windows—had bright lighting that let potential customers admire the merchandise. Many observers were shocked to see streetwalkers on the most prestigious commercial street in the United States; their presence among the wealthy businessmen, shoppers, and promenaders was often interpreted as epitomizing the urban mix of all types. Yet the Broadway streetwalkers were among the most discreet in America. Police discouraged them from speaking with potential customers, so they advertised their purpose through eye contact, flamboyant dress, a peculiar dragging of one foot, and unescorted solitude. Once a man indicated interest, the streetwalker would silently turn onto a darker side street to lead the customer to a brothel or to someplace where an accomplice could rob him. They were almost all gone from Broadway by midnight or 1:00 a.m. Streetwalkers were less restrained in other parts of the city, where they openly negotiated with customers and annoyed neighbors with their noise. One "sporting" paper estimated in 1849 that three thousand prostitutes walked the streets of New York in the hours around midnight.[41]

A lively sexual street culture flourished in midcentury cities, according to leering accounts in the sporting press and horrified descriptions by reformers. Apprentices, clerks, laborers, seamstresses, and servants cruised certain streets at night to find partners for commercial or recreational sex. Part-time streetwalkers, difficult to distinguish from professionals, would take their companions into nearby parks and graveyards. In the 1840s and 1850s, popular cruising grounds in New York included the Battery, the parade ground at Washington Square, and the sidewalks around the public hospital—as well as City Hall Park, which was noted for the availability of prepubescent girls and male "sodomites." These areas had enough public lighting to make faces and bodies visible, yet were near shrubbery or dark alleys that offered concealment. An 1842 article in the *Whip and Satirist of New-York and Brooklyn* describes the scene at Church and Anthony Streets: "We ... stood by the Hospital wall, here countless females passed us by, and boys from the ages of twelve to eighteen, were standing, in loose conversation with harlots of as tender age." Sexual expression became bolder as the evening passed, until in the hours around midnight couples enjoyed intercourse or fellatio outdoors. Similar scenes took place in graveyards and public parks in Boston, Philadelphia, Worcester, New Haven, Newark, Providence, and other cities. The profusion of part-time prostitution (and noncommercial flirtation) resulted in real or feigned cases of mistaken identity, in which respectable female pedestrians were accosted.[42]

Police in the mid-nineteenth century tried to contain the most flagrant displays of immorality in order to preserve public decency. Though they had neither the inclination nor the legal authority to jail all the prostitutes, they sometimes cracked down on streetwalkers who solicited too brazenly, or on legitimately employed women who moonlighted on the streets. Those arrested for vagrancy faced fines or short jail terms. Police occasionally raided the noisiest brothels, particularly if they were in middle-class areas. Outraged neighbors were more effective in chasing out houses of prostitution, since they could ransack, demolish, or torch the offending structure. These actions by police and rioting neighbors helped discourage vice in certain areas, while consumer demand attracted it to others. Places of prostitution gravitated toward the emerging central business districts, drawn by the concentration of theaters, saloons, hotels, and bachelors' boardinghouses.[43]

By the middle decades of the century, clusters of brothels and sleazy "dance halls" lined Almond Street in St. Louis, Fifth Street in St. Paul, and North Street in Boston. These were much smaller than the famous red-light districts that would flourish in the early twentieth century, places such as Storyville in New Orleans or the Levee in Chicago, with boundaries enforced by the police. Mid-nineteenth-century cities tended to have multiple, diffuse clusters of sex-related businesses scattered in and around the downtown. In Philadelphia, which was said to have over two hundred houses of ill fame in the 1840s, the largest vice district stretched along South Street from the Delaware riverfront to Eighth Street; a smaller district was emerging at Twelfth and Pine Streets; and a third cluster of prostitutes occupied some of the boardinghouses west of Washington Square, convenient to the Walnut Street Theatre. A fourth area of Philadelphia brothels was mixed among the homes and boardinghouses west of Franklin Square, Nathan Beekley's neighborhood. Beekley might easily have dropped by on what he called "my usual Saturday evening walk . . . to see the fashions." Similarly, on the eastern edge of downtown Detroit in 1875, a filthy little vice district called "Swill Point" could be found along Larned, Jefferson, and Front Streets near the Michigan Central Depot; about half a mile to the north, a second district called "the Potomac" occupied a few shabby blocks by the Detroit and Milwaukee Depot. Individual homes in more fashionable neighborhoods of Detroit were discreetly turned into brothels. Chicago and New York each had multiple clusters of various sizes and qualities.[44]

With dozens or hundreds of sex businesses in every major city, there was a den of iniquity for every taste and every budget. Near the bottom were the cheap dance halls, filled with "a motley crew of bloated men and painted women, black and white, mixed in the most unreserved sociability," in the words of Boston's Henry Morgan. The customers for these dance halls were sailors on shore, gang members, petty criminals, and other rough working-class types. In impoverished neighborhoods such as the Five Points in lower Manhattan, the worst dance halls occupied cellars and employed women indistinguishable from the desperate streetwalkers of the surrounding slum. "You see the women half exposed at the cellar doors as you pass," wrote Whitman. "Their faces are flushed and pimpled. The great doings in these quarters are at night. Then, besides the prostitution, there are dances, rum drinking, fights, quarrels, and so on." A customer could pay to dance with a woman and could arrange with her for any further services, to be transacted in rooms elsewhere in the building or nearby. Not every dance hall was a squalid hole. Bayard Taylor visited one near New York's waterfront in 1866 and found it surprisingly clean and orderly, almost like a respectable ballroom. He got there early in the evening, just past 9:00, when few customers had arrived. A small band played while twelve women danced a quadrille. There was no obvious indecency, though the women's dresses exposed calves and cleavage. Some German dance halls were more like public balls than fronts for prostitution; although prostitutes worked the crowd, most of those in attendance were working-class couples and young women who just wanted to dance.[45]

Other places of prostitution posed as homes or boardinghouses for young women. The cheapest brothels were in crowded tenements of the slums or in waterfront shanties like those of the "Sands" in 1850s Chicago. These employed prostitutes too old, diseased, or ugly to turn tricks elsewhere, and occasionally a younger girl who had made poor choices. Some were family businesses run by a mother and her daughters. Women loitered outside to accost passersby. They leaned from the windows, exposed their breasts, and quoted prices. Though these places were notorious for robbing customers, they still managed to attract drunken sailors and foolish visitors. Strangers from the countryside often visited city brothels on Sunday afternoon; otherwise, night was the busiest time. Unfortunate neighbors were awakened or kept from sleep by the noise of drunken brothel customers laughing and brawling in the streets outside.[46]

Toward the high end were "parlor houses" that mimicked the appearance of prosperous homes. A visitor entering such an establishment would be welcomed by a servant and ushered into an elegant drawing room just as if he were paying a call on a respectable lady. The prostitutes had the manners and clothing to match. The decor was aggressively opulent: chandeliers, carpets, marble mantels, damask curtains, upholstered furniture, and gilt-framed mirrors. Shy visitors could ask to be taken immediately to a private room, while others gathered in the parlor to meet their hostesses and listen to piano music. Descriptions of such places acknowledged them to be caricatures of wealthy homes. The women's flirtation and low necklines were excessive, the artwork on the walls was racy, the champagne flowed too freely—all this was true, but the parody was close enough to be both reassuring and exciting. Men who carefully contained their passions in domestic settings were invited to throw that training aside. Those who visited were already inclined to accept such an invitation, of course, yet there was no rush. Lonely young men dropped by brothels even when they could not afford to pay for sex, enjoying instead the conversation, music, and card games as if in a respectable parlor. Particularly talented prostitutes might engage clients in protracted flirtations, filled with whispered confessions and love letters. Men did not have to believe this hypocrisy to enjoy it. The parlor house experience was more than physical; it was a playful simulacrum of the deadly serious business of social calling and courtship.[47]

The madams of elite parlor houses even sponsored late night balls to drum up business, drawing some of the same wealthy men whose wives held private balls for respectable society. Some "cyprianic balls" were masquerades; for the first part of the evening, participants appeared in costume with their faces partly concealed by black domino masks. Masked balls were viewed with suspicion in the mid-nineteenth century even when they were not run by prostitutes. The cover of the mask was thought to loosen inhibitions on the dance floor and allow improper people to befriend unsuspecting youths and maidens. Whether masked or not, the cyprianic balls were not orgies. The dresses and dances were much like what would be seen in a respectable ballroom, according to a reporter who attended one such event in St. Louis in 1867, though the hours were later. Prostitutes mingled with a more varied crowd at public masked balls, which saw new popularity in the 1860s and 1870s.[48]

Houses of assignation further blurred the line between the demimonde and respectable society. Renting rooms for short-term guests, they attracted prostitutes and their customers, men and their mistresses, and couples having extramarital affairs. Edward Crapsey claimed in 1872 that houses of assignation had become "the chief danger that threatens the city from the social evil." They were growing in number in New York, and they were far more insidious than brothels. Married women might start going to such places to satisfy their lust but then begin accepting money from their lovers and degenerate into secret prostitutes. Respectable women visited houses of assignation while their husbands or fathers were at work, or they came at night after a visit to the theater. "Private supper rooms" were a variation on the assignation houses. These popular sites of seduction stayed open all night to serve meals in small, private dining rooms; waiters were careful to knock before entering.[49]

SPREES

For young men with money to spend, an evening that began at the theater often continued in the saloon or oyster cellar and concluded at the brothel or gambling hall. Some saloons and eateries operated all night, of course. They could be found wherever men worked night shifts—such as near newspapers and wholesale markets. Others such as oyster saloons stayed open late for the after-theater crowd or, like New York concert saloons, were filled with carousing men through the early morning hours. Brothels did most of their business at night, particularly after the theaters closed. A character in Buntline's *Mysteries and Miseries of New York*, on leaving an oyster saloon after 1:00 a.m., denies that it is getting late: "It's getting early, you mean; but it is just the fashionable hour for the Leonard street ladies, they're in full blast there about this hour." Men often stayed in prostitutes' beds until morning. Disreputable dance halls sometimes operated all night, even past the beginning of Sunday morning church services. By the 1870s, New York had developed a "centre of nocturnal activity" along Broadway between Houston and Bleecker, where concert saloons, gambling halls, and a large restaurant, ablaze with gaslight, did a lively business after 2:00 a.m. A second, cheaper late night district had developed along the Bowery, with all-night eateries, saloons, and private supper rooms.[50]

Packs of young rowdies of all classes reeled through midnight streets on drinking binges known as "sprees" or "skylarks." Controlled and restrained in their daily lives—by employers or social expectations—skylarkers felt exhilarated by freedom as much as by whiskey as they rambled from bar to bar to brothel, getting drunker and louder by the hour. Rarely were these frolics planned; they were celebrations of masculine camaraderie, experienced moment-by-moment in a liquor-fogged rush of impulses. Anything could happen. Skylarkers roared out songs as they stomped through sleeping neighborhoods, broke windows and streetlamps, threw themselves into bloody street brawls on trivial pretexts, fought with police, and smashed up whorehouses for the satisfaction of terrorizing people more powerless than they. The unpredictability and apparent irrationality of this behavior made it especially frightening for any individual accosted by these men. Mousy tailors by day could turn into wolfish predators by night.[51]

Even if no actual crime was committed, moralists asserted, a typical spree followed a downward moral trajectory. In a penitent account of his life written in 1850 before his execution, the murderer Henry Leander Foote described his introduction to New York twenty years earlier. On the boat from his native Connecticut, Foote fell in with a group of wild young men.

> The first night I was led to the Theater, from there to the brothel, and from there to the gambling house. . . . Here we must be fashionable and have a

game of cards and a bottle or two of champane [*sic*]. The cards I objected to, but one said we would be laughed at if we did not follow the fashion of the house.... [The] champane beginning to work, I soon surrendered to their wishes. We played and drank till sometime past midnight, when we concluded it was time to retire. Some one or two proposed to return to the brothel, but that I absolutely refused to do.[52]

Foote shed his moral scruples on subsequent nights and became an enthusiastic theatergoer, brothel visitor, gambler, and viewer of the "model artist" exhibitions of nude women.[53]

GAMBLING

If a night on the town mirrored the moral descent of the sinner, then where did it end? Foote and his companions thought the whorehouse was the fitting conclusion to a spree, the ultimate both morally and chronologically, but many other midcentury Americans considered the bottommost pit of night to be the gambling "hell." Gambling halls were said to keep the latest hours of all urban night spots. There the debauched young man threw away his last dollar and with it the economic virtues that had tethered him to daytime respectability.

Moralists who wrote about gambling were not much interested in games of chance among the poor and the working-class. A lot of this gambling took place in informal gatherings of neighbors, friends, and coworkers at leisure. Boys and young men would pitch pennies on street corners in daylight or play cards in tenements and taverns in the evenings. So little money was at risk that a player caught gambling on the Sabbath might not have even two or three dollars to pay his fine. A small loss in one game could be balanced out by a win against the same opponents the next night. Gambling was a more serious matter when run by professionals, in the view of midcentury writers, but still the cheaper gambling saloons were not worth getting upset over. Games of cards and dice were said to be played amid the same squalor as all the other vices of the poor—if not literally in the haunts of thieves, drunkards, and whores, then in similar places that would not tempt decent men to sin. Surrounding center-city Philadelphia, wrote George Foster in 1849, the districts of Spring Garden, Southwark, and Moyamensing were infested with gambling halls of "an inferior class." These were "generally kept in the back room or garret of some low groggery, and carried on as stealthily as body-snatching or murder. They are vile, filthy dens." A gambling "hole" on New York's West Broadway (a few blocks northwest of City Hall Park) drew loafing clusters of young men who harassed respectable passersby, gambled until 1:00 a.m., "and then [went] away drunk and disturbing the whole neighborhood." The gambling was no worse than the other lower-class pathologies, and certainly not as disturbing as drunkenness.[54]

Blood sports were an exception. Much to the displeasure of moralists, sporting men kept alive the ancient tradition of betting on cockfights, bare-knuckle boxing, and other violent contests. Some of the most popular events pitted dogs against rats, badgers, bears, boars, and each other. Blood sports were portrayed in lurid terms in sensational urban fiction and journalism; descriptions of the fights were often set in horrific cellars crammed with criminals. Yet blood sports drew men of all classes. One 1833 description of a cockfight noted that "sons of the aristocracy" mingled with Boston's lowest elements at a cockfight that lasted until 4:00 a.m. A Saturday matinee at a Boston rat pit in 1856 drew Benjamin Crowninshield and his Harvard buddies.[55]

Most of the contemporaries who described midcentury gambling focused on the more elegant gambling halls. As usual, they feared for impressionable youths from the countryside. In Cincinnati, warned the Reverend Samuel Fisher in 1852, the monster of gambling "lures into her snares thousands of the unwary. More than 500 gaming-houses of various kinds . . . open their doors to seduce the young, the manly, the noble, who enter our city to win a name, and secure a competency. . . . Night after night they are thronged with a motley company of all ages, from all classes of society." Large gambling houses were conveniently near leading hotels and theaters. Tipsy young men were led there by companions on a spree or lured in by touts. Inside, they were awed by luxuries that rivaled those of the most expensive brothels; first-class halls were "fitted up in a style of almost Oriental splendor," with expensive furniture, the finest wines, and a mouthwatering spread of gourmet dishes set out for the free supper. This display of excess made rich men feel at home; those who were seasoned gamblers played until they had lost as much as they were willing to risk and then left the table to converse with fellow gentlemen. Rich men were coveted customers at gambling halls, of course, but the main purpose of the decor was to dazzle ordinary men and evoke their avarice.[56]

A game could be found at almost any hour of the day or night. Gambling halls operated in the middle of the day in the New York's shopping district to serve wealthy ladies and in the neighborhood of Wall Street to serve brokers. But late at night was the preferred time. (We don't have to take clergymen's word for this; ample confirming evidence can be found in news reports of police raids in the wee hours.) Card play at the most elegant houses began near midnight, after a 10:00 supper, and lasted until dawn.[57]

Young men who gambled at night were believed to be robbing their employers—figuratively by arriving at work too exhausted to function, and literally by dipping into the till to cover gambling debts. Captivated by visions of wealth after glimpsing the "regal palaces of vice," the young clerks of Chicago were reported by the *National Police Gazette* to have developed "inordinate ambitions to become what their means will not permit them to be . . . fops, swells, gamblers, and partners of lewd women. The latter two are the most alarming indiscretions in which clerks indulge—indulge to an extent which would astonish their employers, if they were to be made aware of the fact. The gambling saloons—or rather hells—are chiefly supported by clerks, and it is safe to say that every clerk who gambles is a thief." There were enough real examples of embezzlement to make such a claim seem plausible.[58]

Some observers emphasized the similarities between gambling hells and other late night places of amusement. Like dancing, drinking, and whoring, gambling was a vice indulged in under cover of deep night. Like the ballroom, the oyster saloon, and the parlor house, the gamblers' den flashed the bright lights and other trappings of aristocracy to seduce the innocent. It brought the very rich together with the newly destitute. Similarly, it undermined the work ethic. Excessive sexuality, argued the reformer Sylvester Graham, robbed young men of the concentration and self-discipline needed for effective work. Gambling, wrote the Reverend Daniel C. Eddy, taught "the folly of working hard all day, and perhaps all night, for what can be secured in a single game. In this manner, it takes the attention from pursuits of business and industry, and congregates its subjects in saloons and cellars, where they can play at night, and lounge, and smoke, and curse, and sleep, during the day." Gambling halls and brothels even clustered together in some areas, like the Near West Side of Chicago or New York's Greenwich Street; some saloons offered both cards and courtesans.[59]

Yet in certain ways gambling was unique. It was not nearly as visible in the nocturnal city as the other sordid amusements. The extravagant exterior lighting at certain oyster saloons might falsely signify luxury, the light pouring from the windows of cyprianic balls might mimic the festive illumination of a formal dance party, but no gambling hall called attention to itself with Drummond lights or illuminated transparencies reading "Faro Bank." Gambling halls were often hidden in plain sight—occupying prosperous-looking homes on quiet side streets, the upper stories of business blocks, or rooms above shops. Special knowledge was needed to penetrate this mystery of the night. Newcomers first entered accompanied by a friend who knew the way or by a "decoy" covertly employed by the house to bring in potential marks. "There is just enough of secrecy about these houses to give additional zest to the business which is carried on within their walls," declared the editors of the *New York Herald*. By the 1860s, some New York gambling rooms displayed backlit gilt street numbers on their fanlights as signs for those in the know.[60]

The experience of gambling also differed from that of the other vices, whose devotees were encouraged to behave as sybarites seeking all forms of bodily pleasure. Obsessed gamblers instead became almost ascetic, eschewing vices in their monogamous devotion to Fortune. Henry Ward Beecher and lesser-known commentators found this fixation singularly disturbing. Some glimmer of humanity could be found in each of the other vices, but gambling, in Beecher's view, was "dark, malignant, uncompounded wickedness!" Spiritually exhausted by hours of unhealthy passion, the obsessive gambler had no feeling left, no principles, no sympathy for others. On the morning of October 2, 1850, at the moment when Henry Leander Foote was being executed in the yard of the New Haven jail, a game of faro was under way in an adjoining house, according to the reformed gambler Jonathan Green. As the trap opened and the condemned man's neck snapped against the noose, the players heard the sound but "gave no other heed to the horrid solemnities than to turn them into ridicule."[61]

No jokes were told during the heaviest games when men risked huge fortunes. These games, wrote Mason Long, "are conducted with the utmost decorum and amidst the most profound silence." Long recalled one game among gamblers with nerves of iron who continued to play while the building next door burned down. Beecher told of an all-night game of preternatural intensity: "In a room so silent that there is no sound except the shrill cock crowing the morning; where the forgotten candles burn dimly . . . sit four men. Carved marble could not be more motionless, save their hands. Pale, watchful though weary, their eyes pierce the cards, or furtively read each others' faces. Hours have passed over them thus. At length they rise without words." No one would mistake these games for fun, yet neither were they work. Though such gambling was more exhausting than the grimmest desk job, it produced nothing and left participants enriched or impoverished without regard to their efforts. Time passed unnoticed until the players left the table as if waking from a nightmare. Rather than building character like an honest job, gambling destroyed individual integrity, left the player morally if not financially "ruined," and (in some stories) forced him to commit suicide.[62]

All that survived of "play" in these joyless games, all that set them in opposition to work, was the belief in luck—a willingness to rest one's hopes on unseen forces beyond rational understanding. In place of the steady accumulation that rewarded middle-class work, gamblers wagered everything for the chance at sudden wealth. Similar wagers were placed every workday in the great financial markets of Wall Street and LaSalle Street. One of the reasons gambling flourished in cities, the *National Police Gazette* pointed out, was that "it is founded upon the same principle on which many branches of business are based—that of risk, or chance." More dishearteningly, critics such as Green suggested that "luck" was an illusion. Professional gamblers had so thoroughly fixed most games that there was no real chance.[63] Deceived and robbed of their money, suckers left the gambling hall at dawn as another workday was beginning, just in time to return to work.

A REVEALING COMPLICATION

One other late night amusement did not fit neatly into the tale of nightly moral decline. Serenading was at its peak of popularity in the antebellum era. On warm, moonlight nights, groups of perfectly respectable young men could be heard singing under the windows of perfectly respectable young ladies. Tone-deaf serenaders annoyed everyone except, perhaps, the honored targets of their efforts. Others who were more talented or who employed professional musicians could be enjoyed by anyone within earshot. "Late at night, I am often awakened by several voices, singing under the windows of the female Seminary on the corner of Fourth and Vine," wrote a Cincinnati law clerk in May 1843. "It was a pleasant interruption to my slumbers. I have distinguished several favorite airs, 'The Miller's Maid,' 'Midnight Hour,' and 'Near the lake where drooped the willow'—all sung with taste and skill." Enthusiastic young men formed serenading clubs that traveled through the streets after midnight, instruments in hand, to regale their girlfriends.[64]

Not everyone out in the wee hours was a degenerate rake, obviously. Instead, the ritual of serenading reveals a more fundamental feature of late-night leisure: it empowered men and disempowered women. Outside the home were men, actively enjoying themselves; inside were the women, passively listening. The serenade allowed the expression of sexual desire in a highly stylized manner somewhat like parlor courtship, but with an important exception: late at night, the men were firmly in control.

Still dominated by pleasure-seeking young men, the gaslit city encouraged the same habits and values that had prevailed in the preindustrial era:

impulsivity, self-assertion, aggressiveness, and hedonism. These qualities, in turn, continued to shape the public spaces and activities of the nocturnal city. Night entertainment, night work, and night travel expanded interdependently through the late nineteenth and early twentieth centuries. Night entertainment created jobs for actors, musicians, bartenders, and many others; night work meant more demand for public transportation; public transportation made it easier for people to go out at night in search of entertainment. All shared in the common culture of the nighttime street.

CHAPTER 6

Nightmen

Through the era of gaslight and into the age of electricity, more and more people labored at night in arduous, low-paying jobs. "Night life," a term increasingly equated with commercial entertainment, depended on a growing number of stagehands, dishwashers, and "pretty waiter girls." It took more than 68,000 saloon keepers and bartenders in 1880 to fill the glasses of a thirsty America. By 1900 in New York City alone, nearly eighteen thousand men made their living pouring whiskey and drawing beer, while seven thousand men and women struggled for success as performers. In Milwaukee, as many as two thousand bartenders and saloon keepers worked until nearly midnight, and several hundred worked all night. Smaller cities like Omaha and Toledo had their own cohorts of a few hundred bartenders, waiters, actors, and pool hall keepers. Actresses and waitresses made up about a third of their respective workforces, but the other legal entertainment jobs were almost all male. Evening was when these workers were busiest, so they had to live with a schedule that was out of phase with the usual daily cycle of work and rest. Other entertainment workers had jobs that were entirely "countercyclical"; that is, they worked while everyone else relaxed and slept while others started work. Streetwalking prostitutes, for instance, so commonly worked during hours of darkness that they were known euphemistically as "night walkers," a term that appeared in city ordinances and other official documents into the early twentieth century.[1]

Two other types of workers were so closely linked with night that they were named and legally defined accordingly: these were the night watchmen discussed in previous chapters and the night scavengers, sometimes called "nightmen." Both these jobs had to be performed countercyclically; these workers were required to be on the streets at hours when most other people were not. The numbers of nightmen expanded quickly in the mid-nineteenth century even as the growth in recreational nightlife eroded their separation from public activity. Technological innovation in waste removal first increased the scale of night scavenging, then nearly eliminated it. Most other forms of mid-nineteenth-century night work were deliberately scheduled to avoid interfering with the normal rhythm of urban activity. For instance, much of the work of supplying the city's food took place in the predawn hours when streets were cool and empty. Scavengers and marketmen served the urban digestive tract by working on or outside the margins of daily life. Newspaper employees and newsboys could be found at work at any hour, but their busiest times were before and after the normal daylight workday.[2]

COUNTERCYCLICAL WORK: CLEANING THE CITY

Americans in the early nineteenth century deposited human wastes in privy vaults in their cellars or in outhouses behind their homes. When the privies filled up they would be covered over and replaced, or else emptied by scavengers at the homeowner's or landlord's expense. Scavengers in most cities were required to remove the material only at night; hence the euphemism "night soil" was applied to human waste and the terms "night scavengers" or "nightmen" to those who collected it. Privy cleaning was a low-status job performed by African Americans and European immigrants. The introduction of reservoir water into cities in the mid-nineteenth century overwhelmed this system of waste disposal. As thousands of people installed bathtubs, sinks, and water closets before sewers were available to remove the wastewater, their leaky cesspools and overflowing privy vaults saturated the soil and left malodorous pools of standing water in low areas. Cities belatedly responded by creating sewer systems in the 1850s, 1860s, and 1870s. By the early 1890s, New York and Brooklyn together had 844 miles of sewers, Chicago had 525, Philadelphia had 376, St. Louis had 328, and Boston had 291. By 1902 those totals had grown to 1,467, 1,529, 951, 522, and 582, respectively.[3]

Water closets—early versions of the flush toilet—were standard in middle-class urban homes by 1900 and increasingly common even in working-class tenements, but their installation was a prolonged process. By 1880, only about a third of urban households had water closets connected to sewers. The situation varied from city to city. Nearly all the houses in Cambridge, Massachusetts, were so equipped, and officials in Toledo and in Portland, Maine, also claimed that large majorities of local homes had water closets. At the other extreme, water closets were still uncommon in cities with small or nonexistent sewer systems, including such major centers such as Baltimore, Indianapolis, Kansas City, Omaha, and St. Paul. The installation of a sewer main did not guarantee that people would get rid of their privies. Frugal property owners often refused to spend the money for a sewer connection, or in some cases they kept a privy for servants. In all but a few cities, privies remained common in neighborhoods to which sewers had not yet been extended and persisted even in well-sewered areas. As late as 1950, some slum dwellers in Pittsburgh still lacked flush toilets.[4]

Before the water closet and sewer finally replaced the old outhouse, the explosive growth of urban populations turned the once-quiet business of privy cleaning into a huge enterprise employing hundreds of night scavengers as independent entrepreneurs or as employees of major city contractors. In poorly sewered cities where people had connected toilets to cesspools, the flood of wastewater forced more frequent cleanings, sometimes as often as every ten days instead of two or three times a year. It is difficult to find reliable employment figures for night scavengers, but the scale of the operation is suggested by the enormous volume of waste. Nearly one hundred tons of excreta were removed daily from Newark in the 1880s. In Baltimore, which lagged far behind other major cities in building sewers, scavengers removed more than sixty-six thousand tons of human waste in 1905.[5]

City ordinances through most of the nineteenth century required that this work be done during specific hours in the dead of night. In mid-nineteenth-century New York, the work was performed between 11:00 p.m. and 3:00 a.m. in the late spring and summer, or between 10:00 p.m. and 6:00 a.m. the rest of the year. Chicago, Detroit, Milwaukee, Newark, St. Louis, and numerous other cities set similarly specific hours. The work involved scooping filth out of the privy vault with a long-handled dipper or bucket and pouring it into barrels or a tank on a wagon (sometimes euphemized as a "midnight vehicle"). The scavengers then hauled the waste to a convenient dumping spot: on vacant land, on a farmer's field, or in a nearby body of water.[6]

In New York the dumping spot was typically off a pier into the East River or North (Hudson) River. John Griscom, a New York public health official, complained in 1842 that lazy scavengers took advantage of darkness to deliberately slosh much of their load onto the streets or even to dump the entire quantity before they reached the piers. This was a common habit among night scavengers in other cities as well. Unfortunately for New York's air quality, most of the work appears to have been done during warm weather, a trend that may be partly explained by the difficulty of removing frozen waste during winter. In August 1853 well over six thousand cartloads of night soil were dumped off the four designated piers: a full slop cart typically contained close to a cubic yard of material. Roughly the same amount had been dumped in the two-month period of March and April. The stench of the open night soil carts on hot nights was an "abominable nuisance" for neighborhoods where the scavengers were at work, to say nothing of waterfront districts by the piers. The *New York Tribune* urged requiring that privies be emptied only in colder weather, using pneumatic machines and disinfectants, and that the waste be hauled away in garbage scows instead of left to befoul the shoreline.[7]

Major reforms were enacted a decade later. In 1865 the city began contracting for the shipment of its night soil to a buyer in New Jersey. The smell of the city shoreline may have improved somewhat as a result, but the aroma of leaky slop carts continued to annoy New Yorkers and diminish waterfront property values. The annual quantity of night soil hauled through the streets to the piers grew from forty-eight thousand cartloads in 1860 to sixty thousand cartloads by the end of 1866, while additional loads were dumped in vacant lots in the Upper West Side, an area that had already endured the stench of human excrement spread as fertilizer in the construction of Central Park.[8] In 1872, night scavenging came to an end in New York with the city's decision to award a contract for all night soil removal to the Manhattan Odorless Excavating Company. The company's pumping machines and airtight tanks allowed the waste to be removed with a minimum of nuisance at any hour. New York therefore began allowing night soil to be removed during daylight hours, while simultaneously abolishing the old bucket and cart method of emptying privies. By 1880, several other major eastern cities—Boston, Brooklyn, Philadelphia, and Baltimore—also permitted or required the daytime removal of human waste by pneumatic machines. "The work originally was done at night, but the carts made such a racket, lumbering over the cobblestones, that Baltimoreans decided they would rather endure the nuisance by day than have their slumbers disturbed. Since then it has been a 'night soil' business only in name," explained a later observer.[9]

Over the next twenty years, most other major cities began requiring the use of pneumatic machines to empty the shrinking number of urban privy vaults, but they were divided over whether the waste should be removed during the day or at night. Buffalo and Providence opted for daytime removal, but Chicago, Cincinnati, Cleveland, Milwaukee, and Omaha continued to require night removal. Night work therefore persisted in some unsewered slums and working-class neighborhoods, particularly neighborhoods far from the central business district. Moreover, the pneumatic machines' technical limitations often forced scavengers to resort to the old dipper and bucket method, just as nightmen had been doing for generations. Like the lamplighter who continued to walk the poorly lit streets of the urban fringe, the night scavenger remained a familiar anachronism on the technological periphery.[10]

Another type of sanitation worker could be found in the darkened central business districts. During hours when the streets were free of traffic, "broom brigades" in some cities swept up horse droppings and other refuse from the pavement. Every evening in New York except Sunday, according to *Harper's Weekly* in 1884, "nearly eight hundred men, wielding brooms, hoes and shovels, descend upon the dirt in the thoroughfares like locusts upon a field of grain." Workers, mostly immigrants from Italy and other parts of Europe, brushed the cobblestones with thick-bristled brooms, then shoveled the piles of filth into carts to be taken to the waterfront and dumped into barges for disposal at sea. By the 1880s, such work was increasingly supplemented by enormous mechanical street sweepers. Nighttime hours not only kept the sweepers out of the way of traffic but also spared the public from the clouds of dust stirred up by rotating brooms. Snow shoveling in central business districts began whenever the snow stopped falling, often at night. Street cleaning outside the central business districts was largely a daytime job involving men pushing brooms and scrapers.[11]

Some garbage removal was also done at night, particularly in business districts where the cans would otherwise pose smelly, unsightly obstacles to pedestrians and the carts would block traffic. Further, restaurants and hotels sold their kitchen garbage and table scraps to swill collectors who made their rounds late at night. The household rubbish, garbage, and ashes in the residential areas of some cities were collected at night as well. One example was Milwaukee, where public complaints about the stench of garbage wagons in the summer of 1910 forced the collection department to remove household trash between 2:00 and 10:00 a.m.; winter collections took place between 4:00 a.m. and noon. But nighttime collection posed its own problems. The noise disrupted people's sleep, garbagemen could not find the trash cans as easily in the dark, and the cans were sometimes stolen after being emptied. For those reasons, household trash was most frequently collected during daylight hours.[12]

EARLY PHASED WORK: FOOD

Night scavengers worked when most city people slept. In contrast, the work of supplying the city's food straddled the early margin of the ordinary waking day. Provisioning urban America involved a great deal of predawn labor in order to have fresh food ready for early morning shoppers. The hours were dictated partly by the custom of morning market sales, partly by municipal regulations, and partly by the lack of adequate refrigeration; perishable food spoiled quickly when transported or exposed for sale during the heat of the day.

Commercial baking has historically been performed at night, and bakers continued this tradition in nineteenth-century America. They lit a wood fire to bring the brick oven to a high heat; then they raked out the wood and coals, swabbed the bricks clean, and closed the oven for about two hours until it achieved the desired uniform temperature. Meanwhile, the master baker and his journeymen or apprentices sifted flour, prepared a yeasty sponge, mixed the dough, kneaded it, and let it rise. Bread was baked in time to be sold warm to early morning customers. Bakeries in 1850 were still small craft operations. The roughly two thousand bakeries in the United States employed 6,727 wage earners at the time. Household baking declined in the latter half of the nineteenth century. Although small-scale bakeries persisted, new industrial baking companies claimed much of the growing urban market for commercially baked bread. Large-scale, mechanized bread production flourished during the Civil War thanks to government purchasing. By 1899, the baking industry employed over sixty thousand Americans, 80 percent of them men, 17 percent women, and 3 percent children. The numbers of child laborers dwindled in the early twentieth century, but those of women grew quickly.[13]

The nights of preparation and the mornings of selling bread added up to excruciatingly long hours during the era of the small craft bakery, and long hours persisted even as the small shop gave way to the industrial bakery with dozens of employees reporting to foremen. Striking bakery workers in New York and Brooklyn complained in 1868 that they routinely worked sixteen hours a day. Conditions had worsened by the time of an 1881 strike, when workers reported putting in eighteen-hour days and being denied Sundays off. "Extra work has not infrequently been forced upon us, and I myself, within the past 10 days, was required, in a rush of work at my shop, to put in three straight days, during all of which time I had but seven hours for my meals and sleep," said one union leader. A pro-union bakery in 1886 reported that workdays in Buffalo's nonunion shops ranged from twelve to twenty hours, and that men were sometimes asked to work shifts as long as twenty-two hours. This bakery boasted that its own employees were divided into shifts working only ten hours each, the night shift usually ending by 4:00 a.m.[14]

Fresh milk on urban breakfast tables depended on a newly contrived system of night work in and around major cities. Americans had once consumed little of their milk in liquid form, preferring to preserve it as cheese and to drink water, cider, ale, or other beverages that did not spoil so easily. As cities grew larger and more congested in the early nineteenth century, urban consumption was further discouraged by the increasing difficulty of transporting unpasteurized milk to the marketplaces before it turned sour. Dairy cows dried up during the winter, when fresh raw milk could be transported safely, and reached their peak production in summer months when it could not. Large-scale urban dairies arose conveniently near the centers of consumption but became notorious for their dirty, adulterated, low-quality product. The growth of railroads in the 1840s, along with greater use of ice, helped satisfy a growing demand for fresh country milk, now being promoted as a temperance beverage.[15]

Dairy regions developed along the rail lines that radiated from major cities. Farmers carried full milk cans to country railroad stations or later to milk receiving stations, from which the milk was eventually placed on the special "milk trains" that ran toward the city each night. By the 1880s, over one hundred thousand gallons a day arrived at New York's railroad depots and ferry piers. Milk ferries from New Jersey in those years typically arrived between midnight and 4:00 a.m. Milkmen working for dairy companies then distributed the milk between 4:00 and 8:00, pouring it into containers on back porches or inside groceries.[16]

Similar concerns about spoilage influenced the municipally regulated hours of produce markets. Whether located in specially designed halls or in open-air marketplaces, public markets were important centers of perishable food sales in the early nineteenth century and continued to draw thousands of customers through the early twentieth century. Sales typically took place between sunrise and midday, though closing times varied depending on the city, the market, and the perishability of the item being sold. An 1880 US Census inquiry found markets opening as early as 3:00 or 4:00 a.m. in Detroit, St. Paul, Washington, DC, and Allegheny, Pennsylvania (now part of Pittsburgh). Housewives and servants arrived with empty baskets shortly after daybreak on market days, looking to buy meat, fish, eggs, milk, and produce. To secure a good location for their wagons and wheelbarrows, butchers, farmers, and other vendors arrived hours earlier.[17]

This form of night labor extended deep into the surrounding countryside. Late nineteenth-century American cities were ringed by a zone of truck farms and dairy farms serving the urban market. As early as the 1830s, farms in western Long Island focused on producing millions of heads of cabbage, along with other vegetables and fruit. The growth of railroad networks in the 1840s and 1850s provided faster, more convenient links between countryside and city, encouraging a decisive turn to truck farming in New Jersey and parts of rural New York.[18] Some local truck farmers continued to make the trip by horse-drawn wagon in order to deal directly with consumers. The sprawl of the city across former farmland gradually extended the length of the trip. By the early twentieth century journalists observed farmers setting off for the distant city long before midnight—even before dark—to claim good spots in the markets. A reporter for *Harper's Weekly* in 1909 described a nightly migration from Long Island and New Jersey: "Farm wagons enter the city that never sleeps. The drivers take a firmer grip on their lines as they cross the brightly lighted avenues and watch at corners for the swift trolleys that seem to try to pounce on them." Arriving safely at the open-air Gansevoort Market near the Hudson River, some three hundred or more farmers would stop for the night, stabling their horses and sleeping in their wagons. The less frugal ones ate at the late night restaurants, drank in the saloons, patronized the prostitutes who worked the marketplace, or slept in one of the nearby lodging houses.[19]

Nature's Metropolis

"The markets on the water-front are the heart of the city's night life," a New York journalist declared in 1914, using the term "night life" to mean "activity." A similar nocturnal economy served the farmers who arrived at Boston's Faneuil Hall marketplace each evening. Farmers stabled their horses nearby and slept in the haylofts, trusting their wagonloads of produce to the care of the market's night watchman. They were roused at 3:00 a.m. by a wake-up call from a stableman, stumbled off for breakfast or coffee at a nearby restaurant, and were at their wagons when the first wholesale buyers arrived before dawn.[20]

Sales to wholesalers accounted for most of the business at public marketplaces by the early twentieth century, at least in the major cities. Retail market sales were in decline in parts of the urban United States by 1880, as retailing shifted to privately owned groceries and butcher shops. Of the twenty-two hundred businesses selling food in Boston in 1880, only a small minority operated out of the city's Faneuil Hall market. The roughly four hundred marketmen there did most of their business with other merchants rather than directly with consumers. Municipal markets had closed down altogether in many other New England cities, though they still held on to a large share of retail food sales in southern towns and at scattered points in the Northeast and Midwest. The declining importance of direct sales from food producer to consumer can be explained in part by the growing size of the cities; morning trips to the public market became more inconvenient, especially compared with visiting a neighborhood grocery with longer hours. More significant, though, a larger share of the city's food arrived from long distances away by rail or ship instead of coming in on a farmer's wagon. The expansion of the nation's railroad network, and the development of refrigerated boxcars in the 1870s and 1880s, made it possible to transport fresh meat and produce to eastern cities from Chicago or California and to sell it at prices local farmers could not match. This new flood of food arrived at the docks and freight yards and was distributed from privately owned cold-storage warehouses. Some of it ended up being retailed at the public markets, but wholesalers sold much of it to private groceries and butcher shops instead.[21]

The creation of this complex system of middlemen and long-distance shipping did not put an end to night work. Through the late nineteenth and early twentieth century, the produce distribution system still included many small jobbers, grocers, and peddlers who lacked the expensive refrigeration common in large-scale enterprises. Further, produce was often exposed to the open air during the final leg of its journey to market. Whether hauled by a local farmer or rushed from California in a cooled boxcar, fruits and vegetables ended up being carted through the streets of Chicago's Loop to the South Water Street Market. This time outside refrigeration (and the need to avoid heavy traffic) was undoubtedly a strong force for the continuation of predawn marketing hours. Shippers, wholesalers, and retailers seem to have shared a preference for nighttime or dawn transport of produce at every step of the way to the grocery. New York's docks and freight yards in the late nineteenth and early twentieth centuries received produce shipments at all hours, but particularly at night. The Pennsylvania Railroad in June 1893, for instance, ferried two hundred freight cars full of southern produce across the Hudson every night from its Jersey City terminus. The Old Dominion steamship line unloaded ten boats a week, the ships docking about 8:00 in the evening. Freight handlers at the Pennsylvania Railroad's pier began unloading the boxcars about 7:00 p.m., though the consignees were not allowed onto the pier to receive the produce until after midnight. Longshoremen at the Old Dominion pier started carrying out the packaged fruit and vegetables as soon as the ships docked. The day's first sales of southern produce began at a midnight auction on the waterfront, according to a 1911 account. The purchasers carted the produce through the streets for resale at sunrise in Manhattan's West Washington or Brooklyn's Wallabout wholesale market, or for delivery directly to grocers in the early morning.[22]

Retail produce sales continued through the day from peddlers' carts and from inside groceries. In this sense the produce business as a whole was

A SATURDAY NIGHT SCENE IN THE BOWERY, NEW YORK.

FIGURE 8. Night market in New York. Much of the work of provisioning a city took place in the early morning hours, as farmers traveled from the surrounding countryside to be at the wholesale marketplaces before dawn. This engraving shows a different sort of night work—a retail market operating in the late evening. Evening markets did their best business on Saturday nights, when they sold low-grade produce to the poor. "A Saturday Night Scene in the Bowery, New York," *Harper's Weekly*, May 20, 1871.

not truly countercyclical. One final phase of retailing, though, took place at night. As darkness fell, so did street peddlers' prices, as they sought to dispose of wilting lettuce and blotchy bananas. The poor chose this time to do their shopping. Some peddlers specialized in serving this clientele, buying up rotting produce from wholesalers and grocers at the close of the day's business to resell it on the streets. This business reached a climax on Saturday night, as families with the week's wages in hand sought bargains from grocers whose produce would spoil before they reopened on Monday. Washington, Detroit, and other cities set special Saturday night market hours, while New York looked the other way as hucksters occupied certain streets for the evening. New York's Saturday night market in the early 1880s was held on Eighth Avenue at Forty-Second Street; by 1890 it had moved to Ninth Avenue from Thirty-Fourth to Forty-Second Streets, and First Avenue south of Fourteenth Street. On Ninth Avenue, reported *Harper's Weekly*, "The hustling huckster is here in noisy multitude, lining every foot of the curbing with wagon and hand-cart, and encroaching upon the side streets for a distance on either side of the avenue. The wagons, heaped high with the various stocks in trade, are dimly lighted by the red glare from hundreds of smoky naphtha torches. The sidewalks are crowded from curb to house walls with a good-natured, struggling throng, nine-tenths of them women." The hucksters, however, were almost all men.[23]

NEARLY INCESSANT: THE MAKING AND SELLING OF NEWSPAPERS

The islands of nighttime activity centered on the food markets were remarkable but not unique. Similar patterns could be seen in other centers of nighttime work such as the "Printing House Square" (or later "Newspaper Row") district of Lower Manhattan. All-night restaurants and saloons served the many journalists, printers, and drivers of delivery wagons. Near the office of the *New York Tribune,* for instance, midcentury newsboys gathered at "Buttercake Dick's" to swallow hot coffee and greasy biscuits before the first editions came off the presses. Much of the circulation of city newspapers passed through the railroad stations and post offices before dawn for distribution to the suburbs, rural districts, and distant cities. Reporters and printers were regular customers for otherwise underused streetcars and ferries in the pre-dawn hours.[24]

"It is the printing presses and the ovens of New York that keep by far the largest proportion of the Night Workers busied till long past dawn," declared a writer for the *Brooklyn Eagle* in 1901, estimating that 3,120 New Yorkers worked for newspapers at night. These workers included reporters, editors, engravers, compositors, and typesetters, as well as stereotypers, pressmen, and delivery room managers. As the night passed, each group of workers experienced its own successive moment of frenzied work as the process of putting out the morning's paper moved from the scribbling of journalists to the squawking of newsboys in the streets. Only the business offices followed the normal daylight work cycle, processing mail, handling payroll, and interacting with customers concerning circulation and advertising; even there, a few night clerks continued through the evening. Thus, although there was no hour of the day when business paused entirely, the work was largely countercyclical.[25]

The daily (or nightly) miracle of producing a big-city newspaper began at a leisurely pace, according to descriptions from the 1850s through the 1910s. Reporters attended early morning sessions of police courts to gather crime stories and human interest items and continued to work on various news assignments throughout the day. Editors met about noon to sketch out the next morning's news content and to assign the writing of editorials on selected topics. A small early shift of compositors put together extra editions that were issued in the afternoon. The pace of work picked up after 6:00 p.m. The night editor and night reporters came on duty between 6:00 and 8:00. Copyeditors began to revise the articles already submitted and then received new copy as it was written. As soon as a copyeditor completed each article, he handed it to an office boy to carry to the composing room. The night editor decided how best to remake the paper for each edition as events took place in the city during the evening or as important news arrived by telegraph. "After seven o'clock in the evening the composing room becomes a perfect beehive of industry and activity," according to an 1857 description of the *New York Herald*. Once a page was set in type, it was used to form a metal stereotype plate that could be affixed to a cylinder in a high-speed rotary press. The actual printing of the paper took place late at night in the cellar press room. At the *Herald* in the 1850s, enormous steam-powered Hoe presses rumbled to life about midnight and continued until 6:00 or 7:00 a.m., by which time the first editions had already been read by the city's early risers. As papers came off the presses, other workers folded and bundled them, then handed them over to an impatient crowd of mail packers, wholesale dealers, and newsboys who streamed through the distributing department. At the *New York Sun* in the 1870s, the first papers were in the hands of news vendors about 3:30 a.m.[26]

Most of this work took place indoors, but it influenced the public spaces of the city directly through the night reporters who gathered late news and the newsboys who sold it. Though night reporters were few, they were constantly shuttling through the streets from one assignment to another and then back to newspaper offices, often gathering news items along the way. Papers kept a reporter on duty until the final edition went to press and often stationed another at the police headquarters in case some major event occurred. Reporters in the final decades of the century waited at New York's Mulberry Street police headquarters, wrote Jacob Riis in 1901. "The watch now is kept up through the twenty-four hours without interruption. Like its neighbor, the Bowery, Mulberry Street never sleeps."[27] Journalists' familiarity with the nighttime city, and easy opportunities for publishing their impressions, allowed them to interpret the experience of urban night for the general public. Their tendency to accentuate the dramatic, violent, and exotic undoubtedly contributed to the persistent mystique of night in the popular imagination and thus shaped how urban space was perceived and used.[28]

Newsboys were a much more visible and audible presence in the nighttime streets. Many children peddled goods on the streets in the late nineteenth and early twentieth centuries, either to help support their families or to earn spending money for themselves. They clustered outside theaters at night to black boots and sell candy, and they entered saloons in search of profitable drunken customers. Some girls in their teenage years and younger used peddling as a front for prostitution; a bouquet of flowers could serve as an excuse for loitering on the street at night. Of all of these child peddlers, newsboys were by far the most numerous.[29]

The soaring circulation of American newspapers over the course of the nineteenth century created opportunities for hundreds or even thousands of newsboys (and girls, to a lesser extent) in every major city. Most of them sold to morning and afternoon commuters, but newsboys were also familiar figures within urban night. Evening newspapers, which proliferated in the second half of the nineteenth century, relied more heavily on street sales than on home delivery to subscribers. They regularly issued multiple editions, allowing fresh news to be peddled through the evening. Improved lighting on the streets and in homes made reading after dark easier. Morning newspapers in the early twentieth century issued their earliest editions before midnight, particularly Sunday editions that were sold on Saturday nights to crowds emerging from theaters. Any major news event might lead to the printing of "extras" at any hour of the day or night. As each new edition appeared, newsboys cried the headlines along the busiest streets and in hotel lobbies and saloons. On occasion they entered residential areas as well. George Templeton Strong told of being awakened at his Gramercy Park home at 11:30 one night by "a herd of highly excited newsboys" peddling an extra on the welcome arrival of a missing ship. His contemporary New York diarist, Maria Lydig Daly, told how her husband, Judge Charles Daly, heard of the Union Army's triumph in the Civil War: "Last night at midnight he heard an extra called. The Judge rushed to the door. 'Surrender of Lee's army, ten cents and no mistake,' said the boy all in one breath." Thus, the newsboys were one means by which the night world could intrude on the lives of those who normally worked by day and slept by night.[30]

A small but influential minority of newsboys made a more radical break from the daily lives of other children by living on the streets and supporting themselves with their earnings. Not attending school or living in family homes, they were free to adapt their daily schedule to the rhythms of the newspapers and to their own whims. Throughout the late nineteenth and early twentieth centuries, these semi-vagrant children shaped the development of a newsboy subculture that vigorously asserted its members' rights to the city. Experienced newsboys boldly hopped on and off streetcars without paying and went in and out of seedy saloons and brothels. They were notorious for their ploys to elicit sympathy and tips, for crying fictitious headlines, for persistence that bordered on harassment, and for shortchanging customers. They were quick to gang up against any adult who tried to cheat or abuse a fellow newsboy. Having access to small amounts of spending money, they often passed the time between editions in cheap eateries, candy shops, penny arcades, and movie theaters. Newsboys could be found late at night sleeping in out-of-the-way spots near the newspaper offices, typically in groups for companionship and mutual protection. Their numbers swelled on warm summer nights when they were joined by other newsboys who lived mostly with their families. Newspapers were glad to have the boys nearby to peddle the early morning editions or midnight extras.[31]

The children who lived this semi-nomadic existence, known at the time as "street Arabs," were by no means a distinct group limited to night work. Boys floated in and out of this group in response to changes in the weather and to abuse at home. In daylight and through the evening, they shared the streets with newsboys who slept in their families' apartments each night and attended school regularly. Loitering in the news alleys waiting for papers, they also shared their wise guy subculture with other newsboys. Social reformers warned that street Arabs corrupted children from decent homes by teaching them to smoke, swear, gamble, and drink. Early twentieth-century opponents of child labor warned that newsboys late at night were more likely to be exposed to the immoral side of urban life, particularly drunkenness and prostitution. They feared that the sordid culture of the nocturnal streets, combined with the irregular hours, would prepare newsboys for a life of crime.[32]

Except for peddling, most night work in the city took place out of sight and out of mind. A city resident who began the day by reading the newspaper, eating breakfast, and visiting the privy could easily forget that this morning routine was made possible by the night labor of many men. It was indeed men, not women or children, who performed most of this work. The *Brooklyn Eagle* estimated in 1901 that at least twenty thousand people in New York City worked at night, almost all of them men; along with bakers, marketmen, and newspaper workers, the article noted the labor of post office clerks, streetcar operators, wholesale butchers, and telegraphers, among others. Though hundreds of New York women worked into the night sewing garments in their tenements and hundreds of others cleaned the floors of office buildings, department stores, and hotel corridors, these were said to be the exceptions.

"Womenkind play practically no part in the ranks of night workers," the *Eagle* declared; it did not discuss children.[33]

The *Eagle*'s count was incomplete, but it may actually understate how thoroughly men dominated night work in American cities. The paper's estimates did not include shift workers in factories. Shift work was common by this point in sugar refineries, petrochemical plants, gasworks, and electric power stations, all of which could be found in Brooklyn and other parts of greater New York. Few women were employed by such establishments. Outside of atypical Gotham, the situation in the heavy industrial cities of the Northeast and Midwest was even more clear-cut, despite the night labor of some women in textile factories. The ceaselessly running steel mills, glass houses, and other industrial complexes that emerged in late nineteenth century rarely if ever employed women at night; they helped ensure that the nocturnal city would remain a man's world into the twentieth century.

CHAPTER 7

Incessance

New York may have been the capital of night entertainment, but the capital of night work was Pittsburgh. Night work there was obvious to any late nineteenth-century visitor, advertised by the mysterious clanging and roaring behind mill walls and exaggerated by flames from furnaces and coke ovens. A travel writer in 1886 offered the following advice:

> By all means make your first approach to Pittsburg in the night time, and you will behold a spectacle which has not a parallel on this continent. Darkness gives the city and its surroundings a picturesqueness which they wholly lack by daylight. It lies low down in a hollow of encompassing hills, gleaming with a thousand points of light, which are reflected from the rivers. . . . Around the city's edge, and on the sides of the hills which encircle it like a gloomy amphitheatre, their outlines rising dark against the sky, through numberless apertures, fiery lights stream forth, looking angrily and fiercely up toward the heavens, while over all these settles a heavy pall of smoke. It is as though one had reached the outer edge of the infernal regions, and saw before him the great furnace of Pandemonium with all the lids lifted.[1]

Nothing like this had been seen in the United States a generation earlier, despite the presence of iron-rolling mills and glass houses in Pittsburgh since the early nineteenth century. One writer claimed in 1845 that night in the smoky city was so "death-like still" that little could be heard but the cry of the night watchman and "the low rippling music of the waters environing the city."[2]

The spectacle of late nineteenth-century Pittsburgh was created by the

FIGURE 9. Hell with the lid off. Continuous-process industries produced steel, glass, and petrochemicals around the clock in the late nineteenth and early twentieth centuries. This 1931 photograph shows the Jones and Laughlin Steel Corporation's Soho Works at night, with the electric lights of downtown Pittsburgh in the background. F. Ross Altwater, "Industry: Overview of Steel Mill on the Monongahela River at Night," Carnegie Museum of Art, Pittsburgh; Gift of the Carnegie Library of Pittsburgh.

emergence of large-scale industries, operating incessantly, in the decades after the Civil War. Production never stopped in the iron and steel mills, glass houses, rail yards, and oil refineries throughout the Pittsburgh region. Though oil refining soon shifted to New Jersey and Cleveland, the other industries continued to thrive. Just within the city limits of Pittsburgh, the workforce in 1900 included about two thousand glass workers, thirty-five hundred railroad employees, and ten thousand iron- and steelworkers. All but a few workers in each of these industries were men.[3]

The growth of incessant production created a new experience of night for thousands of workers as well as the occasional tourist. Before the late nineteenth century, most night work came only at the beginning or end of a daylight stint. The workday may been stretched out on either end, or shifted earlier or later, but it was still anchored to the normal daylight schedule. Industries that ran twenty-four hours a day attempted to cut that connection, creating spaces dominated instead by the rhythms of production. Yet outside the factory the sun still rose and set, traffic ebbed and flowed, laundry dried on lines, and children yelled as they ran home from school. Workers torn between two schedules found the experience disorienting and unpleasant.

SHIPS, TRAINS, AND STREETCARS

Incessant production was pioneered by the transportation sector. Long-distance shipping had always operated night and day; sea captains institutionalized a prototypical shift system through the tradition of "watches." As suggested by the example of produce shipping, twenty-four-hour work was common in the second half of the nineteenth century on the illuminated waterfronts of American port cities. Night work grew rapidly in the late 1860s in New York, America's largest port at the time. Longshoremen, who performed the backbreaking labor of loading and unloading cargo, had previously worked daylight stints ranging from a few hours to a long day depending on when ships came in. Improvements in lighting in the 1860s allowed shifts to become even more irregular as sundown no longer meant quitting time. One gangway man recalled working a stint in the late 1860s that lasted more than a hundred hours, from Monday morning to the next Saturday afternoon. Men who pushed themselves through such stretches could manage it only with the help of coworkers who covered for them while they snatched quick naps in the hold. Marathon shifts became less common in the late 1890s, but in the early twentieth century many New York longshoremen still worked the occasional thirty-five- or forty-hour shift—two full days, linked by a night under the harsh arc lights of the piers. Thanks to gas and electricity, sunset no longer guaranteed that dock workers could rest; at best, it meant a higher pay rate.[4]

Pressure for these all-night ordeals was stronger in foreign commerce than in the coastwise trade. Owners of large transatlantic ships wanted to minimize unproductive time in port and were willing to pay a premium for night work to ensure that ships were unloaded and loaded without interruption. Longshoremen, their income limited by unproductive hours waiting in waterfront saloons for ships to dock, were willing to extend their shifts into the night even though they considered night work less desirable. In 1872 New York longshoremen secured a pay rate of forty cents an hour for day work and eighty cents an hour for night work on oceangoing ships. Pay rates soon dropped, but workers insisted on maintaining the differential between night and day work in New York, as in Boston, Brooklyn, and other ports. Night work was less common in the coastwise trade, where ships were smaller, arrived in port

FIGURE 10. Yardman on the night shift. Railroads were among the largest twenty-four-hour industries in nineteenth-century America. Many workers were injured in poorly lit rail yards, where tired men labored amid the trains. Thornton Oakley, "South Station, Boston. Looking out from the Station Shed at Signal Bridge and Lights," *Scribner's Magazine*, October 1912.

at more predictable times, and—in the case of New York—tended to dock on the more dimly lit East River waterfront. New York longshoremen serving the coastwise trade did not secure a pay differential until 1898. Longshoremen in Portland, Maine, which served as the winter port for icebound Montreal, struck unsuccessfully in 1911 for a wage scale that would have given them further bonus pay for work between midnight and 6:00 a.m.[5]

Railroads—with their sprawling urban systems of rail lines, marshaling yards and depots—were among the most visible sites of nocturnal work. Shops and factories may have closed, streets might be dark, but lights still glowed in the windows of depots, on lampposts at grade crossings, and in the mysterious colored signals of the yards. In Philadelphia in the 1880s, for instance, the depots were said to be "as bright as day, and every line of track leading into the city is thoroughly lighted to the city limits." People in darkened bedrooms in distant homes could hear the rumble of night trains, the long cry of their whistles. American railroads in their early years preferred operating during daylight hours, though there was some nighttime travel even in the 1840s. New York and Boston suburban trains in the 1860s ran as late as midnight, while long-distance freight and passenger trains ran through the night. Ever larger and busier service facilities developed in and around major cities in the late nineteenth century, employing huge urban workforces of mechanics, yardmen, and stationmen. The 1880 US Census counted nearly 90,000 men employed in the rail shops, and another 63,000 stationmen—together, these made up well over a third of the nation's 419,000 railway employees. The 1900 census found thousands of men working for the steam railroads in each of the nation's largest cities—over 15,000 in Chicago alone. Some railway men worked unpredictable and irregular hours, though not to the extreme of longshoremen. While train crews occasionally put in twenty-four or even forty-eight hours of continuous service, most men worked ten or twelve hours a day Monday through Saturday. Shifts changed at regular intervals in the yards, as befitting an industry that valued precise timekeeping. In the Pittsburgh yards in the early twentieth century, shifts changed every day at 6:30 a.m. and 6:30 p.m.[6]

Streetcar companies also employed large numbers of night workers. Until the introduction of electric trolley service near the close of the nineteenth century, public transportation in most cities stopped between 10:00 p.m. and 12:30 a.m., but the largest cities had later hours. All-night "owl car" service originated in New York City, where by 1856 the Sixth-Avenue Railroad Company was running horse-drawn streetcars every thirty minutes all night. Across the East River, the Brooklyn City Railroad Company announced the start of all-night service in 1859. "This will be a great accommodation to a large number of our citizens who are detained in New York either on account of business or pleasure, to a late hour of the night," declared the editors of the *Brooklyn Eagle*. Late night service began on Philadelphia horsecars in 1857. Municipal pressure and public demand resulted in the introduction of owl cars in some other major cities in the 1870s and 1880s, despite the reluctance of traction companies to bear this expense.[7]

Night service expanded as traction companies throughout the United States created sprawling webs of electric trolley lines in the 1890s and 1900s. The older lines of horse-drawn cars had primarily served the densely populated sections of big cities. Cable-car systems flourished in the 1880s, but their outward spread was limited by the high cost of installation and operation. Most systems shut down at night because the expense of keeping the underground cable in motion far outweighed the potential income from late night fares. In contrast, electric trolley lines were much cheaper to install than cable lines, and even cheaper to operate than horsecars, thus allowing faster transportation over an expanding region at fares many industrial workers could afford. As ridership grew and operating expenses dropped, late night service was no longer financially ruinous, though in most cases it remained unprofitable.[8] Growing numbers of traction companies after 1890 inaugurated owl service, expanded it to additional lines, and reduced the waiting time between cars. (Sometimes they responded only after some determined nudging or threats from the city government.) By 1901, some three thousand streetcar employees worked at night in New York. Most of the lines in Cleveland ran twenty-four hours a day by the mid-1910s, and owl cars could be found even in modest-sized cities such as Des Moines. Sixty of Chicago's ninety-seven surface lines in 1928 ran all night, many with headways of only fifteen minutes. Interurban electric trains in the 1920s also provided expanded night service for suburban commuters and other passengers.[9]

Streetcar companies experimented with different systems of staffing the cars as hours of operation expanded. Companies in the horsecar era clung with surprising tenacity to the practice of keeping the same driver and conductor on each car throughout its daily service. This typically meant that the driver and conductor worked stints of fifteen to eighteen hours with only brief breaks for meals. "No man can work so many hours without shortening his life," an editorial writer for the *Buffalo Express* argued in 1886, "and no man who works such hours can have any hours—or even minutes—left for home life, for social intercourse, for reading, for recreation, for any of the things that make life worth living." A Philadelphia conductor complained, "When I want to see my children, I have to see them in bed. I am off in the morning before they are awake and they are asleep when I come home at night. If they want to see me in the daytime they have to wait for my car." State legislators in Pennsylvania, New York, and Maryland passed legislation in response to the "eight hour" movement of 1886 that limited the hours of streetcar employment to no more than twelve a day. Traction companies complied with the letter of the law by assigning "trippers" to take the early and late runs with an extended break in the middle of the day, allowing the more highly paid regular men to have shifts of less than twelve hours. This arrangement became trickier once the hours of operation expanded during the early years of the electric trolley. Some traction companies attempted to stretch out the workday again, setting off strikes in Philadelphia and other cities.[10]

In addition to the drivers and conductors, street railroads employed large numbers of maintenance and repair workers. Among Baltimore's 1,000 streetcar employees in 1886 were 225 hostlers, 80 carpenters, 80 water boys, 50 blacksmiths, 25 car cleaners, and 10 watchmen. Factory-sized stables in the last quarter of the century held hundreds of horses along with the rolling stock, stores of food, and work space for the blacksmiths, harness makers, and car builders. Stable hands were busy caring for horses at the beginning and end of each rush hour when the largest numbers of horses were on duty. After harnessing and hitching horses for the morning rush, hostlers mucked out the stalls and prepared for the first big feeding about 8:30 or 9:00. The pace of work slackened after the horses returning from the morning rush hour had been groomed and led back to their stalls. Idle men loitered around the stable gambling and drinking. A second feeding took place before the afternoon rush, and a third after midnight. Some large-scale operations divided their stable work into two shifts of twelve hours each. At the Sixty-Fifth Street depot in New York in the 1880s, hostlers arrived for the day shift between 4:30 and 6:00 a.m. as the morning cars pulled out. Some Baltimore streetcar companies alternated the two shifts every week, in a system that let each hostler spend half his life on a daylight schedule but that also forced each shift to endure a regular "long watch" of twenty-two continuous hours of labor.[11] The custom of the "long watch" or "long turn" was becoming common in some manufacturing plants as well.

CONTINUOUS PRODUCTION

Round-the-clock manufacturing facilities were rare in mid-nineteenth-century America; gasworks were one exception, and iron makers' blast furnaces were another. Iron makers, who once located in rural areas such as northwestern Connecticut where charcoal and waterpower were abundant, were concentrating by mid-century in cities. Urban iron makers relied on coal or coke to fire their blast furnaces and run the new steam-powered blast engines. Blast furnaces were likely to crack if allowed to get cold; to interrupt production safely, workers had to "bank" the furnaces in such a way that their reduced rate of burning maintained a certain minimum temperature. A banked furnace continued to consume fuel and thus was an unproductive expense that manufacturers sought to avoid. Some blast furnaces stayed in constant use from the day they were blown in until the day they were abandoned.[12] A range of northern factories adopted shift work during the Civil War to maximize production of ships, gunpowder, rifles, hardtack, and the woolen cloth used in uniforms, but this appears to have been a temporary expedient.[13]

The major growth in shift work began in the decades after the war in capital-intensive factories where a constant flow of production replaced the older style of intermittent "batch" work. Manufacturers of steel, paper, petroleum, chemicals, and glass all developed production processes in the late nineteenth and early twentieth centuries in which any interruption was needless and costly. Good artificial lighting was essential too. The introduction of electric lighting in factories in the 1880s further encouraged nonstop production in these industries, as well as permitting extended night hours in textile plants. At the incessantly operating plants, employers divided their workforces into two or three shifts or adopted more complex staffing systems. By 1927, a survey by the National Industrial Conference Board found that night workers accounted for more than 40 percent of the employees making rubber, sugar, iron, and steel and more than 25 percent of those making paper.[14] Many of these workers experienced continuous production as part of a larger shift in power on the shop floor. Skilled workers' craft knowledge had once given them effective control of the work process. Now, with the installation of costly machinery for mass production, those skills and that power were diminished.

Glass workers saw a drastic change in the decades around the turn of the century. Unionized blowers in the 1880s and 1890s were well paid and set their own working conditions, production quotas, and hours of labor. They established a six-week shutdown during July and August that allowed them to enjoy long, lazy days of fishing instead of roasting in the heat of the shop floor. At this time glass was made on a relatively small scale in flimsy little plants scattered through the towns of Pennsylvania, Ohio, and neighboring states; the biggest concentration was in Pittsburgh, with ninety-three glass factories in 1886. The process typically began in the cool of the night when workers melted sand and chemicals in multiple "pots" inside a furnace, at temperatures around 2,500°F. From this molten material, every bottle, windowpane, and lantern chimney then had to be painstakingly formed by teams of skilled craftsmen assisted by young boys performing less skilled tasks like "snapping up" and "cleaning off." Most of the work of shaping glass was performed during daylight at the preference of workers, even though on a warm day a glassblower would suffer through what one remembered as "terrible heat . . . simply sweating the vitality clean out of his body." The management of the O'Hara Glass Company of Pittsburgh reported in the 1880s that they had once run night shifts but now worked glass only through the day. "This was found to be of great advantage in the character of the work, the health of the employés, and the securing of a better class of boys, the parents always preferring to have their boys home at night."[15]

Efficient, continuous melting tanks began replacing pot furnaces in American glass houses around 1890 and encouraged a shift to larger plants. Glass workers worried that the resulting increase in production would jeopardize their job security or their pay scales, but they found that consumer demand for glass increased as the price dropped. The need for skilled glass blowers was stronger than ever as plants began working glass in two shifts from 6:00 a.m. to midnight, and the pay actually rose. (Most glass houses still used the midnight to 6:00 a.m. period only to melt glass for the next day.) The craft unions' power began to crumble in the late 1890s when semiautomatic machinery made it possible to form bottles with less skilled labor. When the Owens automatic bottle-blowing machine further mechanized production after 1903, glass blowers understood the devastating effect it would have on their craft. Manufacturing interests gloated. The Owens machine "offers complete emancipation" from craftsmen's control of production, exulted the editor of the *National Glass Budget*. The Colburn sheet glass machine had a similar effect on the making of window glass. Instead of being produced by blowing huge cylinders, then opening and flattening them, polished glass now emerged from a machine's rollers in a continuous sheet at a rate of twelve tons a day. Nonunion mechanized shops quickly began running twenty-four hours a day in two twelve-hour shifts.[16]

The remaining skilled glass workers were rapidly displaced by machines in the 1910s and 1920s, as were the remaining boys. The craft unions tried to save their members' jobs by agreeing to concessions the manufacturers demanded: accepting steep wage cuts and giving up their traditional summer stop. Further, they agreed to work glass around the clock in three eight-hour shifts that rotated through the workforce, an arrangement that equitably divided the inconvenience of working at night. William P. Clarke, the president of the American Flint Glass Workers' Union, admitted in 1922 that he personally disliked the three-shift system because it "deprives a married man of being with his family two weeks out of every three." But he and other union officials urged its acceptance in an attempt to help union shops compete against mechanized, nonunion shops. Skilled glass workers nonetheless saw their wages shrinking and their duties reduced to tending machines.[17]

The most spectacular growth of night work was in the iron and steel industry. Through the mid-nineteenth century, iron was made in small batches by skilled craftsmen. Crude pig iron from a furnace was allowed to cool before being transported to a puddling mill to be worked into wrought iron, then rolled to form rails or plates. Technological advances in the production of iron from the 1850s through the 1880s culminated in the creation of integrated steel mills, essentially gargantuan machines that sprawled along an urban waterfront, sometimes for two or three miles. These mills obviated the need for skilled puddlers by producing steel (once prohibitively expensive for many uses) at prices that fell below those of wrought iron. Ore was dumped into huge new blast furnaces and then poured as molten iron into a Bessemer converter or open-hearth steel furnace, from which it emerged as steel, which was rolled into rails or plate without ever fully cooling. Though thousands of men came and went, product flowed steadily through the various stages of the process. Production was partially curtailed on Sundays for repairs in the rolling mills, Bessemer converters, and other parts of the complex, but it always continued at the blast furnaces and usually at the open-hearth steel furnaces.[18]

Night work meant long hours. Pittsburgh steel manufacturers adopted a three-shift system in 1879, since as one industrialist explained, "it was entirely out of the question to expect human flesh and blood to labor incessantly for twelve hours." Steelmakers reconsidered this position in the mid-1880s on the grounds that increased mechanization made steel work less physically demanding. The two-shift system was adopted over the next fifteen years throughout the American steel industry and remained customary until 1923, despite an interruption during World War I.[19] By the early 1910s, about half of all iron and steel employees (and most of the unskilled workers) had a regular workday of about twelve hours. The day shift typically worked from 6:00 a.m. to 6:00 p.m. and the night shift from 6:00 p.m. to 6:00 a.m., seven days a week. More than two-thirds of blast furnace men worked at least an eighty-four-hour week on average. Furthermore, some workers agreed to stay on the job for additional shifts, working thirty-six, forty-eight, and even sixty hours without interruption to make extra money.[20]

In the steel mills, paper mills, glass houses, railroads, and other continuous-process industries, American employers followed a policy of switching workers back and forth between night and day shifts, rather than keeping one group permanently on night work. The policy of alternating shifts produced perhaps the most horrifying aspect of work in the steel mills: the "long turn." Every week or two, the day shift would work an unbroken twenty-four hours, from Sunday morning to Monday morning, at which point the workers staggered home to sleep and returned that same evening as the new night shift. In steel mills and other incessant industries, alternation appears to have been forced by employers' insistence on running only two shifts and workers' aversion to

FIGURE 11. Making Bessemer steel. Steel workers in the late nineteenth and early twentieth centuries routinely worked more than eighty hours a week. Two shifts of workers alternated between night work and day work. The narrow-gauge locomotives and flame-spewing Bessemer converters shown here were among many hazards facing workers. Charles Graham, "Making Bessemer Steel at Pittsburgh—the Converters at Work," *Harper's Weekly,* April 10, 1886. Provided courtesy of HarpWeek.

continuous night duty. When industrial workers were not given alternating shifts, they demanded them. Night workers among the Pittsburgh yardmen of the Pittsburg, Ft. Wayne and Chicago Railroad petitioned management in 1886 to alternate the shifts every two weeks or once a month. Though they were paid an additional ten cents a day for night work, they claimed that continuous night work "has deprived them of many of the pleasures of life." Likewise, several hundred Bridgeport workers at the American Tube and Stamping Company went on strike in 1907 after the company announced that the night shift would no longer alternate with the day shift each month.[21]

EFFECTS ON WORKERS

Descriptions by workers and sympathetic observers in the Pittsburgh region reveal the tremendous burdens night work placed on individual and community life. Some of the best description comes from a book by Charles Rumford Walker, a Yale graduate who in the summer of 1919 posed as an ordinary workingman and took a job on an open-hearth furnace in the Pittsburgh area. Walker describes his weeks on the night shift, when he worked fourteen hours at a stretch, as a time of physical exhaustion and sleep deprivation.

> "What do you do when you leave the mills?" people ask. "On my night-week," I answer, "I wash up, go home, eat, and go to bed." Anything that happens in your home or city that week is blotted out, as if it occurred upon a distant continent; for every hour of the twenty-four is accountable, in sleep, work, or food, for seven days; unless a man prefers, as he often does, to cheat his sleep-time and . . . take a drink with a friend.

Walker claimed that he sometimes slept only two or three hours a day when he was on the night shift. Most night workers were in bed only between 9:00 a.m. and 2:00 p.m. and slept poorly, their bodies struggling to adapt to the alternating day and night cycle of sleep.[22]

The twenty-four-hour long turn was as grim as can be imagined. In *Out of This Furnace*, a 1941 novel documenting life in the Monongahela Valley steel towns south of Pittsburgh, Thomas Bell described it as follows:

> The first twelve hours were much like any day turn except that sometimes, through a break in the mill's rumble, he could hear church bells. If his hands were free he tipped his hat. The second twelve hours were like nothing else in life. Exhaustion slowly numbed his body, mercifully fogging his mind; he

ceased to be a human being, became a mere appendage to the furnace, a lost, damned creature. "At three o'clock in the morning of a long turn a man could die without knowing it."

(This was sometimes literally true: the dazed zombies at the end of a long turn were prone to accidents.) When the 6:00 a.m. whistle blew, workers had twelve hours to eat and sleep before returning for their next shift. Bell writes: "The long turn was bad but this first night turn coming on its heels was worse. Tempers flared easily; men fought over a shovel, or a look, and it was fatally easy to be careless." To conserve energy on the job, wrote Walker, men moved as slowly as possible and sneaked short naps during lulls.[23]

Night work did have some advantages. The summer heat in steel works, glass houses, and cargo holds was easier to endure at night. Streetcar operators had less traffic to negotiate. Above all, work discipline was somewhat relaxed after most of the management went home. "Night work on the whole is not so hurried," observed a writer for the *Brooklyn Eagle* in 1901. "It can be taken more leisurely; in many cases for those who grow to like it, it is more soothing and pleasant than the same toil by day." Working together while the rest of the city slept, some night workers experienced a greater camaraderie. A longshoreman would go fetch food and drink to sustain the rest of his work gang through the hours after midnight, and the gang would pick up the slack to let an exhausted man rest. Enginemen and their firemen on night freight runs developed close friendships that broke down some of the distance created by their master-apprentice relationship. The trust among coworkers, and the relative lack of supervision, also made it easier for employees to steal at night. Professional thieves carried off valuable cargo and equipment in collusion with longshoremen and freight yard men.[24]

Workers who alternated between day and night work could enjoy a semblance of normal life during their periods on the day shift. On their Sundays off, steelworkers enjoyed a twenty-four-hour respite that let them sleep, attend church with their families, or go on a drinking binge. Then they readjusted to a daily schedule, sleeping when it was dark and interacting with friends and family members during their brief hours of leisure. In some early twentieth-century steel mills, like the one Walker worked in, the day shift worked only ten or eleven hours while the night workers were on duty thirteen or fourteen hours. These hours were established at the request of the workers so that they could enjoy free evenings during their day weeks. The shortened hours also let workers rest up from their exhausting night week and gather strength for the coming twenty-four-hour long turn. Some paper mills followed the same practice. The alternating shifts of industrial workers, like the irregular hours of dockworkers and trainmen, meant that these workers were never able to adjust to a fully nocturnal schedule or to develop a separate identity as night workers, but these were costs that workers were quite willing to pay.[25]

Men who always worked the night shift risked becoming alienated from the lives of their families and friends. In one extreme example, a night patrolman in Flint, Michigan, discovered in 1922 that his wife had another husband who worked the day shift in a factory. More commonly, night workers were working at the usual time of the family's evening meal. Those who went to sleep immediately on returning home from work might miss the family breakfast as well, though many streetcar employees on the "owl" shift took additional early morning runs so as to return home as their families woke up. Night workers also had to find creative ways to pursue courtship; steelworkers often had to wait for their weeks on the day turn, while some workers on the trolleys and elevated railroads asked their girlfriends to accompany them in the train cars. "What else can a young fellow do?" one brakeman asked, telling a reporter about a coworker who was conversing with his sweetheart between stops. "He gets no time to visit his girl at home, so she rides up and down with him in the evening. That girl you just saw makes the trip two evenings in the week. I s'pose they'll get married this winter. Yes, that's the way I courted my wife."[26]

Family members suffered many inconveniences in trying to accommodate the schedules of men on the night shift, according to reports by investigators for the early twentieth-century *Pittsburgh Survey*. Women and children were bound to the daylight schedule by the hours of schools, the need to do laundry or play by natural light, and by the light-sensitive rhythms of children's sleep. When the husband was working the night shift, wives had to do housework silently and children had to be kept quiet to let him rest, a challenging requirement in the crowded homes of working-class Pittsburgh. Meal schedules had to change depending on whether the man was working day or night shift that week. When a man and his son worked different shifts, family meals were impossible. Compounding the problem, many immigrant families also took in boarders, sometimes boarders on different schedules so that the same beds could be occupied night and day.[27]

Neighborhood life was also affected by night work, as shown by Pittsburgh's Fifteenth Ward. This ward was about three miles southeast of the central city, inside a bend of the Monongahela River, and comprised the neighborhoods of Hazelwood, Glenwood, and Greenfield. The smoky lowlands by the river were dominated by the furnaces and coke ovens of the Jones and Laughlin Steel Company, as well as by the Glenwood yard and car shops of the Baltimore and Ohio Railroad. Both the steel complex and the railroad complex ran constantly. In the early twentieth century, unskilled immigrants crowded into boardinghouses and tenements near their work, while more affluent families occupied the higher ground. As there was no significant park within walking distance, play space was scarce in this neighborhood, particularly in the crowded lowlands near the mill. The upland terrain was so rugged that much of the land was unusable; steep, eroding hillsides oozed filth from privies.[28]

By necessity children played in the streets of the Fifteenth Ward, their excited voices reaching into adjacent homes, especially during warm months when windows were open. Hazelwood district police records from 1922 show that a number of these conflicts reached the level of formal complaints. Most complaints about noisy ball games came during the afternoon and early evening; complaints about loitering by groups of raucous young men focused on the later evening and midnight hours. One small dead-end street near the Second Avenue commercial strip, lined by the rowhouses of railroad workers, was an especially popular spot for noisy ballplayers. Noise at any hour on a street of railroad workers was sure to keep someone awake and make him grumpy enough to walk the four blocks to the police station. According to the complaint logs, a few Fifteenth Ward complainants explicitly stated that children's daytime play disrupted the sleep of night workers, while a larger number complained about disruptions to sleep at night. The complaints that fill the log create the impression of a neighborhood of sleepless men under siege, unable to stop the racket and facing retaliation if they continued to confront the boys themselves. Delinquents targeted the homes of hostile adults, breaking the windows or pelting the walls with mud, stones, and eggs. Street peddlers were another nuisance for the night workers in the Fifteenth Ward, as well as in other working-class sections of industrial cities. Men selling vegetables and other goods would cry their wares in the streets starting about 10:00 a.m. and continuing all day, keeping men awake and generating more complaints.[29]

Not everyone could endure the strains that night labor demanded. It was young men who had, or were believed to have, the strength and endurance needed to make steel all night, so it was those between the ages of twenty and forty-four who filled the ranks of steelworkers. Young, single immigrant men who were looking to earn money quickly found the steel mills attractive. "One man said to me, 'A good job, save money, work all time, go home, sleep, no spend,'" Walker reported. Some immigrants worked steadily until they had saved enough to return to a comfortable life in the old country. Similarly, neither boys nor old men could handle the exhausting stretches of heavy lifting required on the docks. Longshoremen were routinely expected to carry sacks of sugar, flour, coffee, and salt weighing 250 to 300 pounds. Even strong young men found it difficult to endure this labor for more than a few years. Once they were no longer able to handle cargo, longshoremen found jobs as watchmen or dock cleaners or drifted into less demanding forms of work.[30]

Railroads too preferred to hire what one railroad official called "young men who can bear fatigue and exposure." Nearly half the midcentury workforce consisted of men in their twenties and thirties. Career railwaymen typically began work as laborers in their late teens or early twenties and reached supervisory positions in their thirties. By the early twentieth century, few railroad men were hired past age thirty-five. Street railway companies also found that young men were better able to endure their grueling work schedule. Men as young as eighteen or twenty were hired in some cities to drive the trolleys or collect fares; maximum age limits for hiring ranged from twenty-eight to fifty. Night work often fell to the most recently hired men, whose lack of seniority gave them whatever assignments were not wanted by more experienced motormen and conductors. The newest arrivals worked as "extras," taking undesirable runs trip by trip; others became trippers with split "early-late" shifts. The companies justified these practices by pointing out that inexperienced men could build their skills more safely during hours with less traffic. Nonetheless, some workers with seniority chose to take the early-late shifts, preferring to have time off during the day rather than at night. Drivers, who were stationed on open platforms at the front of the cars, often requested night work in the summer.[31]

"It is true that my hours were long and my work was wearisome," acknowledged one man who had become a streetcar conductor at age eighteen. "I used to get up at five in the morning, hurriedly eat my breakfast, and report for work. I did not get to bed before eleven. I had no time for recreation, and very little time for attending to my private affairs. I used to get home so tired that the effort of writing a letter would sometimes exhaust me, so that I could hardly get to sleep at all. But, in spite of my hard work, I felt better than ever before in my life." Most workers did not feel this good for long. The hours wore men down, and accidents left many incapacitated. Fatigue, inadequate lighting, and relaxed work discipline made night the most dangerous time to work, particularly on the waterfront. Exhausted longshoremen, some fuddled by alcohol, were crushed under falling cargo, fell down ladders, or were knocked into open hatches. Many had fingers nipped off or crushed. Longshoremen lucky enough to avoid being maimed or crippled still had no job security: they were hired only by the day. Railroad workers could count on more stable employment, but most of them chose not to stick with the career for long. The railroad companies were plagued by high turnover. In the paper mills of Holyoke, Massachusetts, workers on the two-shift system were often worn out by age forty-five and were replaced by younger men. Thus night work in these places appears to have been a temporary phase in the working lives of many men rather than a long-term condition.[32]

Children and women were rare in these lines of work, especially at night. Though boys worked at night in the glass houses, their numbers were declining as a result of the same mechanization that expanded the need for night work. In 1880, 23 percent of all glass workers were children; that proportion had dropped to 10 percent by 1904 and under 2 percent by 1919.[33] Virtually no women held night jobs in the continuous-process industries or in transportation; in fact, very few worked in those industries at all, at any hour and in any capacity. Jobs considered proper for females were those that approximated the tasks found within the home, such as clothing manufacture, food preparation, nursing, and teaching, or that reflected such feminine qualities as neatness and cleanliness. The hot, dirty labor in heavy industry was believed to be inappropriate on these grounds and inimical to the delicate female constitution. Pittsburgh's iron and steel industries in 1920 employed 26,975 men and 236 women, most of them probably in clerical positions. Philadelphia's gasworks employed 501 men and only one woman.[34]

Textile manufacturing was one of the very few lines of mass production in which women worked at night in the early twentieth century. Nonstop production had been rare in nineteenth-century woolen and cotton mills; the low quality of lamplight and the high risk of fire discouraged long hours of night work. These technical obstacles vanished with the introduction of electricity about 1880. Night work was now restricted only by human factors such as employers' ability to recruit a competent workforce for the night shift and employees' power to oppose what many saw as exploitation. Workers in the industrial cities of the Northeast and Midwest, where there was a tradition of labor activism, objected to night work and launched occasional strikes to try to stop it. Those in the small mill villages of the Piedmont South lacked the determination and the power to resist. There the night shifts were easily filled with poverty-stricken women and their sleepy children. The difference grew more pronounced in the 1920s, as night work expanded in southern textile mills but not those in the urban North. A study by the federal government's Women's Bureau, which estimated that at least 2.1 percent of all employed women in the United States worked at night, found that women's night work in textile mills was limited mainly to rural southern states that lagged behind the Northeast and Midwest in adopting labor legislation. New Jersey, whose night work law was unenforceable, was an exception. The New Jersey commissioner of labor reported in 1923 that 1,567 women held night jobs in Passaic, and about 400 more in other parts of New Jersey. "Nearly every working woman in Passaic tries night work at some time or other," wrote the labor investigator Agnes de Lima. The women who worked the night shifts in Passaic's woolen mills were mainly impoverished immigrants who were desperate for extra income. Even so, they did not hold the jobs long. Some worked during pregnancy to save up money before the birth of a child. Others started on the night shift once their children were old enough to be left alone at night.[35]

Artificial lighting and continuous production might have permitted city life to be cut loose from the natural cycle of daylight, but that was only in theory. The lives of night workers were still shaped by the dominant schedule. Industrial workers in the cities of the Northeast and Midwest dreaded continuous night shifts so much that they were willing to tolerate the disorientation that came from alternating between night and day work. As long as night workers lived with families, as long as shops and schools followed a daylight schedule, night work could never be the same as day work. It was left mainly to men who lacked the power to avoid it: the young and the poor.

Long hours, disrupted sleep, and personal inconvenience were enough to make anyone leery of night work, and the low status of many night jobs undoubtedly added to the stigma. Being seen walking to and from a night job was a public display of one's modest social rank. Nineteenth-century Americans were acutely aware of differing schedules of different classes, as shown by Walt Whitman's description of the changing character of street traffic on Broadway. Similar processions could be seen around the turn of the new century, though daily journeys had been altered by the physical expansion of cities and by their starker division into zones of production and residence, affluence and poverty. Often traveling greater distances than their predecessors, the daily tide of workers was now linked to the timetables of public transportation. Commuters flowed down Broadway in discrete pulses carried by cable car, as Stephen Crane observed. "In the grey of the morning [the cable cars] come out of the up-town, bearing janitors, porters, all that class which carries the keys to set alive the great downtown. Later, they shower clerks."[36]

The humorist H. C. Bunner wrote in *The Suburban Sage* (1896) of a "time-table test" that supposedly revealed the status of each white-collar suburban breadwinner. From "an upper window in my house that commands an uninterrupted view of the little railway station," he watched the daily procession of poor and rich commuters to New York. Men who rose to catch the 6:00 trains took little care in their appearance and evidently lived grim, meager lives. They were followed by the happily ambitious office boys who arrived to catch the 7:03 train. "But the 7:27 train is quite another affair. The errand-boy has got his promotion. He is really a junior clerk of some sort; and he has the glorious privilege of getting to his office exactly twenty-four minutes later. But, with his first step upward, he leaves light-hearted boyishness behind him and becomes prey to cankering ambition. His companions are men now, but mostly men who have barely escaped the bondage of the 6:38, and in whose breast the hope of ever rising even to the 8:01 is slowly dying out." The status of the commuters rose with each subsequent train, Bunner continued, and so did their concern for reputation. "A commuter's clothes improve from train to train until he gets to taking the 10:17, when he is reputed so rich that he may safely dress shabbily."[37] It remained the privilege of the wealthy to rise late, just as urban elites had done for centuries.

CHAPTER 8

Mashers, Owl Cars, and Night Hawks

Avery D. Putnam seemed an unlikely model of chivalry until the moment the iron rod crushed his skull. Putnam's martyrdom took place in New York City one moonlit Wednesday evening, April 26, 1871. The produce merchant had finished his day's work, but instead of going home to his wife, he paid one of his frequent visits to a lady friend, a milliner named Jeanne ("Anna") Duval who lived on Broadway south of Union Square not far from Putnam's house. Shortly before 9:00 p.m., Mrs. Duval and her fourteen-year-old daughter, Mabelle Virginie ("May" or "Jenny"), had to leave to meet her older daughter, Annie, at a church uptown. Mrs. Duval was in the habit of escorting Annie home after her evening choir practices. Putnam offered to accompany them, so the three boarded a horse-drawn streetcar and rode north on Broadway toward the Church of the Advent.

The car had plenty of empty seats at that hour. Most men who could afford to take the streetcar had already returned home from work, and the theater crowds had not let out. Besides the driver and the conductor, there were only four or five other passengers, including a drunk standing on the front platform with the driver. Near Twenty-Eighth Street, Jenny walked to the front door, hoping to look up through the window at the clock of the newly opened Gilsey House hotel. The drunk was on the other side of the glass, smirking at her. She returned to her seat. He opened the door and stood staring at Jenny and her mother. Mrs. Duval tried to shut the door, but the drunk kept pushing it open. That was when Mr. Putnam rose to defend the honor of womankind. "These ladies are with me; I will not have them insulted," he supposedly declared. The drunk, an off-duty streetcar conductor named William Foster, persisted and eventually threatened Putnam: "When you get out I'll give you hell." Foster attacked as the trio left the car in what is now Times Square. To take them by surprise, he jumped off the front platform and circled around behind the car as they stepped into the street. Jenny glimpsed a bearded man with shining eyes and white teeth, then saw the rod smash down on Putnam's head. Foster tossed the bloody "car-hook" back onto the platform he had snatched it from and ran off into the night. As the car rolled away amid Mrs. Duval's screams, Putnam lay bleeding on the tracks under a gas lamp.[1]

It was another senseless murder in Gotham—atrocious, to be sure, but not entirely out of the ordinary. "Nobody who has any experience of night-cars will be greatly astonished at it," claimed the editors of the *New York Times*. But the press turned it into a symbol of all that was wrong with the city. The murder seemed to the editors of the *Times* and the *Tribune* to be part of a culture of lawlessness in New York that was rooted in the Tammany Hall political organization run by William M. Tweed. The two papers and *Harper's Weekly* were already striving to discredit Tweed but so far had produced little besides indignant editorials and Thomas Nast's scathing cartoons. (That summer, the *Times* would triumphantly publish the solid evidence of corruption that would lead to Tweed's downfall.) "Any one accustomed to travel at night in our street cars will be able to appreciate how completely the respectable inhabitants of this community are at the mercy of our City rowdies," the *Times* declared in its first of many such editorials, which appeared on April 28 while Putnam was dying in the hospital. "Their extensive control over City Boards and local Courts lends to our collective mass of ruffianism a social importance, and a degree of immunity from ordinary restraint, unexampled in any great city of the world." The *Times* implored the better class of male New Yorkers to stand up for law and order, both literally (on the streetcars) and figuratively (in exercising their duties as citizens). "The morbidly cautious passengers of a street-car are but a type of the society to which they belong. We are sacrificing manhood, fair play, chivalry, everything that is worth being proud of, from a slavish fear of the dregs of our populace." In defending ladies against a "rough," the martyred "gentleman" set an example that the *Times* hoped would be followed by an outraged citizenry. Public indignation might thus spur a new spirit of self-assertion in public spaces and public life.[2]

The "car-hook murder" gave the *Times* a useful synecdoche for urban lawlessness over the next several years. Foster turned out to be not just a hoodlum but a former city street inspector appointed by Tweed himself, and the son of a politically connected contractor. Despite his quick conviction for murder, his family's wealth and influence kept him from the gallows through nearly

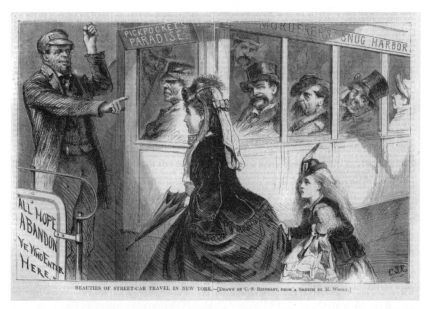

FIGURE 12. "All Hope Abandon, Ye Who Enter Here." This engraving appeared on the front page of *Harper's Weekly*, accompanying an article on the notorious "car-hook" murder in 1871. Though the murder victim was a man, the artist emphasizes the infernal horrors that women faced in the nighttime streetcars of New York. C. S. Reinhart, "Beauties of Street-Car Travel in New York," *Harper's Weekly*, May 20, 1871.

two years of appeals. The father's last petition for the governor to commute the death sentence was supported by local clergymen, lawyers, and public officials. (On reading the testimonies in favor of Foster, Mark Twain observed sarcastically, "I perceive that from childhood up, this one has been a sweet, docile thing, full of pretty ways and gentle impulses, the charm of the fireside, the admiration of society, the idol of the Sunday-school.")[3] Each new step in the proceedings received extensive news coverage both in New York and across the country. Most of the coverage was hostile to Foster and sympathetic to Putnam.[4]

In death, Putnam became an exemplar of the masculine resolve needed to resist urban disorder. Many observers who didn't share the *Times*' fixation on the Tweed Ring still shared its concerns about the dangers of the city—concerns that were reinforced that year by news of the Commune in Paris, the Orange Riot in New York, and a series of strikes that lasted into 1872. If corrupt officials could not keep matters under control, some citizens feared, the propertied classes would have to form a "vigilance committee" to give the scoundrels "the lamp-post treatment."[5]

Women's precarious safety on the streetcars was a particularly disturbing sign of social crisis. It was a matter of national pride that American women had more freedom in public than European women. Unlike hierarchical European societies, the argument went, in America men treated every woman with respect, serving collectively as protectors of the weaker sex. Unescorted American women could travel safely through the streets of major cities, while their counterparts in London and Paris could not. Ladies were even said to be able to travel across country alone, relying on the moral atmosphere of the railroad car and the kindness of strangers. Their safety proved that order and decency could thrive in a democratic society. Middle-class observers were also pleased to see the decorum of the parlor extended into the public space of the streetcar and the intercity passenger car; it seemed to promise that middle-class values would triumph in the modern world heralded by transportation technology. "The people of Brooklyn have long ago made up their minds that a street car shall be just as safe for women and children as their own homes," the editors of the *Brooklyn Eagle* declared in 1880. A few years later they added that "between disorderly conduct in the street or park or anywhere else and similar conduct in a street car there is a great deal of difference.... The street car is particularly sacred to women and children because it is a symbol of our democracy. Consequently, any affront given to ladies in one of these vehicles is a crime committed in a temple."[6]

Like a home or a temple, the streetcar was indeed a furnished interior where women and children could be found, but there the similarity ended. In other ways it more closely resembled the street, in that it was a public space where women were subject to scrutiny by anonymous strangers—women as well as men. Further, it lacked some protections of the intercity railroad car. The fares were so low that they put minimal restrictions on who could enter— drunks, prostitutes, and some workingmen all managed to pay the price of a ride. Large numbers of strangers might enter or leave the car during even a short trip, making it difficult to establish any sense of protective community among the passengers. Police were rarely present.[7]

By the 1860s court rulings had established a law of common carriers that made railroad companies responsible for the safety and welfare of the traveling public; railroads were obliged to set reasonable rules to protect their customers, including from each other. Thousands of ladies and children within cities depended every day on the protection of streetcar conductors and drivers, as a popular magazine observed after the car-hook murder. But Foster himself was a glaring example of how these working-class men could be indifferent or worse, and his fellow streetcar employees had obviously failed to stop him from killing Putnam. The night streetcar was thoroughly infected by the culture of the nocturnal street, where men were rowdier, more aggressive, and more impulsive than during the day. Putnam's widow subsequently sued the Broadway and Seventh Avenue Railroad Company, unsuccessfully arguing that it had neglected its duty to keep order.[8]

Repeated descriptions of harassment on the streetcars show that women's safety remained tenuous through the end of the century and beyond. Yet many women continued to ride the streetcars alone, even after dark. Their right to travel through the city was contested again and again, with no clear resolution either way.

MASHERS AND ROWDIES

Streetcar disorder was a more muted form of the danger that encountered female pedestrians everywhere. Nineteenth-century newspapers are generously sprinkled with brief articles or editorials about men's accosting young women on the streets. The news items are formulaic, often saying merely that the woman was "insulted." The word "insulted" might be taken as a Victorian euphemism, like calling a leg a "limb," but it is worth considering more carefully. The word accurately expressed what people considered the essence of the offense: unwanted sexual advances were viewed not as "harassing" or bothering the woman, but as "insulting" her—showing disrespect by pretending she looked like a prostitute and thus declaring her unworthy of the courtesy gentlemen owed ladies. The insult of being mistaken for a whore was a familiar danger to women both in fiction and in urban reality. A fictional example appears in *The Gay Girls of New York*, an 1853 novel by George Thompson, a leading figure in the "sporting press."

> The illuminated dial of the City Hall clock pointed to the hour of eleven, when a young girl, humbly but neatly dressed, entered the Park from Broadway, and crossed over towards Chatham street. Glancing at the clock, and seeing the lateness of the hour, she increased her speed almost into a run. It was clear that she was not a girl of the town [a prostitute], for she was evidently frightened at her situation, being unprotected and abroad at a time of night when a virtuous woman prefers to be at home, unless she is accompanied by some male relative or friend, to guard her from insult and outrage. . . .
>
> She was passing the rear of the City Hall, when she was accosted by a large, portly and well-dressed man who was seated upon the chain fence of one of

the grassy enclosures, and who, as she approached, stepped directly in her path as if to arrest her further progress.

"Ah, pretty one," said this man, grasping the girl by her shoulders and drawing her towards him, while he admiringly surveyed her pale but surpassingly lovely countenance—"you are looking for company eh? Well, as you seem to be a dainty morsel, I'll go home with you, or take you home with me, just as you like. Where do you live?"

"You mistake me, sir," said the poor terrified young girl, as she trembled in every limb—"I am not what you suppose me to be—indeed I am not! I am only a poor sewing girl, and have just finished work. I live in William street, and am now hastening home to my mother, who is very sick and requires my presence. Therefore do not detain me, sir—let me go, I implore you!"

"Not so fast, sweet lily of the valley," said the man, whose suit of fine broadcloth and massive gold guard-chain announced that he belonged to what is called the upper circles of society, while a certain peculiarity of utterance indicated that he was slightly intoxicated—"not so fast. You're not a common night-walker, then—you are a sewing-girl, eh? Ah so much the better; I will make a fine lady of you, so come along with me, and let your sick mother die in peace."[9]

Besides the improbable language, this scene is atypical for its violence. The evil Mr. Wallingford chokes the struggling maiden until she passes out, then takes her to a house of ill repute with the vilest of intentions. In reality, most insults were only verbal. Yet Thompson managed to compress into this scene several of the themes that appeared in the journalistic and clerical discussions of the problem: the confusion or pretended confusion over the woman's intentions; the overlapping schedules of work and leisure that brought victims and insulters together at night; and the class tensions that inflected such encounters.

Then as now, discussions of sexual advances in public—"mashing," as it became known by 1880—grappled with the problem of ambiguity. Depending on the circumstances and the wishes of the woman, the interaction could fall at any number of points on a spectrum between sexual assault and mutual flirtation. At the extreme, women could literally be raped on a dark street or dragged into an adjoining building or vacant lot. It is impossible to determine the frequency of such rapes in nineteenth-century cities. Newspaper articles and court records reveal mainly the cases in which the woman registered a complaint or in which the crime was interrupted by a police officer. Atrocity stories of gang rapes occasionally appeared in salacious men's publications like the *National Police Gazette*, suggesting that a woman's basic safety could not be assumed even on public streets, or even with an escort.[10]

Further down the spectrum of insulting were incidents in which men physically restrained women, groped them, or stalked them menacingly. In these cases, journalists and public officials declared, decent gentlemen had a duty to intervene by whatever means necessary to protect the honor of American womanhood. One late December evening in 1847, three or four "nocturnal rapscallions" followed a group of ladies as they left the Wesley Chapel in Cincinnati, according to a local newspaper. "These rowdies stepped up to the side of the ladies, and despite their fright and importunities, followed them not only to their own doors, but into the house! One of the ladies, with more presence of mind than the rest, rang the bell, and brought her husband into the room, who, without the usual ceremony of introduction, collared one of the chaps and gave him a sound thrashing." The insulter "got no more than his deserts . . . a black-eye and a smashed countenance," while his friends ran off.[11]

Men often targeted women they considered so far below them in status as to be outside the boundaries of respectability, as in the story of Mr. Wallingford and the sewing girl quoted above. Aggressive sexual advances were among the ways bloods and dandies asserted their claims to status, but the practice extended into the working class as well. Black women were particularly vulnerable, for they were considered to have no rights that any lecherous white man was bound to respect. Eliza Potter, an African American hairdresser in 1850s Cincinnati, wrote in her autobiography that she was saved from attack only by having had the forethought to carry a gun. On one evening when she went to attend to ladies at a party, she recalled, "there were several persons behind me hallooing and hooping, and I could hear them say 'let us frighten her to death,' but I did not feel at all alarmed . . . for I generally had in my basket a good protector." On her return later that night, "there were two men standing on the corner at Fourth and Sycamore—one . . . walked a little distance behind me, until we got to near Race street, when he stopped before me, and he did not speak, but walked close to me. Stepping back, I told him if he took another step he would fall at my feet." Frightened, the man denied he was following her, and soon ran off.[12]

Much more frequently, the insulting behavior consisted of staring or making rude remarks. These incidents took different forms depending on whether the man was in a group or alone. Clusters of men or teenage boys were a common sight on street corners throughout the nineteenth and early twentieth centuries, talking and smoking cigars, particularly in the evenings. These "corner loungers" behaved in fairly predictable ways toward passing women: they told dirty jokes in their presence, loudly judged their appearance, paid extravagant compliments, or called out embarrassing invitations. Men behaving this way had little intention of actually conversing with the women. They were putting on a performance to impress each other, to build camaraderie, and to gain a sense of power from their control of the corner. In a calculated display of misogyny, some corner loungers waited outside churches during evening services to insult ladies as they emerged into the lamplight. Corner lounging was common among men of all classes, though observers often focused on the idle "loafers" at each end of the class hierarchy.[13]

The situation was more ambiguous when an individual man made a pass at an individual woman, or when a woman asked an innocuous question of a stranger. (Was she really asking directions, or was there a sexual subtext?) Every significant American city in the second half of the nineteenth century had hundreds or even thousands of prostitutes who solicited in theaters and streets or worked privately within brothels or their own lodgings. Many of these were part-time prostitutes trying to supplement the meager wages they received from jobs in domestic service or manufacturing. The gray area between the prostitute and the working-class woman was widened by the practice of promenading; a young woman out for a stroll in the street or the park might well be open to flirtation, perhaps more. Police were reluctant to get involved in ambiguous situations, though they would respond to complaints of flagrant solicitation by prostitutes or aggressive harassment by men. "That which was formerly known as Flirtation, is now recognized under the revolting name of 'mashing,'" observed the captain of Philadelphia's Fairmount Park Police in 1882. "While very much is carried on under this guise that is objectionable to persons of refinement, yet it does not amount to a violation of our rules. If many parents could witness the conduct of their sons and daughters in the promenade of public days they would no doubt, very speedily find new rules for their governance."[14]

The presence of prostitutes and other sexually available women created possibilities for mistaken identity, complained the moral reformer John R. McDowall in 1832. "Our mothers, sisters, wives, daughters, friends, and domestics, cannot walk through several streets at any time of day, without being liable to have their morality impaired, and their persons insulted," he wrote. "Vice is so outrageous, that if a female of irreproachable morals is without a protector in the street after dark, she will probably be insulted, even in Broadway." The *Whip and Satirist of New York* claimed in 1842 that men mistaking "decent women" for "harlots" had perpetrated numerous nighttime assaults in the streets. It appears from accounts of actual incidents that the "mistake" was often transparently pretended. Still, the possibility of a legitimate error could be used to excuse or mitigate misbehavior. Thus it made sense at Foster's murder trial for the defense lawyer to challenge the character of the victim's lady friend, Mrs. Duval. He persistently cross-examined Duval about her relationship with her husband and with the victim until she refused to answer any further personal questions.[15]

Nineteenth-century etiquette manuals offered instructions for how women should appear in public in order to avoid "the most severe misconstruction." Women were told to wear conservative street clothes in muted colors, not dresses that belonged in the parlor or ballroom. They were just asking for trouble if they put themselves on display in open-necked gowns and thin stockings. "Those who put on and wear about the badges of infamy, need not much marvel if they are treated as 'infamous,'" declared one New Yorker in a letter published in 1837.[16]

To remain as inconspicuous as possible, respectable ladies were told to pay scrupulous attention to their behavior in the street, even during daylight. They must give the appearance of disengagement from their surroundings, and above all they must never loiter. "Their deportment will be such as to attract the least notice. They will walk quietly, seeing and hearing nothing that they ought not to see or hear," declared an 1892 etiquette manual. "Whether young or old, they will form no acquaintances on the streets, and their conduct will be marked by a modest reserve, which will keep impertinence at a distance, and disarm criticism. The very appearance of evil must be avoided, and she is not a true women who so carries herself in the public thoroughfare that loafers stare as she goes by, and 'mashers' follow her with insulting attentions." A true lady should not stop in public to talk with a friend; the friend should walk beside her if they wished to talk. She should not ask directions of strangers; if lost, she should keep walking until she found a policeman. "She should not turn her head, on one side and on the other, especially in larger towns or cities, where this bad habit seems to be an invitation to the impertinent." Nineteenth-century manuals forbade window-shopping for the same reason. If a lady followed all the rules, she would supposedly have no cause for fear. As one Philadelphian wrote in a 1911 letter to the editor, "The lowest tramp that shambles along the byways of Pennsylvania wouldn't dare approach a true lady. There is something in the appearance, manners and character of a lady that repels the advances of men."[17]

Men sometimes insisted that if any lady were troubled by mashers despite her best efforts she need only call for help. But many women felt uncomfortable doing this. To do so would attract attention in the street, the streetcar, or—worse—the courtroom. Their accusations would likely be met with denials and public embarrassment. If they were behaving as true ladies, after all, they would not have been targeted. Thus, wrote one woman, "We can only find relief for our wounded and insulted feelings in the sympathy of our friends."[18]

MASHERS AND ROWDIES ON THE STREETCARS

Large stagecoaches called omnibuses inaugurated mass transportation in New York in 1827 and spread to other major cities by the mid-1830s. They were soon supplemented by horse-drawn street railways, which offered a smoother ride at a lower cost. The experience of riding in a public vehicle posed special problems for women's strategy of respectability through invisibility. Though women passengers were in motion, as they were expected to be at all times while in the street, they actually remained seated or standing, loitering in the public gaze. The etiquette manuals warned that "there is no position where a dignified, lady-like deportment is more indispensable" than in a public vehicle. Women had to assume that their clothing was subject to public evaluation and that every motion, expression, or gesture was closely watched. They were directed to be silent and motionless, knowing all the while that they remained vulnerable. "No matter how quietly a lady sits in a stage, she is liable to insult," one young lady wrote to the *New York Times* in 1869. "If a young lady in an omnibus puts her hand to her head to arrange her veil or hat, every man's eye is on her and kept on her until she returns her hand again to its place." Some men would continue staring openly, while others might roughly press themselves into a seat next to the woman. "It is impossible to ride down to the ferries without feeling more or less unsafe." Police were well aware that women were frightened, but they felt unable to make arrests merely for "insulting."[19]

Women's experience of riding public vehicles varied hour by hour. The middle of the day was the safest time; with most men at work, middle-class ladies found open seats in streetcars to and from the shopping districts. The evening rush hour presented special dangers for female wage earners who rode crowded omnibuses, ferries, and streetcars home from work. Ladies returning from shopping in downtown Pittsburgh in the early twentieth century, at a time when many skilled workingmen could afford trolley fare, often had their dresses smudged by contact with grimy work clothes. A woman forced to stand on a crowded streetcar might be squeezed indecently against strange male bodies or furtively groped. If she were offered a seat, she might feel a man press his leg against hers or attempt to fondle her under an open newspaper. If she looked up, she might find a masher staring at her. Avoiding eye contact was difficult in vehicles where the seating arrangement forced passengers to face each other.[20]

The press of bodies was less of a concern as the cars grew emptier by midevening, but women faced other dangers. An unescorted woman was now out in public after dark, giving more men the pretext for treating her as a sexual object. Men were likely to have begun drinking as well. Mrs. Duval was acting cautiously by bringing along her younger daughter when she went to escort her older daughter home from choir practice. The authors of etiquette manuals would have approved: women who were obliged to go out at night were cautioned to bring along a servant or another lady "if you cannot procure the escort of a gentleman, which is, of course, the best." In escorting her, Mr. Putnam was acting as the proper gentleman. The situation was risky enough that etiquette permitted women to wear bonnets and hats to shield their faces from view while on a streetcar or omnibus, even though headwear after 7:00 p.m. was otherwise considered a fashion faux pas.[21]

Yet many respectable women took a chance and ignored all this advice. Shopgirls and other women who worked long hours were on their way home from work, while others were out on social calls or theater excursions. Mary Laughlin, a twenty-nine-year-old Philadelphia seamstress who kept a diary in 1888 and 1889, frequently noted the precise time she boarded a streetcar. Laughlin paid visits or made work-related trips around Philadelphia by herself, catching a streetcar home to the outlying Germantown neighborhood in the late afternoon or at 5:45 p.m., at which hour the trip would have carried her through dusk. She also recorded making some evening social calls, but in these cases she was usually accompanied or at least escorted home by her husband, a marble cutter. Not every woman was as fortunate as Laughlin, whose diary mentions no bad experiences on her evening trips. Women were verbally insulted by roughs and mashers and had their pocketbooks grabbed from their hands. Purse snatchers targeted women on the small "bobtail" cars, where there was no conductor who might protect them.[22]

RIDING THE OWL CARS

After the midevening lull in traffic, some streetcars leaving the central business districts of American cities were crowded again when the theaters let out about 11:00.[23] Passengers then grew scarce. The number dwindled to a low point between 2:00 and 5:00 a.m., according to studies of streetcar traffic on the Brooklyn Bridge in 1897 and 1898, and on Broadway in 1901. The timing may have varied slightly from city to city, depending on the liveliness of nightlife and the prevalence of early morning labor. Early twentieth-century traffic studies in Cleveland, New Orleans, and Detroit found the streetcars at their emptiest between 1:00 and 4:00 a.m., a bit earlier than in New York. But as a general rule, 3:00 a.m. was extremely quiet everywhere.[24]

Streetcar service at night was inferior. The horses were decrepit specimens ready for the glue factory. Late-night passengers on one hilly route in Brooklyn often had to get out and push. Even after installing cable car systems in the 1870s and 1880s, traction companies still brought out the slow horse-drawn cars at night rather than running their expensive machinery. The light from flickering oil lamps inside the owl cars was too weak to read by. Riders on a wintry night were chilled despite the coal stoves and the scattering of filthy straw intended to insulate the floor. (The combination of lamps, stoves, and straw also created a fire hazard.)[25] Further, the cars ran much less frequently at night than during the day. Although by 1886 cars on some lines in New York and Chicago ran all night at intervals of fifteen or twenty minutes, headways of thirty minutes to an hour after midnight were typical in other large cities. Hour-long waits between cars remained common into the twentieth century in some major cities, as well as in small cities where owl service was introduced for the first time.[26]

The gaps between cars made it worthwhile for riders to heed the timetables, though the drivers themselves were not punctual. Streetcar schedules were published in city directories, local almanacs, newspaper advertisements, or pamphlets available at hotels and all-night drugstores. Missing one's streetcar meant having to endure a long wait on a street corner, listening for the clatter of hoofs or the jingle of bells in the distance. It was probably quicker to walk than wait, even though the lack of traffic at night allowed streetcars to go a bit faster than their usual sluggish pace (six miles an hour for horsecars, and ten for electric trolleys). Long waits were sometimes unavoidable if one had to transfer from a ferry to a streetcar or from one streetcar line to another.[27] A man waiting on a street corner late at night was an easy target for robbery, and women might well expect worse. "In a city, or in any lonely place, a lady must avoid being alone after nightfall, if possible," warned one etiquette manual. "It exposes her, not only to insult, but often to positive danger." Similarly, unless one's house directly adjoined a streetcar line, the journey ended with a walk through deserted, poorly lit residential streets where crimes could be committed with little chance of witnesses.[28]

Crimes were frequent enough on the cars themselves. Rowdy young men were a common nighttime problem from the beginning of public transportation, when they piled onto omnibuses (and their wintertime counterparts, public sleighs) intent on raising hell. All it took sometimes for a brawl to break out on an omnibus or a streetcar was for the conductor to ask rowdies to stop using foul language or to insist they pay their fares. Conductors and drivers risked being punched, struck with blackjacks, stabbed—or hit by stones if they succeeded in forcing the ruffians off the car. Pickpockets, who relied during the day on the distractions and jostling of crowds, changed their tactics at night to lifting watches and wallets from sleeping passengers. Unusually brazen criminals committed muggings.[29]

These were the exceptions, of course, but relaxed standards of behavior were the rule. The owl car conductors collected many of their fares from late night drinkers and seemed grateful for the business. Passengers on a 2:00 a.m. streetcar, according to a St. Louis newspaper writer in 1881, were "funny, vigorous and frequently quarrelsome. Some are perfectly sober and gentlemanly, the majority are in all stages of intoxication—from the silly and maudlin to the wild man who wants to wake up the city." If conductors were too timid to impose order or expel the profoundly inebriated, rowdies annoyed other riders with loud talk, foul language, cigar smoke, and brawls. The normally boosterish *Street Railway Journal* reported in 1888 that "the late cars out of Boston on the Cambridge division have been made unfit for ladies to ride in by disorderly and drunken persons." The Massachusetts Board of Railroad Commissioners reported in 1909 that the presence of drunks on the late night cars in Worcester and Fall River "creates conditions which at times outrage all sense of delicacy and decency."[30]

Most but not all of the riders on the owl cars were men. Along with the obnoxious drunk was "the tardy 'lodge man'" and many another straggler from an evening of pleasure seeking. Was it truly in the public interest to encourage this behavior? A St. Louis streetcar official argued that owl service would have a "demoralizing" influence that wives and mothers would resent. Advocates of owl service countered—in petitions, editorials, and city council resolutions—that the cars also carried men whose jobs ended late at night: newspaper employees and bartenders, for instance, and shift workers in factories. Some of these claims were questionable, as when an alderman asserted in the 1850s that all-night service was needed by the numerous Brooklyn men compelled by business to be out until 3:00 a.m. Nonetheless, the expansion of night work and the outward growth of the cities in the late nineteenth and early twentieth centuries did create a significant demand for transportation.

In 1910, unions representing thirty-five hundred railroad employees in Harrisburg pushed for expanded night service on the trolleys, arguing that workers at the new Pennsylvania Railroad classification yards in outlying Enola needed to get to and from their homes in the hours between midnight and 5:00 a.m. The city council and the local newspaper supported the unsuccessful request: regardless of its profitability, twenty-four-hour street transportation was said to be a public necessity in any growing, up-to-date city. The newspaper editors insisted that in addition to the trainmen, "there are hundreds of others out after midnight for perfectly legitimate reasons, who have a right to transportation without the expense of hiring a carriage or auto."[31]

The women who traveled late at night were also an eclectic bunch. Rather than walking the streets after midnight, a number of New York prostitutes worked the horsecars, flirting or soliciting openly from the comfort of cushioned seats. Even the marginally safer omnibuses were frequented by prostitutes at night. Waitresses, dishwashers, and actresses rode the owl cars home from work, unaccompanied by a man. Couples returning from an evening's entertainment also rode the first owl cars; they could be seen in the dim light leaning against each other or indulging in further intimacies.[32]

Wherever twenty-four-hour service was established, people adjusted their leisure habits accordingly and eventually came to regard the owl car as a necessity for their work as well. Owl cars allowed night workers to live beyond walking distance from their jobs. So many of the Homestead Steel Works employees were commuters that a 1909 streetcar strike in the Pittsburgh area "seriously crippled" nighttime production.[33] The West Chicago Street Railway Company set off a furor in November of 1896 when it abruptly ended late-night service on its Madison Street line without advance notice. The Daughters of Rebekah had scheduled a ball that night, which ended at 1:00 a.m. with the attendees waiting in evening dress on a cold street corner for a horsecar that never came. On learning that the company intended to keep the line closed to save money, the owner of both the dance hall and an all-night drugstore organized a petition drive. He argued that the line, where cars had run every twenty minutes through the predawn hours, was used by many people attending clubs and dance halls, as well as by night workers returning home. Night workers who traveled the Madison Street line were frequent customers in his drugstore, he said, "and such customers make it worth while to keep open for those who need medicine at night for emergency cases." West Side residents who joined his cause argued that it was unsafe to make them ride the parallel lines along Van Buren and Lake Streets, because that meant walking through dark side streets where criminals lurked. When the City Council responded by ordering the streetcar company to resume night service, company president Charles Yerkes denied it had the right to so but agreed to comply anyway.³⁴

HACKS AND HELPLESSNESS

Hacks, the late nineteenth-century equivalent of taxicabs, were not a viable alternative for most women at night. They were expensive, for one thing. In 1871, the minimum fare for a night hack ride in Boston was fifty cents before 11:00 p.m. and a dollar afterward. People in every city grumbled about the high cost and the drivers' tendency to overcharge.³⁵ Further, hacks were scarce. Milwaukee in 1897 had about two hundred "night hawks"—hack drivers who worked after midnight—for a city of about 250,000. Night hawks made their money from well-heeled men who enjoyed the more disreputable amusements: drinking, gambling, and whoring. Some owl car conductors refused passengers who were obviously drunk, and some drunks were too incapacitated to board a streetcar anyway, so hack drivers found many fares by waiting outside saloons. Rich men on a spree, or successful gamblers, would hire a hack as a celebratory gesture. A john and a prostitute, or a couple of lovers, would take a hack to travel in privacy to a place of assignation.³⁶

Some hacks rolled through the nocturnal streets looking for customers, but most preferred to await their next fare in a promising location and rest their horses. Hackmen stationed themselves wherever they thought they could find business: at railroad stations, outside restaurants, near theaters that were letting out. Certain central locations became known as places where hacks could be found, such as Madison Square in New York, or Broad Street near Chestnut in Philadelphia. After midnight, night hawks waited outside oyster saloons, billiard halls, and gambling dens, hoping for customers too drunk to count their change. Liveried cabmen, who had contractual arrangements with expensive hotels and clubs, could be found outside those establishments.³⁷ A solitary woman who entered a hack was taking a risk. She was delivering herself into the care of a working-class stranger who had the power to take her anywhere with no one watching. Some drivers were known to abuse this trust. According to one advice manual, it was better to take a streetcar or omnibus "where there are plenty of people" than ride alone in a hack at night.³⁸

There was no safe way to travel through the city at night without a man's protection. For all the rhetoric about the American woman's freedom in public, the fact remained that at night her freedom grew more limited with each passing hour. Unescorted women persisted in walking through the streets, boarding a streetcar, or hiring a hack, but this did not mean they traveled on equal terms with men. No sensible woman was oblivious to the possibility that something very unpleasant could happen to her. No matter how carefully she dressed and behaved, she was taking a risk, and she knew it.

A 1903 short story in *McClure's Magazine* provides a glimpse into the emotional trauma that women experienced in navigating the city by night. In this story, "A Little Surprise" by Mary Stewart Cutting, the protagonist Anita Gibbons feels increasingly vulnerable after she ventures out to her suburban New Jersey train station to greet her husband on his return from work. A series of miscommunications leaves her alone and worried on a commuter train and ferry to New York, hoping to join her husband for a dinner party. She is shocked to find that her husband is not there to meet her at the ferry slip. She asks helplessly for him at the newsstand, when to her great relief is rescued by a neighbor couple who offer to escort her to the restaurant. "I didn't know it was so dark at night when you were out alone by yourself, until I came off the ferry-boat," she confides to them. But her husband isn't at the restaurant either. Mrs. Gibbons is watched closely at each step along her route through the city—by the clerks at the newsstand, by men on the streetcars, by waiters and diners at the restaurant. Rather than abandoning her, the neighbors invite her along for an evening at the opera and then supper with their friends until nearly midnight.

A delay on the elevated train makes them miss their ferry, causing further delays. Eventually she and her two escorts find themselves on a crowded, lurching trolley in New Jersey.

> Men sat with their heads on their sweethearts' shoulders, in true early-morning trolley-car fashion, and every inch of standing room was packed too thick for the eye to penetrate with a singing, drunken, cat-calling, indecent crowd, the last scum of a great city. It was an offense to delicacy to be there. The lights flared wildly up and then went out at intervals. When they went out, Mrs. Gibbons felt a cold terror. She had always been afraid of drunken men, and she was so used to the protection of love!

She begins to cry, attracting the attention of men nearby. Matters get even worse as she attempts to transfer to another trolley; she is shoved aside by a fleeing pickpocket as she tries to board after her friends, and she is left all alone when the trolley pulls away.

Of all the chances and changes of this wild Walpurgis night, there could be nothing stranger than this, that she, Nita Gibbons, should be sitting alone amid the dark marshes, in front of a Jersey "gin mill" at half-past two o'clock in the morning. It was so entirely past all imagining that frenzy had left her. She would probably never get home again, but she had ceased to struggle against fate.

Here, at the nadir of feminine helplessness, Mrs. Gibbons is saved by the unexpected arrival of her husband. "You were just crazy to do such a thing," he scolds her. "It makes me wild to think of it. You don't know what might have happened. I'll be afraid to go off and leave you home alone. I don't know what you'll do. You ought to be looked after like a child." The story concludes with Mr. Gibbons kissing his weeping, chastened wife as the first light of dawn appears in their empty suburban street, then leading her home where she belongs.[39]

CHAPTER 9

Night Life in the Electric City

Until the 1880s, gas lighting gave the nocturnal city a landscape of light and shadow. Brightly lit thoroughfares, with glowing shop windows and illuminated signs, intersected dark side streets that led through the slums. Public lamps were too weak and too few to illuminate a whole block. Unless supplemented by privately owned lighting on businesses or homes, gas lamps left dark gaps where faces could not be recognized and shapes were only dimly discerned. Uneven lighting reflected and in some ways encouraged the human qualities of the night city: the dramatic contrasts of safety and danger, wealth and poverty, work and leisure, virtue and vice.

Electricity produced a different visual experience of urban night. Far brighter than gas, electric streetlights could fill a city block with light instead of just the circle around the lamppost. Shadows shrank back, revealing facial expressions and house numbers. By the late 1920s, a wash of electric brightness had covered the old chiaroscuro landscape of the gaslit city. Major streets could now be seen at night with a detailed clarity once possible only in gaslit rooms. The glow even in dimmer side streets was usually sufficient to make faces recognizable. The big city might still be an anonymous world of strangers, but not because of darkness.

In other ways, one could argue that electricity merely updated the qualities of the gaslit city. Disparities in lighting persisted. Brightness remained the marker of conspicuous consumption. Nighttime entertainment still concentrated in the most brilliantly illuminated streets, and incandescent displays in shop windows gave an aura of luxury to retailing districts. Chicago's Clark Street at night, observed a newspaper reporter in 1892, was "resplendent with

FIGURE 13. Saturday night on an electrified street. Electric streetlamps did not illuminate the entire city equally, but by the 1920s they did make most streets bright enough that faces could be recognized and obstacles easily avoided. The improved lighting helped encourage more pedestrian activity. John Sloan, "Bleecker Street, Saturday Night" (1918), Encore Editions.

flashing lights and gaily dressed promenaders from all quarters of the city. Here are grand hotels, fine theaters, variety shows, dime museums, shooting galleries, railroad ticket offices, dance-halls, and ready made clothing stores, all ablaze with light and filled with pleasure seeking humanity." Ordinary citizens rubbed shoulders with gamblers and prostitutes inside glittering billiard parlors and "gin palaces."[1] All this seemed to echo the descriptions of gaslit cities fifty years before.

The effect of new lighting technology on the nocturnal street, as it turned out, was less powerful than the effect of the social and economic changes occurring at the same time. After 1880 many more people found the opportunity to go out at night, and they enjoyed a greater array of attractions. Hours of labor shrank from 1880 to 1910, giving city residents more free time in the evenings. A tenfold increase in clerical jobs gave more people incomes that could support small luxuries. Even working-class men enjoyed modest increases in spending power, while unprecedented numbers of young women entered wage labor for the first time.[2] Electricity, as we have seen, enabled some of this economic expansion to take place during hours of darkness: it encouraged night work in factories and night travel on trolleys. These changes, more than just the exciting brightness of the electric streetlamp, were what put crowds on the streets at night.

Night life in the electric city drew a different type of crowd: younger, more open to working-class participation, and not quite so heavily male. The emerging youth market welcomed an explosion of cheap mass entertainment. A 1911 survey of recreation in Milwaukee estimated that 52 percent of evening movie audiences consisted of people from fifteen to twenty-five years old; another 14 percent were under fifteen. Two-thirds of vaudeville audiences and dance hall crowds, too, were twenty-five or younger. Studies of Detroit, Kansas City, and Providence reached similar conclusions.[3] Though traditionalists still condemned late night activities for destroying morals and work ethics,[4] perceptive observers saw more insidious threats. The new commercial leisure activities did not follow the model of propriety set by the reformed theatrical performances and concerts that enticed middle-class Americans in the mid-nineteenth century, nor were they as flagrantly immoral as the old gambling halls and brothels (which continued to flourish as well). Instead, vaudeville shows, movie theaters, and dance halls reflected shifting standards that critics feared were undermining respectability without flouting it outright. Entertainment entrepreneurs and social reformers proved unable to keep nocturnal street culture from infecting these amusements or the young people who attended them. Moral ambiguity now suffused all "the watches of the night," muting stark contrasts much like a streetlight whose dull glare never matched the brightness of day. The old moral chronology of night was falling into chaos.

ELECTRIFICATION

Electric light first filled American streets at 8:00 p.m. on April 29, 1879, with the demonstration of a dozen high-voltage Brush lights in downtown Cleveland. A band played, and ten thousand Clevelanders gawked as the two-thousand-candlepower arcs flooded Public Square with the light of noon—or at least the light of 350 gas lamps. "It turned every gas light within a half dozen squares green and yellow with envy," gushed the *Cleveland Herald*.

> It floated out Euclid avenue and paled the erstwhile radiance of the Opera House triple illuminator. It was wafted either way on Ontario street, almost

kissing the dark blue waves of the lake on the one hand, and just peeping over into the dense darkness of the flats on the other. Westward on Superior street it penetrated far enough to make the big high bridge stand in somber splendor by contrast, and toward the east it threw great shadows up against the City Hall and old Trinity. It shimmered and sparkled in and through the playing, bubbling water of the fountains, and reflected like sunlight on the gilded vane that surmounts the tall centennial flagstaff.[5]

In early arc lights, a dazzling bolt of current hissed between carbon electrodes, stabbing painfully any eyeball that looked at it directly. Approximating daylight in its intensity and spectrum, arc light let people see colors and read newspaper type. The innovation quickly spread to the streets and waterfronts of other cities and was applied also in theaters, circuses, mills, hotels, department stores, and saloons—not always to great acclaim. Casting a glare of up to four thousand candlepower, arc lights were too bright and too hot to be used as if they were gas. It was simply unacceptable to place them on the old gas lamp posts, which rose only a few feet above head height. Instead, cities found it best to install arc lamps on tall poles so that the arc would be outside pedestrians' field of vision and the expensive brightness would cover more ground. Minneapolis installed an "electric moon" in the center of its business district, consisting of a cluster of arc lamps atop a steel mast 257 feet tall. Detroit tried to illuminate most of the city from 122 arc light towers. Those in the downtown rose 175 feet above the streets and stood between 1,000 and 1,200 feet apart; those in outlying neighborhoods were twice as far apart, shorter, and not quite as bright. Arc light towers proved an unsatisfactory way of lighting cities. They were so tall that trees and fog interfered with illumination. Cities more commonly installed the arc lamps about twenty feet above intersections, in globes hanging from pole arms or wires, but this arrangement was not ideal either. Rather than bathing the streets in a uniform glow, the low-altitude arcs threw pools of stunning brilliance directly below, and from there the light tapered off into dimness. Cities often spaced these lights 250 to 400 feet apart, leaving dark gaps like those between the old gas lamps.[6]

Even as public squares shimmered with "piercing light-blue sparks of electricity," gas and oil lamps continued to sputter throughout American neighborhoods. Fewer than 20 percent of urban streetlights in 1890 were electric; more than 60 percent burned gas, and the rest were even more primitive. Given the limitations and expense of arc light, cities continued through the early twentieth century to install more gas lamps, now brightened by new

Welsbach mantles. New York City's streets were lit by 25,483 gaslights in 1890 and 42,777 in 1903. St. Louis lit its streets with 8,709 gas lamps in 1890, 13,260 in 1903, and 17,079 in 1907, by which time the city still had only 4,538 incandescent electric streetlights and 1,082 arc lights. The nocturnal streetscape in the early twentieth century was a crazy quilt of different forms of illumination: arc light towers, arc light globes near street level, incandescent lights, gaslights with and without mantles, lamps burning gasoline or kerosene. Interior lighting lagged even further behind; some affluent homeowners did not upgrade from kerosene to gas until the 1890s, and electric light remained a luxury.[7]

Incandescent lighting provided an appealing alternative to arc light. Incandescent lamps did not hurt the eyes when placed close to street level and did not require as much maintenance. They were mainly used indoors at first but also proved well suited for advertising: colored bulbs could be arranged to spell out messages. Seedy working-class thoroughfares like New York's Bowery looked surprisingly cheerful when lit with electric lamps, signs, and display windows; bright light drew the eye to storefronts and places of amusement at sidewalk level, while the looming tenements and elevated railroad tracks receded from view. Incandescent bulbs, clustered for greater brightness, increasingly replaced gas streetlamps and arc lights after 1900, especially with the introduction of improved tungsten filaments after 1907. Instead of the variegated pattern of glare and shadow created by widely separated arc lights, the street could be washed in even light from multiple incandescent lamps. By 1903, Pittsburgh, Providence, and Columbus already had more incandescent streetlamps than arc lights.[8]

As electric lighting spread through the city in the early twentieth century, merchants and public officials eagerly replaced older electric lamps along commercial streets with the newest and brightest models, creating what were called "great white ways." They aimed to do more than just help people see better. Electricity represented energetic modernity and signaled that the city was a forward-looking, prosperous, exciting place. By imitating New York's Broadway, merchants and city officials hoped to bring to their humbler centers some of the same bustle—or at least to draw more people downtown after dark. Minneapolis officials, for instance, began in 1908 to install brilliant new tungsten filament lamps on ornamental poles along Nicollet Avenue, claiming to make it one of the best-lit cities in the world. In Hartford, a crowd of twenty thousand turned out one December night in 1911 to see the first block of Asylum Street become another white way. When the mayor threw the switch on the new clusters of globes, wrote a reporter, "Night was turned fairly into day, and the block can now raise itself head and shoulders above any other section of the city as far as lighting is concerned and say, 'Where are you fellows, anyway? I can't see you down there in the dark.'"[9]

Besides drawing more people out at night, improved lighting technology seemed to promise greater safety from crime, as it had in the early years of gas. "Known criminals shrink from a glaring light in which they may be recognized, while criminals of all classes are kept in restraint by an illumination which makes their movements so clearly seen," claimed a writer for the *Boston Evening Transcript* in 1887, five years after electric streetlights were introduced in Boston. The weak statistical support for such claims did not shake people's faith that light discouraged crime. St. Louis officials confidently inflated an old saying by asserting that a single streetlight was as effective as five policemen. In Minneapolis, a sudden loss of electric lighting following a 1911 explosion in a generating plant raised fears about crime, much as happened during the gaslight blackouts of the nineteenth century. Minneapolis's business streets got their electric lights back first, and the remaining gas and gasoline lamps kept other districts from being totally dark; the crime wave never came. In Chicago in 1920, just before the outbreak of Prohibition gang warfare, a local politician promised that "with more lights in our dark alleys and streets, crime, such as murder and holdups, will be minimized to a great degree." Similar rhetoric could be heard throughout the twentieth century, though it remains debatable whether improved lighting significantly reduces crime or merely pushes it into comparatively darker areas.[10]

The spread of electricity seemed initially to follow the pattern of early gaslight. It first appeared in businesses and wealthy homes and along the most prestigious commercial blocks. In turn-of-the-century New York, electricity heightened the contrast between the bright zones of affluence and the dark slums that still relied on gas and kerosene. In Minneapolis, electric streetlights were first installed along Hennepin, Nicollet, and Washington Avenues and other important downtown streets, then along the outbound streetcar lines. Minneapolis officials slowly added individual lamps between streetcar lines in response to petitions from neighbors, until in 1908 the city launched a more far-reaching program of lighting the outlying neighborhoods.[11]

The distinctions between slums and wealthy neighborhoods faded in the 1910s and 1920s as electric rates dropped, working-class income rose, and cities replaced gas streetlamps with electricity even in ordinary residential neighborhoods. By 1925, 89 percent of Chicago's streetlights were electric. The electrification of urban America was nearly complete by the end of the 1920s, at least in the cities of the Northeast and Midwest. By the time Commonwealth Edison installed its one millionth meter in Chicago in 1929, few Chicago dwellings lacked wiring. Thereafter the rising consumption of electricity in cities depended largely on more intensive use, as homes and businesses installed more and more lights, appliances, and electric-powered machines. The remaining untapped markets were mainly in rural areas, where as late as 1935 barely 10 percent of farmers had electricity. It took federal intervention, through the Rural Electrification Administration, to bring this urban amenity to the countryside in the late 1930s.[12]

Within the city, though, one sharp distinction in lighting persisted: the much brighter illumination of the central business district and particularly the blocks devoted to nighttime entertainment. In part this was the result of city governments' determination to continually upgrade streetlights to keep the white ways brighter than their surroundings. The effect was enhanced by display windows, ornamental floodlighting of facades, and especially electric signs. Broadway led the way in the 1890s and 1900s with advertisements composed of hundreds of colored lights, some flashing or simulating motion. The monumental signs first clustered between Herald Square and Madison Square, noted for its garish Heinz pickle made of green lightbulbs against an orange background. By 1910, electric advertisements lined more than twenty blocks of Broadway and were becoming a tourist attraction in their own right. Times Square emerged as the brightest node of Broadway in the early twentieth century, with brilliantly lit theater marquees and towering ads for coffee, cigarettes, and toothbrushes. "Times Square can truly be called New York's twenty-four-hour corner," declared a 1929 article in the *New York Times*, which was headquartered there. "When its offices close its theaters open, and after these are the night clubs and cabarets."[13]

The accepted wisdom was that bright lighting could "improve the moral tone of even the most vicious neighborhoods," as a St. Louis official put it in 1908. But what about the "bright lights" districts? Could it be that the lurid illumination heightened the moral danger? The leaders of Chicago's Juvenile Protective Association thought so. "The bright lights and open doors of cheap theatres and pleasure resorts urge a constant invitation upon the girls and boys whose dreary home surroundings drive them into the streets for recreation," the association warned in its 1911 report. Drawn to the electric glow, their desires morbidly stimulated, huge numbers of young people might fall under the demoralizing influence of urban vice.[14]

FIGURE 14. Unveiling night's secrets. Guidebooks promised male travelers a shortcut to local sporting men's knowledge of urban entertainments. This cheaply printed example from 1892, intended for men visiting Chicago at the time of the World's Columbian Exposition, was subtitled "The Pleasure Seeker's Guide to the Paris of America." The word "night" is decorated with what appear to be shining red lightbulbs, suggesting both the modern illumination and the open prostitution that could be found on Chicago's streets.

HOME LIFE AND NIGHT LIFE

Concerned citizens worried that the excitement of night life, now exaggerated by electric lighting, was competing more successfully with the dull decency of home life. They feared, moreover, that the home itself had grown inadequate. Earlier concerns about failed domesticity had focused on bachelors living in boardinghouses. Bachelors continued to draw attention in the decades around the turn of the century, as their numbers rose sharply and they constituted an unusually large proportion of the population. In Chicago in 1890, 13.8 percent of the total population consisted of bachelors aged fifteen to thirty-four, and 44 percent of all men fifteen and over were unmarried. Single young women were also very numerous. Though most of these unmarried adults lived with relatives, many others lived on their own in boardinghouses, lodging houses, and apartments. Furnished-room districts filled with single adults arose in cities throughout the country; the residents often dined at nearby restaurants, cafés, and saloons and spent much of their free time away from their rooms.[15]

Family living spaces, too, were said to be inadequate around the turn of the century. As cities grew larger and denser, wrote Michael M. Davis in a survey of New York City recreation in 1911, people were forced into cramped apartments in crowded neighborhoods and used them only for eating and sleeping. This complaint was more than just nostalgia. The urban dwelling was losing many of its functions to public and semipublic spaces. Adult men had long since grown accustomed to leaving home each day to work, and children now spent their days sitting in classrooms and playing on the street. Working-class people often bathed in public bathhouses. On warm evenings, they went into the streets in search of fresh air and even slept outside on rooftops and along sidewalks. "In a crowded city there is human pressure upon the street hardly less great than that within the home," Davis wrote. "Offshoots from the street arise to meet this pressure,—the candy shop for the children, the ice cream and soda parlor, the moving-picture show, the vaudeville, the dance hall, the saloon. To these places people pay to go, partly to seek positive pleasure, partly because to remain within the straits of the home or the moil of the street means positive pain or discomfort." Settlement house workers and other social commentators feared that the family was being torn apart as a result and that the values associated with domestic life could survive only if somehow extended to public space.[16]

Privacy, a core value of American domesticity, was scarce in apartment buildings where families shared rooms with boarders and voices carried through walls. William Dean Howells's 1889 novel *A Hazard of New Fortunes* provides a more disturbing example of public scrutiny, in which a married couple gazes into lighted apartments at night from their seat in an elevated train. Mrs. March enjoys the fleeting sense of "domestic intimacy" she gets from her voyeurism, while her husband declares that "it was better than the theatre, of which it reminded him, to see those people through their windows: a family party of work-folk at a late tea, some of the men in their shirt sleeves; a woman sewing by a lamp; a mother laying her child in its cradle; a man with his head fallen on his hands upon a table; a girl and her lover leaning over the window-sill together. What suggestion! what drama! what infinite interest!" Whereas the private dwelling was in public view, a sense of privacy paradoxically could be found in public: in the anonymity of the public street, the distant park, the darkened theater, the crowded saloon, or the elevated train. The vastness of the city allowed people to enjoy an evening's pleasure without the constant sense of being recognized and watched. It was a freedom unavailable either in the rural community many city residents had come from or in the crowded neighborhoods where they now lived.[17]

The relaxation of the middle-class opposition to commercial entertainment also helped the growth of urban nightlife. Leisure activities in general had been regarded with some suspicion through the first half of the nineteenth century, but now they were advocated by clergymen and physicians as antidotes to overwork. Middle-class Americans increasingly shared working-class people's enthusiasm for "fun," even in such unproductive forms as watching a slapstick routine on the vaudeville stage. The culture of self-improvement and moral uplift, to which P. T. Barnum had been careful to genuflect, was losing its power at the turn of the century.[18] Entertainment entrepreneurs cheered on this cultural shift, seeking to build a mass audience by promoting attendance among respectable women.

THEATERS, NICKELODEONS, AND MOVIE PALACES

Playhouses were so thoroughly tamed by the turn of the century that legitimate theater was considered a woman's entertainment. Women predominated in matinee audiences at theaters as early as the 1860s. Though men continued to attend the theater in the evening, they were often presumed to be chaperoning their wives and daughters to performances aimed at flighty females. The legitimate theater's high ticket prices and pretensions of high culture left it with a dwindling share of the theater audience in the early twentieth century. Separate, low-priced theaters offered melodramas and foreign-language performances for working-class men and their families, but these theaters grew scarce as vaudeville and movies provided cheaper alternatives.[19]

When physical humor and broad jokes withdrew into the male preserves of the concert saloon and minstrel show after 1860, some variety houses in New York resisted the trend. Tony Pastor and other theater owners offered sanitized versions of the variety acts found in concert saloons. Pastor advertised "ladies' nights" in which escorted women were admitted free, and by the 1880s he was successfully drawing the men and women of the middle class. An 1884 report in the *New York Clipper* described one evening's crowd as "decidedly a 'family' audience, a large number of ladies, many of them coming in twos and threes, without escort, showing that it is politic to manage an establishment of this description in such a manner that no gentlemen need fear to bring his wife, sister or mother to 'see the show' or even allow them to go by themselves."[20]

In the 1880s Benjamin F. Keith and Edward F. Albee tinkered with the respectable variety show in theaters they owned in Boston and Providence and developed the format that gained popularity as "vaudeville." They offered a continuous rotation of short acts: musicians, gymnasts, sideshow freaks, blackface minstrels, trained animals, celebrities, comedians, and other performers. Curious passersby could join the audience at any time and stay as long as they liked. While promising "something for everyone," vaudeville theaters particularly sought to attract women, whose presence signaled respectability. Keith, Albee, and their imitators (at least at first) suppressed language or innuendos on stage that might be considered offensive. They banned smoking, whistling, stamping, and peanut crunching in the audience and hired ushers to enforce the rules. They opened as early as 11:00 a.m. to attract middle-class ladies on shopping excursions. As the format proved successful, vaudeville entrepreneurs bought theaters in multiple cities and created circuits through which lineups of performers traveled, changing weekly. The vaudeville palaces of the 1890s and 1900s courted an affluent audience while holding prices below those of legitimate theaters. Cheaper imitators on the "small-time" circuits brought vaudeville to working-class neighborhoods and small towns. According to Michael Davis's 1911 survey of New York City recreation, men made up a slight majority of the vaudeville audience; adult women accounted for 29 percent of those in attendance, and children another 19 percent. Davis estimated that 60 percent of the audience came from the working class and 39 percent from the "clerical" middle class. By the 1910s, vaudeville was the second most popular theater experience in America, drawing 20 or 25 percent of the theater audience, according to studies of several cities. Only the movies drew more people.[21]

Despite the presence of middle-class women at vaudeville shows, Davis complained that the moral tone of the performances had slid to the point "where the impropriety and obscenity were no whit less offensive than they are at a so-called burlesque show." The comparison was plausible mainly as an expression of outrage against the danger of exposing women and children to inappropriate material. In contrast to vaudeville, burlesque in turn-of-the-century America was an overtly salacious genre of variety theater consisting of leg shows, raunchy skits, and off-color jokes. It had developed a reputation as a crude, working-class entertainment for leering young men and teenage boys. Nonetheless, some burlesque theater operators attempted to present tamer alternatives to the sexual humor and "coochie dances" that characterized "hot" shows. Further, the audiences were more diverse than critics believed: middle-class men and even women attended—many of them no doubt drawn by the same voyeuristic spirit that took the more adventurous on "slumming" expeditions into vice districts and Chinatowns. Davis's study estimated that a slight majority (54 percent) of the New York burlesque audience consisted of people in the "clerical" class—a higher proportion than in either vaudeville houses or movie theaters—and that 19 percent of the burlesque audience was female.[22]

Davis considered movies the least objectionable of the leading forms of theatrical entertainment. Carefully rating the moral content of the shows they witnessed in New York, Davis and his investigators concluded that five-sixths of burlesque shows were "demoralizing" (the worst grade), and the rest were "lowering" (the next to worst). One out of five vaudeville shows were "lowering," three-fourths were "not objectionable," and the small remainder was "of positive value." In contrast, half of the movies were "not objectionable," and the other half were "of positive value." A study in Kansas City concluded that 79 percent of movies were "good," compared with 72 percent of stage performances and 23 percent of dance halls. Regardless of whether they liked drama or dancing, anyone planning an evening out had to choose carefully.[23]

The motion picture business had begun unpromisingly in the mid-1890s with the installation of peep show machines in Kinetoscope parlors in central business districts. By dropping a penny in the slot, an individual could peer into the eyepiece at a short film in which images miraculously moved. Even such ordinary sights as waves on a beach seemed fascinating at first, but the novelty quickly wore off and the Kinetoscope parlors closed. Peep shows survived in a slightly different format, the Mutoscope machine, in downscale amusement arcades. Young men and boys enjoyed watching slapstick skits, boxing matches, and such sexually suggestive films as "How Girls Undress" and "Getting Ready for the Bath." Davis deplored the arcades in his 1911 study. "By construction, and usually by situation, these places invite the listless 'hanging-round' which is one of the worst characteristics of the street; and they tend to create that most dangerous mixture—a minority of hardened street children scattered among the majority of normal children who have drifted in to the show." By then, however, the peep shows had long since lost much of their remaining audience to competition from movie theaters.[24]

Short motion pictures were projected first in vaudeville houses in the late 1890s and then in nickelodeons—cramped storefront theaters that got their name from their five-cent price of admission. "Nickel madness" swept through American cities starting in 1905. Nickelodeons sprouted up almost overnight throughout central business districts, drawing women and men of all classes. They were even more numerous on commercial streets in or near working-class, immigrant neighborhoods, where people of all ages enjoyed the silent films regardless of whether they spoke English. A *Harper's Weekly* article in 1907 dismissed the criticism that the dark, stuffy rooms harbored pickpockets; thieves would have found slim pickings in audiences composed of "workingmen . . . tired drudging mothers of bawling infants [and] the little children of the streets, newsboys, bootblacks, and smudgy urchins." In 1907 there were 200 nickelodeons in Manhattan alone and 158 in Chicago. A 1911 recreation survey estimated a weekly attendance of over 210,000 in the movie theaters of Milwaukee—more than half the city's total population. Vaudeville theaters drew only 75,000, burlesque shows 24,000, melodrama 17,500, and legitimate theaters under 22,000.[25]

Not everyone shared the popular enthusiasm for movies. Starting in the winter of 1906–7, social reformers joined journalists in denouncing the tawdry nickelodeons and the vulgar one-reel films shown in Chicago. "Most [nickelodeons] are evil in their nature, without a single redeeming feature to warrant their existence," declared the *Chicago Tribune*. An investigation in 1909, wrote Louise de Koven Bowen of the Juvenile Protective Association of Chicago, found that "the pictures not only showed crime of all kinds, but scenes of brutality and revenge calculated to arouse coarse and brutal emotions." A movie called *Black Hand's Revenge* showed a bombing, a killing, a robbery, and an "attack on a woman." In another egregious film, titled *The School Children's Strike*, students get revenge on a stern principal by burning down the school. "Such films could not fail to have an injurious effect upon young people," Bowen observed. The imposition of local police censorship soon put an end to the showing of the most demoralizing films, and a follow-up investigation in 1911 found that "the motion picture shows of Chicago are now very decent," Bowen wrote. The moral quality of films rose nationwide after 1909, when movie producers decided to submit their films to the review of a National Board of Censorship.[26]

Social investigators worried that the diversity of the audience posed another moral danger. Low admission prices let large numbers of children and working-class adults attend the shows, usually not in family units. Crowds of unaccompanied children milled outside the theaters, gawking at posters and wondering how to gain admission. Some children stayed very late in the evening until the admission price dropped near closing time. "These crowds were often worked by evil-minded men, who are generally to be found where little girls congregate," Bowen wrote. "The boys and men in such crowds often speak to the girls and invite them to see the show, and there is an unwritten code that such courtesies shall be paid for later by the girls." Prostitutes also loitered in some theater lobbies, along with immoral young women looking to pick up men, the Bridgeport Vice Commission reported in 1916. "'Charity girls,' young, wayward girls, who perform immoral acts for gifts or entertainment, but with no thought of financial reward, were 'picked up' by young men and boys who make it a practice to attend shows for this purpose." Once inside the theater, amorous couples, casual pickups, and child molesters all used the darkness as cover for "liberties." Vice commissions and other reform organizations warned that depraved men prowled the unsupervised rooms, seeking the opportunity to "sit beside the innocent boys and girls without a question or suspicion until irreparable harm is done." Intimacies beginning in a theater supposedly started many girls down the path of immorality that led to the brothel.[27]

The reputation of movie theaters improved in the 1910s and 1920s with censorship of the films and with the opening in downtown areas of new "movie palaces," which were distinguished from the older nickelodeons by their size, garish architecture, exuberant decor, and tolerable ventilation. Potential rowdies were awed into silence by the setting or shushed if necessary by ushers. As in the earlier developments of reformed theater and vaudeville, the new movie houses succeeded in drawing affluent people to what had been considered a low-status amusement. The architect George Rapp called the movie palaces "shrines to democracy, where there are no privileged patrons, [where] the wealthy rub elbows with the poor—and are the better for this contact." Meanwhile, some older melodrama and vaudeville theaters switched to showing movies. Movies now dominated theatrical entertainment. Worcester, Massachusetts, had 9,300 theater seats in 1910, 44 percent of them in movie houses; in 1918, the number of seats had nearly doubled to 17,600, of which 82 percent were in movie theaters. By the mid-1920s, movies were shown in more than 20,000 theaters across the United States.[28]

Movie theaters did not structure leisure time the way playhouses or concert halls did. Like the vaudeville shows with which they remained intertwined into the 1920s, movies did not require audiences to arrive or leave at any particular hour. In the 1910s and 1920s, live performances of all sorts started at 8:00 or 8:15, but movies ran continuously, according to *Chicago Tribune* advertisements. Expensive theaters in and around the Loop advertised shows in the early 1910s that started at 10:00 or 11:00 a.m. or noon and did not end until 11:00 p.m.; smaller neighborhood theaters typically ran continuously from 6:45 p.m. to 11:00. "Stay as long as you like," urged the Colonial theater downtown. This was literally possible at the Lyric theater, the only one to run nonstop for the benefit of night owls, insomniacs, and anyone else who liked attending movies in the predawn hours. Late at night, homeless men would pay a dime for a comfortable seat. Bums snored in the darkness as movies flickered in rotation hour after hour; they were roused at 5:00 a.m. and moved to the balcony while scrubwomen cleaned the main floor, then fell back to sleep. Two other all-night theaters had opened by 1916 and, in the opinion of the Juvenile Protective Association of Chicago, posed a menace to children. By the early 1920s, many other theaters had expanded their hours, though rarely past 1:00 a.m. The Orpheum ran continuous shows from 8:00 a.m. to midnight, while the Castle and the Randolph each ran from 8:30 a.m. to 12:30 a.m. Many neighborhood theaters now started in the midafternoon but still closed at 11:00 or 11:30. The State-Lake (downtown) and the Bryn Mawr (in the North Side Edgewater area) were unusual in listing separate evening showtimes.[29]

DANCE HALLS

Though public dance halls had existed before the nickelodeon craze, they gained greater popularity during those same years. The most common form of public dancing in the second half of the nineteenth century was the public ball, a special "affair" in a rented hall to raise money for charitable or civic organizations. Affluent people concerned about their social status avoided these gatherings, which instead drew working-class couples and immigrants. Some hall owners held public balls for their own profit, but these had an unsavory reputation. "There were at least 200 young girls at the dance, mingling with a lot of roughs and behaving in a manner that was revolting," declared a Pittsburgh-area reformer in 1886 after visiting a Saturday night ball at the Allegheny Coliseum; he reported seeing girls in their early teens drinking with tough-looking men and boys. In addition, young people in the 1890s established new "pleasure clubs" whose main purpose was to hold small weekly gatherings for members and invited guests as well as annual or semiannual public balls for which tickets were sold. As these grew in popularity, they were joined by new commercial dance halls, dance academies, and ballrooms.[30]

Most public dance halls were improvised spaces in rooms upstairs from or adjacent to saloons. Some of these were merely excuses for consuming alcohol: men and women drank at tables during prolonged intermissions between badly played piano tunes; four minutes of dancing would be followed by a fifteen-minute break for refreshments. The worst of these dance halls were fronts for prostitution, but more commonly they attracted teenagers and young adults of varying degrees of chastity. It is impossible to determine the numbers of dance halls with any certainty. Manhattan business directories listed 195 in 1910, but this is undoubtedly an underestimate: many smaller dancing rooms would not have been listed at all. Social investigators in 1914 estimated that the true number in Manhattan was about 600. In Chicago, wrote Louise de Koven Bowen in 1917, 440 city licenses for dance halls understated the true number by about half; saloon keepers often did not seek licenses for their adjacent halls. Cabarets in the 1910s provided opportunities for wild public dancing late at night, in addition to risqué dance acts and performances of ragtime music.[31]

Dancing academies and ballrooms offered expanded opportunities for dancing and succeeded in attracting young teenagers and middle-class young adults who shied away from the saloon dance halls. Some academies were just dance halls with minimal instruction a few nights a week; some even sold liquor on the public "reception" nights. But more commonly they offered regular classes in an alcohol-free environment; unlike the case in the dance halls, dancing proceeded with few interruptions. "The average attendance at the dancing academies of New York in a week is one hundred thousand young people, ninety percent of whom are under twenty-one years of age, and forty-five percent under sixteen," wrote the social investigator Belle Israels in 1910. Well-managed academies prohibited the "tough dances" seen in many dance halls, dances such as the "bunny hug" and "shimmy" whose suggestion of sexuality appalled social reformers. They closed by midnight, unlike dance halls that stayed open into the predawn hours.[32]

Specially designed new ballrooms opened in central or secondary business districts in the 1910s and 1920s. Among these were luxurious "dance palaces" such as the Trianon in Chicago, billed as the "world's most beautiful ballroom." Couples there waltzed in a huge, high-ceilinged hall inspired by the Grand Trianon palace in Versailles, amid trappings of marble, velvet, and brocade—a far cry from doing the "grizzly bear" in a fetid backroom with creaking floorboards. Dance palaces that cared about their reputations in the 1910s and 1920s put restrictions on tough dancing and on the lascivious music that accompanied it—jazz. Dance palaces were not exclusive: young adults from working-class and immigrant backgrounds mingled with those from the middle class. Rather than feeling intimidated by the setting, working-class men and women flocked to the fancy dance halls for an affordable taste of elegance. Dancing grew so popular by the mid-1920s that total dance attendance in Pittsburgh was estimated to be approaching two million a year, over triple the city's population. More than two-thirds of those attending were from sixteen to twenty-five years of age.[33]

Social reformers from settlement houses and vice commissions feared that dance madness would destroy public morality. They were troubled by the drinking, the tough dancing, and the late hours in many dance halls, particularly those connected to saloons. A 1910 investigation of Cleveland's dance halls found that teenagers as young as fourteen attended dances where liquor was sold, where prizes were offered to girls who could drink the most, and where "immoral women" were allowed to solicit. Some of these dances continued until 3:00 a.m. In these situations, other investigators warned, young people could easily lose their moral bearings. Pressed close to the bodies of their dancing partners, they were "rather like children who, with blood aroused by liquor, their animal spirits fanned to flame by the mad music, simply threw caution and restraint to the winds." Teenagers paused from their obscene gyrations only to indulge in "brazen petting parties" during intermissions.[34]

Even if the dancing was decent and no liquor was sold, a basic problem remained. Like antebellum theaters, public dance halls and cabarets in the early twentieth century were feared to be places where the depraved corrupted the innocent. Any man had the freedom to approach any girl without introduction and ask her to dance, regardless of their difference in age, class, or moral station. "The dance problem is the problem of the girl," wrote Belle Israels in 1910. A working-class girl who had reached the legal working age of fourteen but had not yet attained her full legal majority typically viewed the experience of going to the dance hall as a rite of passage to adulthood, a statement that she was "a free and independent being," Israels wrote. Such girls supposedly were easy prey for lechers.[35]

After the dance, and sometimes during it, women were besieged with offers to escort them home. A few dance halls posted signs warning girls not to accept rides from strangers, and some stationed employees in the hallways to prevent pickup attempts. But even a conscientious dance hall operator could not control what happened outside, where teenagers clustered on the sidewalks. Men and boys accosted girls they had seen inside and tried to make dates with them; others called out invitations from the windows of automobiles along the curb. When the invitations were accepted, couples or small groups drifted off in search of other entertainment: late night soda fountains at best, sleazy hotels or suburban roadhouses at worst. The possibility of sexual immorality worried reformers even if the frequency was low.[36]

Contrary to what the social investigators feared, most interactions between strangers at a dance hall were distant and impersonal. Young women did not hesitate to turn down dance invitations. They could afford to be choosy since there was often a surplus of men over women. When they did agree to dance, they did so with a detachment that bordered on rudeness, then left their partners as soon as the music ended to return to their own groups of friends. Some regulars danced with each other on many evenings without ever introducing themselves. At the end of the evening, the girls usually went home as they had come, in a group of girlfriends. Nonetheless, an evening's excursion to the dance hall demanded skill and persistent vigilance if a woman hoped to fend off unwanted advances.[37]

LODGES, SALOONS, AND VICE DISTRICTS

Even if mostly innocuous, the new cheap amusements flourished amid a nighttime milieu that was not. Movie theaters and dance halls put throngs of men, women, adolescents, and children into the street at night, where they encountered the same moral diversity that had troubled earlier commentators on urban danger. The late night street was still unquestionably a male-dominated space, occupied by loiterers and debauchees and by the disproportionately male population traveling home from night jobs and night entertainment. Men outnumbered women in the evenings even in the dance halls and movie theaters (though women were the large majority at matinees). Decent women who attended cheap amusements at night took moral risks comparable to those of decent men who had attended concert saloons fifty years before. Meanwhile, older male-oriented leisure activities persisted—even thrived—amid the boom in commercial entertainment. Thousands of saloons, clubhouses, and Masonic meetings kept men out late while their wives, daughters, and sisters stayed home—a situation that Brooklyn's Reverend T. DeWitt Talmage called "an assault upon domesticity."[38]

Millions of men in turn-of-the-century America attended meetings of fraternal organizations, secret societies, and other clubs. Some of these met twice or even three times a week. Secret societies such as the Free and Accepted Order of Masons and the International Order of Odd Fellows expanded quickly in the decades after the Civil War, a period described by a contemporary as the "Golden Age of Fraternity." They were joined by newer orders such as the Knights of Pythias and the Improved Order of Red Men. By the mid-nineteenth century secret societies had stifled their traditions of drunken revelry, filling their evenings instead with elaborate rituals. Middle-class and many working-class men responded enthusiastically. Fraternal lodges boasted five and a half million members by the 1890s. In Philadelphia alone, listings for the hundreds of lodges filled page after page in the 1890 city directory. Millions of other American men throughout the late nineteenth century joined semi-social organizations with strong elements of ritual, such as the Grand Army of the Republic and the Knights of Labor. These organizations made a public point of building Christian character, though critics pointed out that lodge meetings interfered with attendance at church events. The hours varied enormously. George Watson Cole's Masonic meetings in 1875 Bridgeport ended as early as 8:00 p.m., but rituals in other lodges extended meetings long after midnight.[39]

No matter how late a lodge meeting ended, there was sure to be a saloon still open to receive anyone not yet ready for bed. There were over 200,000 licensed liquor dealers in America at the end of the nineteenth century, plus an uncounted legion of unlicensed "blind pigs" and kitchen barrooms—a drinking spot for every taste, budget, and schedule. An 1895 study found that there was a licensed saloon for every 317 people throughout the ninety-five cities surveyed. A 1911 study put the total number of saloons at over 10,000 in New York and over 7,000 in Chicago. Milwaukee had 2,265, or one for every 166 people. The liveliness of saloons varied hour by hour, with different schedules prevailing in different parts of the city. Saloons were heavily concentrated in and around the central business districts, where one could be found on nearly every street corner, but these typically drew most of their trade during and immediately after working hours. In the early evenings, saloons were at their liveliest along major thoroughfares and near streetcar junctions, where men stopped in for a drink or two on their way home from work. Some downtown saloons closed as early as 8:00 p.m., after the last commuters left. Activity focused then on the taverns in the "bright lights districts" that competed with theaters for customers and drew additional business from audiences before and after the shows. Late night drinking spots could be found in the bright lights districts, as well as in downtown hotels. The door of one popular tavern in Chicago, described in an 1892 guidebook, was "tightly closed at midnight, but the initiated may gain ready admittance by learning the pass-word of the night and roaring it . . . through the key-hole." The pretense of secrecy was so flimsy that hacks lined the curb outside the tavern to wait for the after-hours customers.[40]

All-night saloons operated openly in some cities. In Philadelphia, reported the *Public Ledger* in 1890, "They are not numerous, and are usually situated near railroad depots or at junctions of night lines of cars. There are some thousands of men who are habitually out after midnight on legitimate business." Saloons and restaurants selling alcohol could be found near train stations, at the junctions of owl car lines, and in areas with many night workers, such as near newspaper offices. The existence of these places helps explain the prevalence of homeward-bound drunks on the owl cars. A more specialized class of all-night dives allowed impoverished drunks to sleep on the floor. Some saloons never closed, or (as in Pittsburgh) they attempted to evade local laws by "closing" at midnight and reopening ten minutes later. Others opened at 5:00 or 6:00 a.m. to fortify workingmen before their morning labors.[41]

It is next to impossible to determine how many men spent their evenings in saloons. Boston's Committee of Fifty for the Investigation of the Liquor Problem tried to estimate the patronage of the city's 606 licensed barrooms in the summer of 1895. Relying on policemen's guesses about the saloons on their beats, the committee calculated that a daily average of 226,752 people entered the bars—a figure that exceeded the adult male population of the city. Obviously this is not a credible estimate, even accepting that it includes repeat customers and residents of other towns. At the other extreme, a 1913 study of New York workingmen's leisure activities found that less than a third of the men surveyed would admit to visiting a saloon even once a week. It is easier to generalize about the number of women spending their evenings in saloons: very, very few. Saloons served an almost exclusively male clientele before Prohibition. The female presence was felt only in the lyrics sung by sentimental barflies, in pictures of nudes over the bar, and in some places by the women of dubious virtue who drank in curtained "wine rooms" at the back. An occasional housewife might slip into the saloon through a side door in the late afternoon, but only long enough to buy a growler of beer to take home. At night, any woman in a saloon was assumed to be a prostitute.[42]

Undisguised prostitution flourished in the decades around the turn of the century in "red-light districts" where police tolerated it in exchange for bribes. Nearly every major city had at least one of these areas, typically adjacent to the central business district, within easy walking distance of streetcar lines and train stations. In theory, segregating prostitution in carefully bounded districts was supposed to limit its immoral influence on the rest of the city, but many working-class families lived inside the districts, and prostitution always sprawled outside the boundaries as well. Nor did the practice succeed in making commercial sex less visible. In the Levee district south of Chicago's Loop, for instance, "The lighted street, the sound of music, the shrill cries and suggestive songs of the inmates and entertainers, all of those features tend to bring the business to the attention of the public," the Chicago Vice Commission reported.[43]

SEX AND DANGER IN THE MAN-FILLED CITY

Though men continued to dominate the nighttime street, women and children were there in growing numbers; their presence was visible enough to suggest a serious social problem but not large enough dispel it. "One of the most disturbing phases of the present situation in Minneapolis, and an alarming social symptom, is the large number of young girls in the streets at night in the downtown sections, and in the business districts of the outlying sections," declared the Vice Commission of Minneapolis. The Minneapolis police had actually attempted a census of minor-aged girls on the streets after 10:00 p.m. one night in June 1911. They counted 1,646 young girls unaccompanied by adults.

> They may be found in numbers loitering about the fruit stores, drug stores and other popular locations, haunting hotel lobbies, crowding into the dance halls, the theaters and other amusement resorts; also in the saloon restaurants and the chop suey places and parading the streets and touring about in automobiles with men. It would not be fair to charge that all or a large proportion of these girls are prostitutes. It is perfectly plain, however, that many of those who are not, are on the direct road.[44]

Obviously, social investigators at this time tended to exaggerate the sexual dynamic at play in night life, and to conflate sexual expression with deviance, danger, and crime. But we do not have to share their preoccupation with female virginity to see that the late-night city was suffused with sexuality.

FIGURE 15. Privacy in public. The public spaces of the nocturnal city helped individuals escape the supervision of family and neighbors. Here a man and a woman take the opportunity for a private conversation; other couples enjoyed closer physical intimacies. Edward Hopper, *Night on the El Train*, 1918. Wadsworth Atheneum Museum of Art/Art Resource, NY.

Most of this public eroticism was playful and consensual. It took such mild forms as the flirtation outside a dance hall, or women's ambiguous acceptance of men's offers to treat them to ice cream. In these cases, the power relationship between the sexes was relatively equal and the sexuality was mostly just a titillating possibility that added spice to the interactions.[45]

Overt public sexuality in the late nineteenth century most commonly took place off the streets. People continued to use city parks for flirtation and pickups, just as their parents and grandparents had done. Though the possibility of seclusion allowed consenting couples to carry the interaction as far as they liked, it also led to situations where men forced their attentions on women. Brooklyn's Prospect Park superintendent John Culyer declared in 1884 that "it has always been my desire that young ladies and children should visit the Park unattended without the slightest danger," but he acknowledged that mashers boldly accosted ladies during the day and that "rowdies and dangerous characters" prowled about at night.[46] Philadelphia's Fairmount Park police struggled to suppress sexual activity by arresting dozens of park visitors every year for such offenses as fornication, "indecent conduct," and "open lewdness" as well as for rape, child molesting, indecent exposure, and homosexual liaisons in the bushes and bathrooms.[47] The captain of the park police, Louis Chasteau, complained in 1882 that although the park covered thousands of acres and drew over five million visitors a year, he had only seventy-nine policemen to "exercise surveillance by night and by day." Among the more troublesome spots in the 1880s were the restaurants licensed to operate within park limits. One of these, at Strawberry Mansion, was little more than "a badly kept Beer Garden with the addition of a degrading dance house." Park guards in that area were kept busy "<u>very</u> properly driving couples from isolated and dark places." Drunkenness and immorality spread into the park after midnight from roadhouses nearby, Chasteau reported in the 1890s, making it necessary to assign more officers to all-night service. The nighttime park remained a haven for illicit sexuality into the 1910s.[48]

In the early twentieth century, Omaha's teenagers gathered at Hyland Park on pleasant nights to drink beer and have sex. New York's servant girls and factory workers found a precious semblance of privacy in Central Park at night, even when they were surrounded by other amorous couples. Young lovers could be seen lying together on the Central Park lawns, kissing on the park benches and so forth. "Sing a song of lovers, / Lovers o' Central Park– / Happy lads and lasses / Hidden in the dark," wrote one observer in 1923. By 9:00 p.m. on a pleasant evening, he noted, the secluded spots in the park had long since been taken, and even the most conspicuous benches were occupied. "It is downright shocking, for by now the ardent young men and women have cast aside all pretense at modesty. Now they pay you no attention as you stroll along the darkened paths, but continue their love-making flagrantly and unabashed." In evenings in the 1910s and 1920s, gay men filled the benches in the southern corners of the park and enjoyed a lively cruising scene.[49]

Women who did not want to see or experience sexuality could certainly choose to stay out of the parks at night, but the streets were harder to avoid. Late at night, sexual expression on the streets took on an aggressive edge. Prostitutes and clients were out seeking each other, looking closely at prospective partners as they walked along the sidewalks. Unescorted women on the streets could expect to be accosted by strangers. In center-city Philadelphia in

1900, according to the *Public Ledger*, women walking along Chestnut Street in the late evening were "ogled and insulted by crowds of men."[50] The 1922 log of Pittsburgh's Hazelwood police precinct is filled with complaints about young men loitering and drinking on street corners until after midnight. Men and boys shot craps in alleyways and lit fires in back lots. Foul-mouthed teenagers clustered on the front steps of apartment buildings, harassing female passersby. A homeowner on a neighborhood commercial street complained that frisky couples were using his porch swing while he slept. "Dos not want his poarch used as a place for men and women to do thair Didling," the desk sergeant recorded.[51]

This was the vulgar and often frightening world that women and children entered when they ventured into the public spaces of the city at night. Despite the lighting that made it possible to identify pedestrians on the street, and despite the increased activity produced by the expansion of night work and night entertainment, nocturnal street culture still stood in opposition to the values of respectability and domesticity. It was possible for women and children to traverse public spaces with reasonable physical safety in the early watches of the night if they avoided certain locations. But merely avoiding rape is not the same as experiencing the city on equal terms with men. Whenever they walked outside the mixed-gender theater or dance hall at night, women and children were stepping onto adult male turf that could be as crude, drunken, and violent as the roughest saloon.

"If children are to be in the streets, the streets must be made a safe place for children," declared the Philadelphia Vice Commission in 1913.[52] It was a fine sentiment that few people took seriously, particularly as it applied to the nighttime. Even as electricity swept away the shadows of the gaslight era, social reformers and public officials took steps to remove children and women from the public spaces of the city.

CHAPTER 10

Regulated Night

In his 1915 book *Street-Land*, Philip Davis wrote of Tommie, the well-brought-up young son of a Boston schoolmaster:

> Like his parents, Tommie was taught from infancy to retire soon after sundown. One evening, while in his "nightie" and in the midst of his prayers, he heard the fire alarm. He ran excitedly into the front-room, flung the window open—just in time to catch sight of the fleeting shadows of little children running madly behind a clanging, hissing fire-engine. "Mama," he asked in great surprise, "are these the night children?"

Contemplating the phrase "night-children," Davis used a revealing analogy:

> Did you ever hear of night birds? There are owls, to be sure. Those on the Boston Common, known to sight-seers the country over, are as cosmopolitan by nature as our city children. But their reputation is rather low in birdland. Their reputed wisdom is of a doubtful sort, mostly derived from a knowledge of things which well-bred babes of birdland close their eyes to before nightfall. Nevertheless, the owl, in its nocturnal habits and dark wisdom, strikingly resembles the "wise-guy" of Street-Land. The alarming thing about city children is that they are becoming more and more owlish.[1]

Davis's vignette of Tommie and the night children reflected a widespread concern among the American middle class at this time. From about 1880 to 1930, children's access to the nighttime city became a subject for public hand-wringing and for intervention by middle-class social reformers and government officials. Middle-class Americans feared that children's development would be undermined. The orderly process of learning and acculturation that took place in schools and homes would be tainted by the "dark wisdom" to be gained in urban public spaces—knowledge of the world that was unsuitable for young people.

This concern revealed cultural conflicts between middle-class and working-class Americans over the meanings of modern night, and over the raising of children. Middle-class city dwellers, like Tommie's parents, already ensured that their own children stayed indoors at night and went to bed early. Working-class parents did not. Confident of their own cultural superiority, middle-class Americans searched for ways to remove all children from the streets after dark. Their desire to shield even poor children from premature exposure to adult knowledge lay behind efforts to create supervised recreation centers such as boys' clubs, to regulate child labor in the streets, and to impose juvenile curfews. These three reform campaigns won major victories in American cities in the decades of the 1880s through the 1920s. By 1930, a new institutional framework was in place to restrict children's access to urban night.

Curfews and child labor laws restricted girls far more than boys. The widespread belief that females needed protection contributed also to an effort to restrict adult women's presence in the nighttime city; women's rights activists and progressive reformers in the early twentieth century sought state legislation forbidding night work by women. This campaign did not achieve the same success as the effort to get children off the streets at night. Nonetheless, it did force women out of some night jobs, and it ensured that men would get the lion's share of night work in the decades ahead. The campaign further reinforced the idea that women were not full citizens of the nocturnal city.

THE NIGHT CHILDREN

Working-class children participated in both nightlife and night work. Children and teenagers loitered in the shopping and entertainment districts, listening to the music that spilled out of concert saloons and waiting for excitement. Some would take in a vaudeville show or a movie. Newsboys and child peddlers worked the nighttime crowds. In working-class neighborhoods, children and teenagers flocked to cheap storefront movie theaters. Others could be seen on the streets returning from work in the factories and sweatshops or running errands.[2]

Most working-class children, though, spent their evenings playing on the streets of their neighborhoods. As population densities soared in central cities and open space dwindled, those who lived in cramped apartments had nowhere else to play; their parents often wanted them outside so they would not interfere with the housework. The sidewalks and streets of immigrant districts such as New York's Lower East Side or Chicago's Near West Side were filled with children from the moment school let out until late in the evening. According to an 1891 article in *Scribner's Magazine*, "Every doorway pours forth its little quota, and it is sometimes with difficulty that one can thread one's way through the crowds that literally swarm about the sidewalks. Some are playing quietly; some are fighting; some are 'passing' ball when the policeman on the beat is not in sight." Girls were greatly outnumbered by boys but could still be seen skipping rope under the streetlamps.[3]

Even as electric lighting spread through the streets and homes of the city, night had nowhere been turned into day, either in the literal sense of illuminating the street as brightly as sunlight or in the figurative sense of creating an island of virtue in a sea of moral darkness. The increasing use of urban public spaces ensured, rather, that far more people—particularly working-class people—came in contact with the dubious elements that had long characterized the night. Children on even the most respectable streets after dark encountered all sorts of people, good and bad, rich and poor. On the well-lit sidewalks, one could see ladies and gentlemen in evening dress emerging from restaurants and theaters; middle-class and working-class couples out for a stroll; factory hands returning home; shopgirls rushing to an evening's entertainment; German families headed for respectable beer halls; young "bloods" starting a debauch; peddlers, newsboys, matchgirls, beggars, drunks, pickpockets, and—conspicuously—streetwalkers. Rough young men loitered on the street corners, while the darker side streets might well harbor muggers and rapists.[4]

Modern urban night was not an extension of day; it was a liminal new world in which conflicting moral values mingled uneasily. This spectacle of moral diversity, even chaos, unmistakably heightened the excitement of the nighttime street. The "white ways" struck many observers as the essence of the urban experience, buzzing with energy and barely suppressed eroticism. The effect on adolescents could be devastating, wrote Jane Addams in 1909. The teenage boy, at an impressionable age when he was first confronting powerful urges, walked unsupervised through the streets and into places of commercial entertainment. "It is nothing short of cruelty to over-stimulate his senses as does the modern city. This period is difficult everywhere but it seems at times as if the great city almost deliberately increased its perils. The newly awakened senses are appealed to by all that is gaudy and sensual."[5]

Addams was drawing on new ideas of child development that had emerged in the late nineteenth century and had been given clearest expression in G. Stanley Hall's famous 1904 book *Adolescence*. Hall, the foremost child psychologist in the United States, argued that the child recapitulated the stages of evolution of the human race, from presavagery to civilization. To become a happy adult, the child must successfully pass through each of these stages. Adolescence, between thirteen and eighteen years of age, was particularly crucial. "The dawn of puberty," Hall wrote, "is soon followed by a stormy period when there is a peculiar proneness to be either very good or very bad."[6]

Child experts and reformers viewed the city as an unnatural environment that threatened to upset the schedule of development. They urged parents and teachers to shelter children from influences inappropriate for their age. The proper childhood was thought to take place in a single-family home, with a private bedroom for the parents and separate bedrooms for boys and girls, to shield the child from premature exposure to sexuality and sexual difference. Children were supposed to spend their days in a schoolroom with their peers and in specially designated play spaces such as private backyards and playrooms. The sheltered child should also be kept to a regular schedule for meals and sleep, with an early bedtime that differed from that of adults. Dr. F. S. Churchill, a Chicago pediatrician, wrote in 1912: "We should apply not only the principle of regularity but we should stick to natural hours in our regularity. For countless ages the young of all animals have naturally slept and rested at night. They have not been careering around cities."[7]

As Davis suggested in his story of Tommie and the night children, a sheltered childhood was still far more common among the middle class than the working class. The seclusion of children, like the domesticity of women, was a luxury that many working-class Americans neither desired nor could afford. Working-class parents admired the industrious child who helped provide for the family and the spontaneous, self-reliant child who could hold his own in the streets. Less concerned about strict scheduling either of daily activities or of child development, they granted their children freedom. As a result, Davis warned in his 1915 book, the street remained at least as powerful an influence on the working-class child as the home and the school. Most disturbingly, the street was where children learned about sexuality, the most carefully guarded part of adult knowledge. "For a decade, this country has hotly debated the where and when and how of teaching sex hygiene. During this same period, the street has been teaching it at all hours, under all sorts of conditions, to thousands of children regardless of age or sex." Contaminated by illicit knowledge, working-class children had crossed the line into a dangerously precocious adulthood.[8]

These beliefs about child development, coupled with the increased use and moral uncertainty of the nighttime street, were what motivated the middle-class concern about "night children." In the late nineteenth and early twentieth centuries, social reformers and public officials strove to bring the blessings of a proper childhood to the urban poor and working classes.[9] The boys' club, child labor, and curfew movements were part of this larger project of cultural evangelism.

BOYS' CLUBS

The boys' club movement originated in New England from a Protestant sense of Christian stewardship. The earliest charitable club for boys was the Dashaway Club of Hartford, Connecticut, created in 1860 by women volunteers at a Congregational church mission in the slums. It fell apart during the Civil War but was followed by a succession of three other boys' clubs in Hartford between 1867 and 1876, all of them affiliated with Congregational churches. These early clubs attempted to mimic the atmosphere of a respectable home, to compensate for a lack of domesticity in the lives of poor children. Typically, a club was furnished with a piano, books, and pictures to make it look like a middle-class parlor. As in a parlor, the evening's activities included playing games and singing songs. Boys were strongly encouraged to wash their faces or to bathe if bathtubs were available, in keeping with the middle-class fetish of personal cleanliness. The supervisors, many of them women, attempted not only to remove the children from the influences of the streets but also to provide moral training. Virginia Smith, organizer of the Hartford Boys' Club, explained: "We hoped to make the place one of entertainment, instruction, comfort, and good, homelike Christian influences for those boys who for any reason needed such a place of resort, or had fallen or were in danger of falling into evil ways in unemployed hours." By the mid-1870s, the Hartford clubs had been joined by similar organizations in Providence, New Haven, and Salem.[10]

The early Hartford clubs were small, drawing nightly attendance of no more than a few dozen children, many of them newsboys. As the boys' club movement expanded throughout the Northeast in the late 1870s and 1880s, local philanthropists formed much larger clubs, starting with the Boys' Club of

New York in 1876. The mass boys' clubs differed from the small clubs by offering a wider range of activities. In addition to providing books, board games, and magic lantern shows, the Boys' Club of New York opened a gymnasium with trapezes, horizontal bars, and boxing equipment. It offered classes in singing and art, and instruction in carpentry, bookkeeping, and printing. The Boys' Club of Fall River, Massachusetts, had its own bowling alley, indoor running track, and swimming pool. By 1923, the Boys' Club Federation counted 180 boys' clubs in the United States, mostly in the cities of the Northeast and Midwest. Dealing with hundreds of unruly boys, the organizers of mass boys' clubs had to use methods different from those of smaller clubs. The leader of a mass boys' club could not know all the boys individually or serve as a surrogate parent. Rather, he was what one advocate called "a moral policeman," who would patrol the rooms breaking up fights and preventing thefts of the equipment. The Fall River superintendent was so insistent on order that he trained boys to fall silent at the blast of a whistle.[11]

Club officials regretted that the mass boys' clubs could not provide the personal attention found in smaller boys' groups, such as those that opened at settlement houses in the 1890s. But they pointed out that their organizations reached more children every night, often tougher children who would otherwise be getting into trouble on the streets. Like the early Hartford clubs, the mass boys' clubs targeted "street Arabs"—wandering ragamuffins, presumed to be homeless. John F. Atkinson founded the Chicago Boys' Club in 1901 specifically to serve homeless newsboys and bootblacks who slept in the alleys of Chicago's Loop. The Chicago Boys' Club initially combined the functions of a social club and a homeless shelter for children, providing free lodging, meals, baths, and clothing. As late as 1911, the club's publication was titled *Darkest Chicago and Her Waifs*, though the typical club member was acknowledged to live with his family in an immigrant neighborhood.[12]

Even if the boys were not literally homeless, boys' club advocates argued that they might as well be. According to a club leader in Boston, "It is a true and trite saying that a good home is a better place for a boy at night than a boys' club. If all homes were perfect homes, then would the boys' club be useless." The tenements of the poor were far from perfect, club leaders contended. In response to complaints that the Chicago Boys' Club would draw boys away from their homes and into the "bright light section," John Atkinson took his critics on a tour of the hovels that he feared were crushing the spirits and character of Chicago's youth. His point was obvious: the boys would be better off in a supervised clubhouse downtown, where they might even pick up civilized habits that could be spread to their parents. In any case, the tenements

FIGURE 16. Childhood and night. Boys' clubs made it their mission to get city boys off the streets at night, where they might develop bad habits or criminal tendencies. Here the two meanings of the "Danger Hour" are vividly displayed, in the juxtaposition of the vulnerable boy and the menacing nocturnal street. *The Danger Hour* (Chicago: Chicago Boys' Club, 1918), Chicago Historical Society.

failed to hold the boys inside. Turned loose, boys would amuse themselves shooting craps, watching prostitutes, and learning the ways of the street. Children of low morals taught the others to drink, smoke, swear, and steal.[13]

The clubs' most important benefit was thought to be their power to save boys from a life of crime. In Binghamton, New York, the superintendent of the boys' club asserted that juvenile crime dropped by 60 percent in the year the club opened; the activities at the club so engaged boys' interest that they forgot all about delinquency. The Chicago Boys' Club tried to garner support by hinting that its clients were potential muggers and gangsters. The club's 1911 annual report warned: "The slums of Chicago are places where crime breeds. They fester with filth, squalor, sin, vice and corruption of all kinds. Amid such conditions as this thousands of helpless children are born each year." Such children, the club pointed out repeatedly, could be more cheaply entertained by the Boys' Club than incarcerated after the dark street had worked its influence. The Chicago Boys' Club had initially focused on children between eight and thirteen years old. Because of its increasing concern with juvenile crime, it expanded its work in the 1920s to include more boys in their later teenage years, when the turmoil of adolescence and the temptations of the gangs were strongest. Adolescence and night were both troubled times, the club suggested in a cleverly designed fund-raising pamphlet, showing a boy on a menacing nocturnal street under the words, "The Danger Hour."[14]

CHILD LABOR LAWS

As the boys' club organizers had hoped, thousands of children from the toughest neighborhoods were persuaded to spend their evenings indoors playing checkers or basketball. But since attendance was voluntary, children could escape supervision whenever they wanted. Even though hundreds of supervised playgrounds, boys' clubs, outdoor gymnasiums, and vacation schools had opened throughout the urban United States in the early twentieth century, the street was the favorite play space. One playground advocate estimated in 1913 that 90 percent of play in Boston took place on streets. The freedom from supervision was both the greatest attraction of the street and, from the perspective of child care experts and social reformers, its greatest danger. Children could not be kept inside at night without some kind of coercion, reformers realized. Two such efforts at coercion came from the child labor reform and juvenile curfew movements.[15]

Child labor reformers in the early twentieth century asserted that childhood should be a period of natural development and education, free from economic interference. Reformers criticized factory and mine labor for enslaving small children and denying them the stimulation of uninhibited play in the open air. They attacked the street trades for nearly opposite reasons: children who worked on the streets as newsies, bootblacks, peddlers, or messengers were given too much freedom and stimulation. Reformers paid surprisingly little attention to the dangers children faced in traffic. Physical danger was secondary in their minds to the danger of disrupted development and moral corruption.[16]

Of all the street trades, selling the news drew the most children. In the late nineteenth and early twentieth centuries, every major American city had hundreds or even thousands of newsboys. Most were under sixteen years old, and many were under ten. There were a few newsgirls among them, but only a few. In Chicago in 1905, girls were estimated to total perhaps 5 percent of all newsies.[17] The busiest time of work was from 3:00 to 8:00, when newsies peddled evening editions to people leaving work. Many continued late into the night. In metropolitan centers like Chicago and New York in the early twentieth century, the publication of multiple editions ensured that fresh news could be sold around the clock every day of the week. Saturday night was especially busy; newsboys in the entertainment districts sold theatrical papers and early Sunday editions long after midnight.[18]

"The main characteristic of street work is its unwholesome irregularity," wrote settlement house worker Ernest Poole in 1903. In contrast to the disciplined pace of the factory, the street trades retained a premodern mixture of work and leisure that—in Poole's opinion—made the boys unsuited for later jobs. The hours were defined only by the availability of newspapers and the presence of crowds. Between editions, newsboys would loiter in the streets or in the "news alleys" outside the pressrooms. They passed the time smoking cigarettes and shooting craps. Whereas proper children dined regularly, the newsies gobbled snacks of doughnuts or hot dogs washed down with large bowls of coffee.[19]

Selling newspapers allowed boys to spend their evenings free of adult control. Since their earnings were unpredictable, newsboys found it easy to keep money for their own use instead of turning it over to their parents. Their profits and their freedom from supervision allowed them to enjoy movies, cheap plays, shooting galleries, and peep shows. Boys who lost their money gambling or attending shows would often sleep out to avoid punishment at home; they would huddle together for warmth in stairwells and over steam grates. Eventually they might grow so independent that they would join the homeless "street Arabs," supporting themselves with newspaper sales and petty theft. Even boys who did not go to this extreme found that the late night excitement of the street made it hard to pay attention to their boring schoolwork the next day.[20]

"Street Arabs" and juvenile delinquents mingled with boys from decent homes, child welfare advocates warned, teaching them bad habits and worse morals. According to Chicago activist Florence Kelley, "There is a tradition among the boys themselves that in order to be a 'wise guy' he must know the greatest possible amount of evil, and what he does not know he must invent, and he must tell the last newcomer everything he knows or can invent." The nocturnal street was the source of this illicit knowledge. While selling papers at night on the streets and in the saloons, boys learned about prostitution, gambling, and drinking by watching the adults engaged in them. Further, young boys were known to be sexually assaulted at night or prostituted by the newspaper distribution workers and strangers who loitered in the news alleys. Some developed venereal disease as a result.[21]

These concerns were expressed even more strongly in descriptions of the night messenger boys, who were the targets of a reform campaign between 1909 and 1915 led by the New York–based National Child Labor Committee (NCLC). At this time, in every significant city, telegraph companies such as Western Union and American District Telegraph employed dozens of boys as young as twelve to carry telegrams and notes, and sometimes to run errands for customers. After about 9:00 or 10:00 at night calls to legitimate businesses dropped off, wrote Owen Lovejoy, an NCLC leader. "The service at night plunges us at once into another world. As day is given up to industry in most cities, so the forces of pleasure and recreation hold chief sway at night." Night messengers were called on to deliver some messages to respectable hotels, theaters, and ballrooms, but more often they glimpsed the darker side of urban recreation. According to another reform article, night messenger work left the boy "exposed to temptation at the time when he is least able to resist—just as he is passing through those bodily changes which accomplish adolescence, and when his mind is swayed by the natural curiosity of youth toward the other sex."[22]

NCLC investigators reported in horrified detail the moral atrocities boys witnessed late at night. "It is almost sickening to think of this boy still on duty at this hour [2:00 a.m.], and on duty that takes him into the worst and most degrading environment that the hours of night in an evil city could afford," wrote one investigator, describing a sixteen-year-old Louisville messenger boy. At the brothels they were called to, messengers would see women in various states of undress or even engaged in sexual activity. Boys eventually gained a thorough knowledge both of sexuality and of the prostitution business, including its kinkier specialties. Some boys were drawn into active participation in the vice trade. They were sent to buy condoms, liquor, cigarettes, opium, and cocaine. They received commissions for sending customers to certain prostitutes and tips from men who wanted to be shown the red-light districts. According to the NCLC, boys as young as fourteen or fifteen became customers in the brothels, sometimes being initiated by women who thought they could get rid of venereal disease by giving it to a virgin. Others attended the "circuses" in which prostitutes performed sexual stunts.[23]

Charity workers and reformers were particularly disturbed by the presence of girls on the streets. About sixty school-age girls, mostly the daughters of immigrants, sold newspapers in Chicago about 1910. They did a brisk business in the saloons, where their incongruous presence appealed to kindly drunks. Others sold gum, shoestrings, and other small items until late at night, even sometimes in the red-light vice districts. Newsgirls could also be found in Detroit, Hartford, New Haven, Manchester, New Hampshire, and Wilmington, Delaware, but they were rare in many other cities. In Cincinnati, reported a child labor reformer in 1908, police vigorously discouraged any girl who attempted to sell papers. Nevertheless, a few advocates of women's equality insisted that girls had just as much right as boys to be on the street, at least in the daytime and evening. Instead of forcing girls off the street, suggested Mary Hall of the Hartford Equal Rights Club, "My remedy would be [to] make the streets clean morally, so clean that a girl who wishes to sell newspapers can do it in safety."[24]

Thanks to efforts by the NCLC and local child welfare advocates, most of the urbanized states of the Northeast and Midwest began regulating the street trades in the early twentieth century. The laws took so many forms as to defy easy generalizations, but they typically established permissible hours and permissible ages for boys and girls in the street trades. Ohio, for instance, prohibited children under eighteen from serving as messengers between 9:00 p.m. and 6:00 a.m. It prohibited newspaper sales on the streets of Cincinnati at any hour by boys under ten and girls under sixteen. Girls in Chicago could not legally sell papers at any hour until they were eighteen; boys under fourteen could sell papers between 5:00 a.m. and 8:00 p.m., and older boys could sell at any time. Such laws were essentially attempts to keep the most innocent and vulnerable children out of the street trades and to impose curfews on the older boys. By 1912, child labor reformers had persuaded state legislatures to adopt laws keeping boys under twenty-one out of the night messenger service in the cities of Massachusetts, New Jersey, New York, Utah, and Wisconsin. Five other states allowed no one under eighteen to do this work, and eleven more allowed no one under sixteen. By 1915, thirty states restricted newspaper sales and other street trading by boys and girls in the major cities; most of these states either set higher minimum ages for girls or prohibited girls altogether. Other regulations were imposed by municipal ordinances. The enforcement of these laws and ordinances left much to be desired. Police rarely enforced the Chicago municipal ordinance, although they were occasionally persuaded to crack down on flagrant violations, including sales by newsgirls.[25]

JUVENILE CURFEWS

The establishment of permissible ages and times for street work reflected the era's obsession with scheduling human activities. The age consciousness that shaped school systems and ideas about child development around the turn of the century was paralleled by a stronger consciousness of precise times within the day. As described in previous chapters, American work had become increasingly defined by time rather than by the task, and by clock time rather than by the passage of the sun. As Americans grew more accustomed to planning their daily activities by the clock, they also grew to accept a stricter scheduling of evening entertainments such as stage shows and movies. Time became not only scheduled but standardized. Railroads divided the continent into time zones in 1883, effacing local variations in order to simplify train schedules. Most city dwellers soon scheduled their lives by "railroad time." Everywhere, man-made schedules were replacing natural rhythms and variations in the temporal order of modern life. The restrictions on children's street labor were one way Americans consciously tried to shape this new temporal order in the city. Juvenile curfews were another.[26]

In contrast to the boys' club and child labor movements, which were strongest in the nation's urban industrial core, the most direct attack on children's access to night started at the periphery. The juvenile curfew movement appears to have originated as an import from small towns in Canada. The town council of Waterloo, Ontario, adopted a resolution in the late 1880s that the town bell should be rung at 9:00 every evening, and that children under fifteen could be arrested after that hour if they persisted in loitering on the streets without parental supervision. Encouraged by enabling legislation in Ontario's 1893 "Children's Law," many other towns and small cities adopted curfews by 1896.[27] Curfews, which dated back as far as the Middle Ages, had not previously been directed so narrowly against children. They had often been used to control potentially unruly subordinate groups, such as Saxons under William the Conqueror. In the United States, general curfews were used as temporary emergency measures in the aftermath of disasters such as the Great Chicago Fire of 1871. Some eastern towns such as Portsmouth, New Hampshire, also maintained a long-standing tradition of a general curfew, but this custom was regarded as a sign of backwardness inappropriate for any ambitious city.[28]

After originating in Ontario, the juvenile curfew movement leaped the border in the mid-1890s to emerge in the towns of the midwestern United States. There it quickly spread under the leadership of an eccentric named Alexander Hogeland, who claimed for himself the titles of "father of the curfew law" and "the newsboys' friend." Hogeland had grown interested in the plight of newsboys in the early 1870s, apparently while employed as a superintendent of newscarriers for the *Louisville Commercial*. "The boys of many families roam the streets early and late, away from home restraint. They fall victim to vicious habits and practices, and thus go rapidly to swell the army of criminals in our prisons," he wrote in his 1883 book *Ten Years among the Newsboys*. In this sentimental and self-congratulatory tract, Hogeland presented himself as delivering street-corner sermons to crowds of guttersnipes. After moving the boys to tears, he would try to bring them to Christ, or at least to an evangelical night school where they could wash their faces. He would use his connections with local businessmen to find jobs for promising young lads.[29] Hogeland expanded his activities throughout Kentucky and the Midwest. In 1886 he organized the National Youths' Home and Employment Association to help street boys find homes and jobs. From his new home base in Lincoln, Nebraska, Hogeland traveled throughout the Midwest under the auspices of this shadowy organization, supporting himself from charitable donations and by selling publications boasting of his work.[30]

Hogeland appears to have launched his curfew campaign in the mid-1890s, when he persuaded city officials in Lincoln to impose a curfew on children under fifteen. Under Lincoln's widely copied 1896 ordinance, children could not be on the streets after 9:00 p.m. in the spring and summer, or after 8:00 in the fall and winter, unless they were accompanied by a parent, were running an errand, or held a job that required them to be on the streets. This last exemption was evidently intended for Hogeland's friends the newsboys. In his speech supporting the ordinance, Hogeland promised that it would help parents regain control of incorrigible children, would improve the morals of Lincoln's youth, and would diminish crime and vagrancy. He also claimed that the measure had been successfully tried in many villages and towns, particularly in Nebraska. He said it had done wonders for the towns of North Platte and Grand Island.[31]

Hogeland's allusion to these precedents revealed one of the major liabilities of the curfew movement in its early years: curfews were associated with small towns, not thriving cities. The Nebraska towns at the forefront of the curfew movement in the 1890s were far from thriving. Nebraska was suffering from both a severe drought and the nationwide depression. Towns that had boomed in the 1880s saw their urban pretensions shattered—their banks ruined, their businesses closed, and their people fleeing to more promising places. North Platte's electric light company collapsed in 1895, forcing townsfolk to resort to kerosene lanterns again. Grand Island's streetcar system was failing and would soon end service. In Lincoln in 1895, two hundred houses and numerous downtown storefronts were vacant. The curfew would initially be viewed as an attack on modern urban life, but urban life had limited appeal in such towns; citizens feared crime but could not enjoy the bustling activity that prevailed elsewhere.[32]

The news of Lincoln's curfew inspired similar ordinances throughout the struggling towns of the Midwest, Pacific Northwest, and New England. Hogeland promoted the movement by writing to newspapers and public officials nationwide and by addressing municipal councils. Hundreds of villages and small towns adopted the curfew in the late 1890s, as did some of the major cities of the central states, notably St. Paul, Kansas City, Des Moines, Indianapolis, Fort Wayne, Denver, Dallas, and Omaha. Except for the two Indiana cities, all of these were places that had at least doubled in population in the 1880s but saw only slow growth or decline during the hard times of the 1890s. In Omaha in 1896, Hogeland successfully persuaded the city council to adopt the curfew despite the mayor's veto and the hostile editorials in a local newspaper. The *Omaha Evening Bee* argued that "a curfew ordinance may be all very good in little villages, but does Omaha want to put itself in the same class with every country cross-roads town?"[33]

Hogeland's idea was scorned or ignored in larger, more prosperous metropolitan centers in the 1890s. Chicago alderman and saloonkeeper "Bathhouse John" Coughlin introduced a parody curfew proposal that would have forbidden anyone of any age from being on the streets after 9:00 p.m.; he called for hiring a thousand bell ringers to enforce the measure. The leader of the curfew movement in New York City reported that her proposal encountered "derision, and unbounded incredulity." The *New York Times* mocked the curfew as a needless restriction worthy of Prussian autocrats, on the rare occasions when it deigned to notice the movement at all. "Much literature from the 'curfew' cranks has reached this office, and enough of it was read to justify its wholesale deposition in—or around—convenient waste baskets. The documents sent gave enough evidence that by shutting children up in the house at nightfall juvenile crimes and misdemeanors had been made less numerous in several towns, but what of that? Even more effect in this direction would be produced by cutting the throats of everybody under age."[34]

Far from cutting throats, police and courts seemed unwilling even to make curfew violators pay small fines. Lincoln Police Court records show that not a single child was arrested during the first two months the curfew was in effect. (Supporters of the ordinance, however, claimed that no arrests were needed because the children were off the streets.) In Indianapolis, an average of four children a year were arrested in the first six years of the curfew. City officials in Wilkes-Barre, Pennsylvania, made repeated threats to enforce the curfew, without result.[35]

The curfew slowly spread to more cities in the early twentieth century, including two of the largest. In 1915 the Philadelphia police department imposed a curfew order that allowed officers to apprehend any children out unattended after 11:00 p.m. and to put them in the House of Detention until their parents came to get them and explain their presence on the streets. After the start of Prohibition in 1920, even Chicago joined the movement. Concerns about the independence and immorality of young people coincided with a perceived breakdown of law and order at the time. Illegal liquor was flowing into the city, gangsters were killing each other, and police were rumored to be too corrupt to do anything. An attack on delinquency was not demanded by real public safety needs—juvenile crime was actually dropping—but it was included for symbolic reasons in a campaign to reform the police department and end the crime wave. A grand jury investigation in the fall of 1920 called attention to the problem of unsupervised youths, and the *Chicago Daily Journal* ran a sensational series on teenage promiscuity. As in the boys' club movement, teenagers were feared to be both vulnerable and menacing. Grand jury members urged imposing fines on the parents of delinquents and forcing all children under sixteen off the streets after 10:00 p.m. The *Chicago Tribune* used a familiar metaphor to support the proposal: "The street corners and vacant lots of the city are the kindergartens of a school of crime. The primary and intermediate classes meet in vicious poolrooms. Cabarets and tough saloons are offering advanced lessons, and post-graduate instruction is available in the jails and penitentiaries. Parents who provide their children with clean entertainment and interests in their own homes . . . will keep them out of the path to a criminal education." The Chicago city council passed the curfew unanimously in April 1921, but enforcement was never vigorous and soon lapsed entirely. Five years later, in an attempt to protect teenagers from sexual assault, the Women's Protective

League and other Chicago women's groups proposed a midnight curfew on all unaccompanied girls under eighteen. The police then began for a while to enforce the nearly forgotten curfew ordinance, particularly on girls.[36]

None of these three reform efforts succeeded in forcing children off the streets at night in the period before 1930. The boys' clubs never matched the popularity of street play, while the street trades laws and curfews were not well enforced. The larger significance of the new framework for controlling the night was educational and symbolic. It advertised middle-class belief in the value of scheduling children's lives and made it a matter of public policy to give each age group its appropriate activities, spaces, and daily routines. It emphasized also that girls did not have the same right to the nocturnal city as boys. By developing the boys' clubs, child labor laws, and curfews in the decades leading up to 1930, middle-class city dwellers tried to limit the social and moral problems raised by modern urban night without destroying its appeal for adults.

NIGHT WORK LAWS FOR WOMEN

Somewhat different motives lay behind the early twentieth-century wave of legislation concerning night work by women. Compared with the child labor reformers, advocates for women's night work laws did not devote nearly as much energy to documenting the evil effects of the nighttime city. Their expressions of concern about women's night work seemed at least in part to be calculated strategies in pursuit of a larger goal: reducing the length of the workday. Nonetheless, they routinely warned of dangers lurking on the nighttime streets, and they made it clear that women did not belong there.

Women entered the paid workforce in increased numbers after about 1880, as technological advances both eased the burden of housework and created new employment opportunities. Daily household chores were made somewhat easier by the wider use of coal-burning cookstoves and running water; women's efforts to feed and clothe their families were assisted by the mass production of inexpensive bread, soap, canned vegetables, and ready-made garments. All these things required money, so many married women with children, as well as unmarried teenage daughters, found they could best help their families if they worked for wages. As the tasks of housekeeping moved out of the home, women were hired to do similar work in canneries, garment shops, and commercial laundries. The mechanization of production, assisted in many cases by electricity, allowed employers to replace male employees with unskilled women who could be paid less and who lacked strong traditions of labor activism. The total number of female wage earners soared. The number of females in manufacturing rose from 631,000 in 1880 to 1.77 million in 1910. In 1880, 14.7 percent of all women and girls at least ten years old worked for wages, a total of 2.6 million in all lines of work. By 1910, 23.4 percent did so, and they numbered more than 8 million.[37]

Comparatively few women worked at night. They remained shut out of the incessantly running steel mills, glass houses, and railroads that operated large and growing night shifts, as discussed in chapter 7. Nonetheless, a few of the industries that did employ women asked them to work long hours that extended after midnight. American bakeries, already notorious for long hours of night work, employed 2,210 women in 1879 and 38,901 in 1919. An 1892 investigation found that women pie makers in the bakeries of Chicago usually started work at 3:00 in the afternoon and continued until 2:00 a.m., but on Saturday nights and nights before holidays they continued until 8:00 the next morning and then took the rest of the day off.[38] Bookbinding was another branch of manufacturing in which a great deal of night work took place. Women and teenage girls tended to hold the less skilled jobs, which grew in number as the industry mechanized in the late nineteenth century. By 1900, women outnumbered men among the thirty thousand bindery workers. Half of this workforce was located in the four major publishing centers: Boston, Philadelphia, Chicago, and especially New York. Manhattan alone had some 280 binderies in 1910, with about six thousand female employees binding books, magazines, and other printed materials. The center of the industry was the printing district of Lower Manhattan—a semicircle east of Broadway within a mile of City Hall. Unlike the steel mills of Pittsburgh or the cotton mills of the rural South, very few of New York's binderies ran separate night shifts. Instead, during busy times of the year they routinely prolonged the daily hours of their workers long into the night. During the pre-Christmas rush, women employees reported working "long days" once or twice a week that began as usual about 8:30 a.m. and sometimes lasted until midnight or even a full twenty-four hours until the next morning.[39]

The long hours and low wages of women industrial workers drew the attention of labor activists in the 1890s and early 1900s. Union men in American Federation of Labor had ulterior motives in calling for reforms; any new laws limiting women's hours and increasing their pay would also discourage employers from hiring women to replace men. More vigorous reform campaigns in the early twentieth century were launched by the American Association for Labor Legislation, led by economists at the University of Wisconsin, and the National Consumers' League, a predominantly female organization headed by Florence Kelley, the former chief factory inspector of Illinois. Kelley had drafted and lobbied for passage of an 1893 Illinois law that forbade factories to employ women for more than eight hours a day. That reform experience helped inspire similar laws in other states but offered a warning: it was overturned by the Illinois Supreme Court in 1895 on the grounds that it interfered with the "freedom of contract," a right of employers and employees to negotiate their own labor agreements that was supposedly guaranteed by the Fourteenth Amendment.[40]

The Illinois labor law proved to be just one of many pieces of protective legislation to face hostile court reviews. In 1905 the US Supreme Court struck down a New York State law that limited the workday in bakeries to ten hours and the workweek to sixty hours, for both males and females. In *Lochner v. New York* the court held that the state had failed to prove that the long hours in bakeries harmed either the bakers or the public; therefore no compelling issue of public welfare justified the state's inference with freedom of contract. Then, in a 1907 case concerning bookbinderies, New York's highest court used similar reasoning in rejecting a law more specifically banning the night work of women and children. The state had failed to demonstrate that night work was more harmful to women than to men, thus it could not properly deny women the same right to work.[41]

Despite the 1907 setback, advocates of shorter hours thought that laws focused specifically on women would be easier to defend than laws restricting the hours of all workers. Some state supreme courts had already signaled a willingness to consider women workers a separate, weaker class, unable to bear the same workload as men. The challenge therefore was to provide courts with evidence of these unique female disabilities and their connection to the larger public welfare. By so doing, labor reformers could firmly establish one form of protective legislation that might then serve as an opening wedge to break through the wall of judicial hostility.[42] In 1908 the labor attorney Louis Brandeis succeeded in persuading the US Supreme Court to uphold an Oregon law that set a ten-hour maximum workday for women in laundries and factories. His famous "Brandeis brief," compiled with the assistance of Josephine Goldmark, presented a collection of quotations from various authorities to show that long hours weakened women's health and thus the public welfare. Brandeis and Goldmark approvingly quoted medical opinion about the host of dangers resulting from women's overwork, especially during their "periodic semi-pathological state of health." The Court agreed. "As healthy mothers are essential to vigorous offspring, the physical well-being of woman becomes an object of public concern," wrote Justice David Brewer in the

Muller v. Oregon decision. "Her physical structure and a proper discharge of her maternal functions—having in view not merely her own health but the well-being of the race—justify legislation to protect her from the greed as well as the passion of man."[43]

It was a peculiar argument, but peculiarly effective. At a time when women were increasingly working outside the home and gaining a stronger voice in public affairs, their advocates presented them as shrinking violets needing special protection. The argument worked beautifully. Persuaded by appeals to their prejudices and cowed by reams of data modeled after the Brandeis brief, state assemblymen approved an outpouring of protective laws in the two decades following the *Muller* decision. Courts now routinely upheld them. (During the same years, courts began accepting laws limiting the hours for men as well as women, thus incidentally undercutting one of reformers' original justifications for seeking special protections for women.) By 1922, almost all states had imposed at least some restriction on women's workweek, ranging from a forty-eight-hour maximum in Massachusetts to a seventy-hour maximum in Illinois. A fifty-four-hour maximum was the most common. States also passed restrictions on women's night work in particular, mainly to make the hours laws easier to enforce. Without night work laws, employers might legally have had women work two full shifts consecutively, one ending and the other beginning at midnight. If certain hours were prohibited entirely, then it was a simple matter for a factory inspector to catch a manufacturer in the act of breaking the law. The industrial states of New York, Massachusetts, Connecticut, and Pennsylvania all had night work laws by 1921, as did Indiana, Nebraska, Kansas, and Wisconsin, among others. The laws typically concerned work between 10:00 p.m. and 6:00 a.m. No night work laws applied to men.[44]

Arguments against women's night work cobbled together disparate points about women's anatomy and household responsibilities. Testifying before the labor committee of the Ohio House of Representatives in 1919, Dr. Emery Hayhurst of the Ohio State University argued that night work was unhealthy for everyone, especially females. Unnatural hours disturbed the biological rhythms of sleep and digestion, artificial lighting strained the eyes, and fatigue led to a much higher accident rate, particularly between 2:00 and 5:00 a.m. Women's additional need to attend to housekeeping and child care kept them from getting the rest they needed. Night hours in themselves were detrimental to the female body, because "menstruation, pregnancy, and nursing (lactation) are all phenomena of natural origin and coördinated with the natural arrangement of the order of day and night." In *Radice v. New York* in 1924, the US Supreme Court accepted the state's right to protect women from threats to "their peculiar and natural functions." Given the difficulty in busy cities of resting properly during the day, women's "delicate organism" might soon be damaged by night work.[45]

Another argument against women's night work dealt with the danger of sexual assault in the streets. Many working women walked the entire distance to and from work to save carfare or because there was no convenient streetcar service at the hours they needed to travel. Those who took the trolley still had to walk at least a short distance at either end of the ride. According to a 1908 report by the Nebraska Bureau of Labor Statistics, women workers had to "face the wiles and insults of loafers and mashers when out alone late at night. Attacks on young girls returning from work late at night are not infrequent." The labor investigator Mary Van Kleeck warned of similar dangers to the New York bookbinders. She cited the example of a girl who was "employed at the age of seventeen to stitch programs for opera houses and theaters. During the theater season she worked overtime until 11 or 12 o'clock at night, a day of fourteen and a half hours. She walked home alone, past the closed business houses downtown. 'Only bums are down there at that hour of the night,' she said." Merely the fear of what might happen to them in the streets late at night was enough to jeopardize women's health, Dr. Hayhurst testified in 1919.[46]

The prospect of moral corruption was thought to be only slightly less alarming: young women and teenage girls who walked through the night streets might be ruined by the unsavory influences surrounding them. Campaigning in the 1910s to reduce the Saturday evening hours of women store clerks, the Consumers' League of Connecticut warned that the clerks of Hartford left the department stores at an hour when the downtown was bustling with theatergoers and other amusement seekers. "Every Saturday night at ten o'clock, crowds of young men and boys gather in front of our department stores in long lines on the edge of the sidewalk, waiting to see the girls home." Many a foolish girl accepted a stranger's invitation to escort her home. Within the workplace, men and women with loose morals, coarse speech, and a taste for liquor might corrupt younger girls who were still innocent. Waitresses who worked in late night restaurants, such as those in theater districts and railroad stations, had to deal with flirtation and harassment from customers. "The men are always fresher to girls at night than in the day time," complained one waitress. "Perhaps it's because so many of those gamblers come in drunk."[47]

The workers affected by protective legislation were not always as grateful as reformers might have hoped. Anna Schmidt, a waitress in a Buffalo restaurant, was trying to support herself and her two children by working from 6:00 p.m. to midnight while her husband was serving in the army during

World War I. Evening hours allowed her to care for her children during the day, and they paid well; the after-theater crowd in particular was known for its generous tipping. But since those hours violated New York State law, her employer was arrested and fined twenty dollars in a case that finally ended in the US Supreme Court's *Radice* decision.[48]

Anna Schmidt's situation was not unusual. Most women who took night shift jobs in factories were married with children, and they found that night work interfered less with their household responsibilities. Mothers could leave their children to sleep, with relatives or unattended, then head to the factory. "I don't want to have to work days, as then my children are alone," explained one woman who worked the night shift at a cordage factory in Auburn, New York. Widows with children showed a similar preference for night work, often cleaning office buildings. Some wives sought night work as a way to adapt their schedules to those of their husbands who held night jobs. These arrangements were hardly ideal, as the National Consumers' League pointed out: children were neglected at night and poorly attended during the day as mothers tried to catch up on their lost sleep. Household duties ensured that women could not possibly get enough rest before it was time to return to work. But in the absence of any better alternatives, women continued to choose night work.[49]

Feminists in the National Woman's Party opposed the special protections on the grounds that they were a form of discrimination that interfered with women's ability to earn a living. As a result of New York's night work laws, women could no longer work at candy counters and soda fountains in the evenings. One employer fired more than eighty women and hired men in their place, the organization contended in a pamphlet issued in the 1920s. Women lost night jobs as linotypists and proofreaders for the newspapers, conductors for the street railways, and pharmacists in drugstores. A newsstand chain replaced its women employees with men. "The Woman's Party contends that if night work is undesirable for women because the streets are not safe for women, as is claimed, the solution is to make the streets safe instead of barring women from night occupations." Laws aimed at promoting the public health should apply to both sexes, not just women. Special protections, even if well intended, prevented women's progress toward true equality, the Woman's Party argued to no avail.[50]

Some private employers took a paternalistic interest in the safety and well-being of women who continued to work at night. The telephone industry, which employed young women to operate its twenty-four-hour switchboards, took great care to ensure that workplaces approximated the cleanliness, comfort, and decorum of a proper family home. It scrutinized the family background and educational attainment of applicants so that only those of "the right class" would be hired. Telephone companies in cities provided dormitories for the night operators so they would not have to walk through the streets at night. In Connecticut's cities, operators were not allowed to leave or enter the building between 10:00 p.m. and 7:00 a.m. except in emergencies.[51] Similarly, when Connecticut's munitions factories began running night shifts of women during World War I, there was so much public concern about immoral influences on the shop floor that companies established welfare departments to investigate and fire women of questionable virtue. During World War II, when labor shortages forced American heavy industry to hire women on the night shifts again, companies had the streets, bus stops, and rail stations carefully patrolled at the time of shift changes.[52]

CONCLUSION

"We do progress in a century," marveled Edward Hungerford in a 1910 description of New York at night. "Any well-born citizen of 1810 might rub his eyes in wonder—the cold white iridescence of a myriad sputtering arcs makes the city street at night as brilliant as midday." A viewer from above would see the entire city outlined by light, with brilliant strips distinguishing the major streets, waterfronts, and bridges.[53]

People's behavior had changed as well. Railroads, trolleys, newspapers, and steel mills now ran twenty-four hours a day, ensuring that somebody had to be traveling through the streets at any hour. Saloons, restaurants, and theaters stayed open later, and a few never closed. Teenagers and young adults stayed up late reading, studying, and socializing. City people had stayed up later than country folk even in the mid-nineteenth century; by the early twentieth century, wrote an advocate of daylight saving time, "there is an increasing tendency to shift the day farther and farther forward. Without a doubt the number of people who sleep till noon is constantly increasing. If the matter is left to adjust itself, one naturally wonders how far we shall go in the substitution of night for day. Will day and night finally be wholly transposed?"[54]

City people and country folk agreed that the nighttime city represented a startling intensification of the urban experience. True, the sidewalks were emptier and the street traffic less frenetic, but the nocturnal city conspicuously exaggerated some of the distinctive social aspects of urban life: the anonymity of the individual, the relative weakness of community oversight, and the encounter with a diverse mix of strangers. A young man walking through strange streets, face shaded by his hat brim, was free of the informal constraints he might feel in a village where he was likely to be recognized by neighbors who knew him and his family. He was free also of the discipline of the family home, of the structure provided by his daily work routine, and of the natural limitations imposed by the cycle of the light and darkness. From the 1820s until after the turn of the century, the partial illumination of the city provided enough light to get around, but not so much as to lift "the cover of darkness" from those who sought concealment. Further, the growing number of legitimate activities at night provided a useful camouflage for everything else that went on. Was that shadowy figure innocently on his way to a prayer meeting or a job on the night shift? Was he sneaking off for a night of debauchery and crime? Or, more disturbingly, was he trying to make up his mind as he walked along? As the English clergyman James Hervey had warned long ago, sinners felt free to follow their impulses at night, "Each soothing himself with the fond Notion, That all is safe; That no Eye sees."[55]

This freedom was not absolute, nor was it equally available. The right to the nighttime city belonged especially to men, above all to young men, subject to the inconsistent limitations imposed by male police officers. Young men had dominated American streets at night since the colonial era, whooping, carousing, and generally being obnoxious to anyone else who had to be out. Gradual improvements in lighting and policing in the nineteenth century helped a broader range of people feel somewhat secure on the streets after dark, though never to the same extent as young men. Women, children, and older men had fewer reasons to be out late anyway. They remained uncommon in the types of work that prevailed at night, and they were underrepresented in the heavy industries that saw the greatest growth of shift work in the late nineteenth century. Commercial entertainments, despite efforts at reform, were often at odds with the domestic values associated with the proper middle-class woman. Women who defied these values by going out could do so only under male protection, or they risked harassment and physical violence.

Public policy, in the form of curfews and labor laws, inadvertently reinforced the old nocturnal culture. Well-intentioned legislation discouraged the presence of women and children on the streets at night, thus perpetuating an imbalance that might otherwise have subsided. This response echoed the simultaneous attempt to mitigate the problems created by widespread prostitution. In the same way that the vice district boundaries (and later zoning regulations) were intended to resolve conflict by dividing the city spatially, the effort to remove children and women from the streets at night aimed at dividing the city temporally. Vice districts were not places for decent women and children; night was not a proper time for women and children to be in public.

Technological change undoubtedly contributed to a change in the life of the city, but that change remained incomplete through the 1920s.

In some ways the 1920s may seem an odd point at which to close our story. That decade is often remembered as one of accelerating change in American urban life: a self-consciously modern era filled with jazz, flappers, and bootleggers. Historians who peer beneath this glittering surface find that the cliché contains important truths. City life really was changing. The outward expansion of the metropolitan area in this decade, combined with the sudden rise in the use of automobile, produced new configurations and experiences of urban space. Growing numbers of people were whisked across long distances without having to set foot in the streets and streetcars that made urban travel an intensely public experience. Women in particular appreciated the automobile's potential to free them from the hazards of public space. Relations between men and women were also changing. Affluent young women felt somewhat more free to venture out at night. The increasingly popular ritual of dating shifted courtship from the private home to places of commercial amusement: restaurants, theaters, and night clubs. Prohibition's closing of the saloons destroyed an institution whose boisterous male camaraderie had extended into public drunkenness and street brawls that could frighten passersby. In its place, cabarets and speakeasies nourished a new public drinking culture that included women as well as men. The Great Migration of African Americans to northern cities was inaugurating a demographic shift that, over the next half century, would alter white Americans' perception of the city's pleasures and dangers. Its effects were already being felt in the surge of white interest in the nightlife of New York's Harlem and Chicago's Bronzeville.[56]

But these changes belong to a different story. Alterations in public policy, gender relations, and urban demography from the 1920s onward are difficult to place within a history framed by artificial lighting. To the limited extent that they were shaped by technology, it was by the automobile, not the gas jet or lightbulb. Moreover, though many people were now able to go out at night with minimal public interaction along the way, those who walked or rode streetcars encountered the same old problems of disorder and unpredictability. Women continued to rely on the protection of male companions. Unescorted women after 1920 continued to risk being harassed, mistaken for prostitutes, or worse. The concern was strong enough that a new female-focused taxi company formed in New York in 1923, offering women the reassurance of riding with women drivers.[57]

Even in the electric era, night was not turned into day in any sense of the phrase. The codes of behavior that prevailed in the dark streets of preindustrial America proved remarkably resilient, preserving the night as an incompletely civilized realm within the modern city. Whether the human mind and body could ever fully adapt to a nocturnal schedule is a question that cannot be answered by historical evidence and has proved challenging even for researchers in medicine and psychology. Researchers have found that some of the body's circadian rhythms—notably the daily variations of body temperature—can adjust almost completely to night work if the worker remains on an uninterrupted night schedule. Of course, it is almost impossible for anyone to remain on such a schedule except in highly contrived situations. The major problems with daylight sleep appear to be social and cultural rather than inherent in the human body. As in industrial Pittsburgh long ago, daytime sleepers still complain of being distracted by ambient noise and household activity, or they force themselves to get by on less sleep in order to socialize or attend to family duties.[58] Night survives as a potent influence on human life.

However commonsensical this mixed result might seem, nineteenth-century Americans had foreseen at least the possibility of something different: a technological future so thoroughly cut loose from nature and the human past that, in the words of Ignatius Donnelly, "Night and day are all one . . . and the business parts of the city swarm as much at midnight as at high noon."[59] Today, as visionaries continue expecting technology to produce a utopian (or dystopian) transformation of mankind, it is worth peering back into an age where such expectations were unfulfilled: the century when humans gained the power to walk with seeing eyes throughout their nocturnal environment. Today, for better or worse, the brilliant products of human innovation remain shadowed by the natural rhythms that shape all life, and by the cultures we inherit from our dim past.

ACKNOWLEDGMENTS

Over the past decade of work on this project, my research materials have piled up while my memories of the research process have faded. An archaeologist examining the middens of notes in my offices and attic could discern several generations of data storage in the accumulated CDs, zip drives, 3.5-inch disks, shoeboxes of index cards, and cartons filled with photocopies of print and microfilm documents. These remain as readable as the day I created them. Yet nowhere in this mass of information is an adequate record of all the people who helped me along the way. Those records I kept only in my head, and now, having finished the book, I'm embarrassed to see how fragmentary they have become.

I recall the origins of the project in summer of 1999, at a National Endowment for the Humanities institute at the University of Illinois's Chicago campus. Organized by Bob Bruegmann around the topic "The Built Environment of the American Metropolis," the institute was a stimulating gathering of scholars who have gone on to produce some of the most remarkable recent work in urban history. Among all those present, I remember Robin Bachin as especially helpful in shaping my thinking about what was at the time a somewhat off-beat topic: the history of night. The discussions that summer gave me a lot to mull over on my nocturnal walks around Chicago's Lincoln Park and Lakeview neighborhoods.

As the project got under way, I benefited from the comments of colleagues and friends who read or listened to the early papers I spun off. I particularly thank Michael Ebner, Ellen Eslinger, Adam Rome, Mark Rose, Peter Stearns, and Joel Tarr for their suggestions and encouragement. Pat Cohen provided thoughtful answers to my research-related questions and, along with several of the colleagues mentioned already, helped me obtain funding. Conversations and e-mail with Randy Burgess, Ray Weissman, Alan Petigny, Karen Renner, Cindy Lobel, Roger Ekirch, Edward Gray, Nancy Shoemaker, and the late Susan Porter Benson alerted me to issues and sources I might otherwise have neglected. As always, Howard Chudacoff was a reliable source of encouragement and assistance.

A year at the University of Connecticut's Humanities Institute provided both an intellectual community and an invaluable release from teaching duties during a productive research year in 2003–4. At the UCHI, some of the most perceptive questions and critiques came from two scholars who were already my colleagues in UConn's history department: Dick Brown and Sylvia Schafer. The University of Connecticut provided additional aid through a Provost's Scholarship Development grant in 2009.

The American Antiquarian Society proved to be my most frequent research site. Especially in the early years of the project, before so many materials had been digitized, I relied heavily on their extensive collection of nineteenth-century newspapers and books. A Kate B. and Hall J. Peterson Fellowship in the summer of 2004 gave me some pleasant and productive weeks of "history camp" in Worcester, Massachusetts, along with fellow campers Sean Adams, Laura Pruett, and Chris Phillips. Joanne Chaison, Caroline Sloat, and the rest of the AAS staff were exceptionally helpful and supportive.

Though my research has taken me again and again to the AAS, most of my other research trips have been fleeting, typically no more than a few days or at most a couple weeks in any one place. It is for these trips that my memory fails me. I regret that I cannot recall the names of the many people who assisted me at historical societies, university special collections, and public libraries across the Northeast and Midwest. The best I can do is to list the sites by name and to emphasize that, almost without exception, I was given exceptional service. My thanks to the librarians and archivists at the following institutions (pause for deep breath): Boston College, Boston Public Library, Brown University, Carnegie Library of Pittsburgh, Chicago Historical Society, Chicago Public Library, Columbia University, Cornell University, DePaul University, Free Library of Philadelphia, Harvard University, Historical Society of Pennsylvania, Indianapolis Public Library, Library of Congress, Louisville Free Public Library, Massachusetts Historical Society, Massachusetts Institute of Technology, Nebraska State Historical Society, Newberry Library, New York Public Library, Omaha City Hall, Omaha Public Library, Philadelphia City Archives, Rhode Island Historical Society, Senator John

Heinz History Center, Toledo Public Library, Trinity College, University of Connecticut, University of Illinois at Chicago, University of Louisville, University of Nebraska at Lincoln and at Omaha, University of Pittsburgh, and University of Toledo. The friendly people at the American Flint Glass Workers' Union in Toledo were unusually accommodating not only in providing materials and research space, but also in allowing my wife and infant daughter to use the conference room for nursing and naptime.

Robert Devens, Alice Bennett, and Tim Gilfoyle at the University of Chicago Press have been of immense help in preparing the manuscript for publication. Tim provided perhaps the most perceptive and helpful set of comments I have ever received on any manuscript. The anonymous readers also made valuable suggestions and saved me from some errors.

It has often been a challenge to balance the demands of this research project, my teaching and service duties, and my family responsibilities. But in this last area I have no regrets. I am deeply grateful for the love and companionship of Vicki Magley and our two wonderful children, Anne-Marie and Sammy. Instead of taking walks through the streets of Hartford, Providence, and Chicago, I now enjoy the evening ritual of bedtime stories and "night-night songs."

ABBREVIATIONS

AAS: American Antiquarian Society, Worcester, Massachusetts
AGLJ: *American Gas-Light Journal* (and variant titles)
AIS: Archives of Industrial Society, University of Pittsburgh
BE: *Brooklyn Eagle*
BG: *Boston Globe*
BS: *Baltimore Sun*
CBC: Chicago Boys' Club Papers, Chicago History Museum
CIO: *Chicago Inter Ocean*
CLC: Consumers' League of Connecticut Papers, Schlesinger Library, Radcliffe College
CT: *Chicago Tribune*
ERJ: *Electric Railway Journal*
GMLC: Geography and Maps Reading Room, Library of Congress, Washington, DC
HTC: Harvard Theater Collection, Harvard University
HW: *Harper's Weekly*
JPAP: Juvenile Protective Association Papers, University of Illinois at Chicago
KCCU: Kheel Center for Labor-Management Documentation and Archives, Cornell University
LO: *Lowell (MA) Offering*
LVP: *Lowell (MA) Vox Populi*
MHS: Massachusetts Historical Society, Boston
NCLC / LC: National Child Labor Committee Papers, Library of Congress
NPG: *National Police Gazette*

NYH: *New York Herald*
NYT: *New York Times*
NYTr: *New York Tribune*
PHS: Pennsylvania Historical Society, Philadelphia
PI: *Philadelphia Inquirer*
PPL: *Philadelphia Public Ledger*
RIHS: Rhode Island Historical Society, Providence
SRJ: *Street Railway Journal*

NOTES

CHAPTER ONE

1. Fanny Trollope, *Domestic Manners of the Americans* (1832; London: Penguin Books, 1997), 209–21.

2. The book focuses on the cities of the Northeast and Midwest, though it frequently stretches this region to include discussion of border cities such as Baltimore, Washington, Louisville, and St. Louis. The northeastern quadrant of the United States was the nation's urban-industrial core in the nineteenth and early twentieth centuries. This was where the nation's largest cities were concentrated and where cultural and technological changes tended to originate. Focusing on this region helped me keep the research process manageable. A book that fully examined the experience of night nationwide would have to deal to a much greater extent with regional differences. In particular, any consideration of the southern states and California would have to explore in depth how the distinctive racial composition and racial etiquette in those regions inflected their experience of urban night. During the period covered by the book, the cities of the Northeast and Midwest had comparatively small nonwhite populations, typically no more than 5 or 10 percent. More than 97 percent of the people in New York City were white during the period from 1850 through 1920, according to the US Census (Campbell Gibson and Kay Jung, "Historical Census Statistics on Population Totals by Race, 1790 to 1990, and by Hispanic Origin, 1970 to 1990, for Large Cities and Other Urban Places in the United States," Population Division Working Paper 76, US Census Bureau, February 2005: http://www.census.gov/population/www/documentation/twps0076/twps0076.html.

3. Harold L. Platt, *Shock Cities: The Environmental Transformation and Reform of Manchester and Chicago* (Chicago: University of Chicago Press, 2005); Basil Hall, *Travels in North America in the Years 1827 and 1828*, vol. 1 (Edinburgh: Robert Cadell, 1830), 161; Rudyard Kipling, *American Notes*, in *Selected Works of Rudyard Kipling*, vol. 3 (New York: Peter Fenelon Collier, 1900), 276, 277.

4. Ignatius Donnelly, *Caesar's Column: A Story of the Twentieth Century* (1890; Middletown, CT: Wesleyan University Press, 2003), quotations at 10 and 111.

5. Harold L. Platt, *The Electric City: Energy and the Growth of the Chicago Area, 1880–1930* (Chicago: University of Chicago Press, 1991), xvi. For a fuller exploration of these issues, see the introduction and the essays in *Does Technology Drive History? The Dilemma of Technological Determinism,* ed. Merritt Roe Smith and Leo Marx (Cambridge, MA: MIT Press, 1994). See also Thomas P. Hughes, *Human-Built World: How to Think about Technology and Culture* (Chicago: University of Chicago Press, 2004); Ruth Schwartz Cowan, *A Social History of American Technology* (New York: Oxford University Press, 1997); John F. Kasson, *Civilizing the Machine: Technology and Republican Values in America, 1776–1900* (1976; New York: Hill and Wang, 1999).

6. On the effects of the 1776–79 blockade of Narragansett Bay, see Welcome Arnold Greene, *The Providence Plantations for Two Hundred and Fifty Years* (Providence, RI: J. A. and R. A. Reid, 1886), 61–62. On the *Argo*'s 1779 exploits, see *An Historical Sketch, to the End of the Revolutionary War, of the Life of Silas Talbot, Esq.* (New York: G. and R. Waite for H. Caritat, 1803), 65–97; Henry T. Tuckerman, *The Life of Silas Talbot, a Commodore in the Navy of the United States* (New York: J. C. Riker, 1850), 76–77; *Providence Gazette,* July 10, 1779; *Providence American Journal,* Sept. 23, Nov. 11, 1779; On the *Argo*'s 1780 service as a privateer, see *Providence Gazette,* June 3, July 22, 1780, and *American Journal,* July 6, 1780; Zuriel Waterman, "Journal of an Intended Cruize against the Enemies of the United States in the Privateer Sloop *Providence*, Capt. James Godfrey Comander," in Richard Waterman Family Papers, RIHS, entry dated Jan. 26, 1780.

7. Waterman journal, especially entries for Sept. 15–17, Oct. 4, 9, Nov. 13, 19, 1779; Jan. 1, 25–26, Feb. 12, Apr. 18, 1780; Dr. Zuriel Waterman Memorandum Book, 59, 127, Richard Waterman Family Papers, RIHS; Rhode Island General Assembly, "Instructions Intended for Capt. Talbot Commander of the *Argo* . . . ," Dec. 1779, Silas Talbot Papers, RIHS; Manfred J. Waserman, "Dr. Zuriel Waterman: An Itinerant Surgeon in the Revolutionary Era," *Proceedings of the American Philosophical Society* 117, no. 5 (Oct. 1973): 388–403 (Waserman provides a brief biography of Waterman but appears to have misunderstood the *Argo*'s discharge); David Swain, ed., "Diary of a Doctor-Privateersman 1779 to 1781 by Zuriel Waterman," in *Rhode Islanders Record the Revolution: The Journals of William Humphrey and Zuriel Waterman,* ed. Nathaniel N. Shipton and David Swain (Providence: Rhode Island Publications Society, 1984), fol. 3, p. 80; privateering license dated Apr. 14, 1780, Silas Talbot Papers; Robert Grieve, "The Sea Trade and Its Development," in Field, *State of Rhode Island and Providence Plantations at the End of the Century,* ed. Edward Field (Boston: Mason, 1902), 2:427. The shoreline village of Pawtuxet is part of the township of Cranston, southwest of Providence; it is not the same as the city of Pawtucket to the north.

8. Waterman journal, Jan. 9, 15, 26, 29, 31, 1780; *Providence American Journal,* Sept. 16, 1779; Clarence Saunders Brigham, "Rhode Island in the Revolution," in Field, *State of Rhode Island,* 1:246; Blake McKelvey, *Snow in the Cities: A History of America's Urban Response* (Rochester, NY: University of Rochester Press, 1995), 13–14.

9. Waterman journal, Jan. 26, 1780. This description of Providence is based on the following: Marquis de Chastellux, *Travels in North-America in the Years 1780–81–82* (New York, 1828), 19; Nancy Fisher Chudacoff, "The Revolution and the Town: Providence, 1775–1783," *Rhode Island History* 35, no. 3 (Aug. 1976): 71–89; *Names of the Owners or Occupants of Buildings in the Town of Providence from 1749 to 1771* (Providence, RI: Sidney S. Rider, 1870);

Greene, *Providence Plantations*, 54-60; John Hutchins Cady, *The Civic and Architectural Development of Providence, 1636-1950* (Providence, RI: Book Shop, 1957), 27-43, 56; Powderhorn engraving of Providence, 1777, reproduced in Hope S. Rider, *Valour Fore and Aft: Being the Adventures of the Continental Sloop "Providence," 1775-1779, Formerly Flagship "Katy" of Rhode Island's Navy* (Annapolis: Naval Institute Press, 1977), 108-9. About 4,300 people lived in Providence at the time, a respectable number for a colonial seaport. Carl Bridenbaugh, *Cities in Revolt: Urban Life in America, 1743-1776* (New York: Alfred A. Knopf, 1955), 216-17.

10. Waterman journal, Jan. 26, 1780. On Russell, see Gertrude Selwyn Kimball, *Providence in Colonial Times* (Boston: Houghton Mifflin, 1912), 327-29. On the practice of charivari and other mob action against prostitutes, see A. Roger Ekirch, *At Day's Close: Night in Times Past* (New York: W. W. Norton, 2005), 223-24, 253-54; Bridenbaugh, *Cities in Revolt*, 316-17.

11. Waterman journal, Jan. 26, 27, 1780; Tavern license for Nathaniel Jenckes, 1780, Providence Town Papers, book 5, no. 2211, RIHS. On wartime inflation, see Greene, *Providence Plantations*, 62; table titled "Philadelphia, June 30 1780," in Waterman memorandum book. On street pavement, see Greene, *Providence Plantations*, 54, 58. American towns before the mid-nineteenth century did not attempt to clear snow from the streets; see McKelvey, *Snow in the Cities*, 6-9, 30-32. On drunken pranks played against the military sentries in wartime Providence, see Chudacoff, "Revolution and the Town," 75-76.

12. Ekirch, *At Day's Close*, 48-56; Carl Bridenbaugh, *Cities in the Wilderness: The First Century of Urban Life in America, 1625-1742* (1938; New York: Capricorn Books, 1964), 364-72; Bridenbaugh, *Cities in Revolt*, 98-105, 298; Chudacoff, "Revolution and the Town," 74.

13. Ekirch, *At Day's Close*, 7-47; Peter C. Baldwin, "How Night Air Became Good Air, 1776-1930," *Environmental History* 8, no. 3 (July 2003): 412-29.

14. James Hervey, *Contemplations on the Night* (New York: James Rivington, 1774), 17-19, 24.

15. William Otter, *History of My Own Times*, ed. Richard B. Stott (1835; Ithaca, NY: Cornell University Press, 1995), 42-50; Richard Stott, *Jolly Fellows: Male Milieus in Nineteenth Century America* (Baltimore: Johns Hopkins University Press, 2009).

16. "City and County of New York, Court of Session-May Term," *Poulson's American Daily Advertiser*, May 23, 1815; Josiah Quincy, *A Municipal History of the Town and City of Boston, during Two Centuries* (Boston: Charles C. Little and James Brown, 1852), 102, 104.

17. Bryan D. Palmer, *Cultures of Darkness: Night Travels in the Histories of Transgression from the Medieval to the Modern* (New York: Monthly Review Press, 2000), 141-43; Samuel L. Mitchell, *The Picture of New York, or The Traveller's Guide through the Commercial Metropolis of the United States by a Gentleman Residing in This City* (New York: I. Riley, 1807), 211; Jill Lepore, *New York Burning: Liberty, Slavery and Conspiracy in Eighteenth-Century Manhattan* (New York: Alfred A. Knopf, 2005), 148-49; Ekirch, *At Day's Close*, 155-68.

18. David Hackett Fischer, *Paul Revere's Ride* (New York: Oxford University Press, 1994), 22; Rider, *Valour Fore and Aft*, 3-4; Bridenbaugh, *Cities in the Wilderness*, 388-89; Bridenbaugh, *Cities in Revolt*, 316-17; Edward H. Savage, *A Chronological History of the Boston Watch and Police, from 1631 to 1865; Together with the Recollections of a Boston Police Office, or Boston by Daylight and Gaslight* (Boston: Edward H. Savage and J. E. Farwell, 1865), 12; Timothy J. Gilfoyle, *City of Eros: New York City, Prostitution, and the Commercialization of Sex* (New York: W. W. Norton, 1992), 76-91.

19. Arthur H. Hayward, *Colonial Lighting* (Boston: B. J. Brimmer, 1923), 70; Sidney Irving Pomerantz, *New York, an American City, 1783–1803: A Study of Urban Life* (New York: Columbia University Press, 1938), 305–6; Selden D. Bacon, "The Early Development of American Municipal Police: A Study of the Evolution of Formal Controls in a Changing Society" (PhD diss., Yale University, 1939), 402–3, 504, 593; Allen Richmond, *The First Twenty Years of My Life* (Philadelphia: American Sunday-School Union, 1859), 121, quotation at 132; Wolfgang Schivelbusch, *Disenchanted Night: The Industrialization of Light in the Nineteenth Century* (Berkeley: University of California Press, 1995), 82–96; Thomson King, *Consolidated of Baltimore, 1816–1950: A History of Consolidated Gas Electric Light and Power Company of Baltimore* (Baltimore: Consolidated Gas Electric Light and Power Company, 1950), 8; Warren S. Tryon, ed., *A Mirror for Americans: Life and Manners in the United States, 1790–1870, as Recorded by American Travelers*, vol. 1, *Life in the East* (Chicago: University of Chicago Press, 1952), 12; quotation from Osgood Bradbury, *Mysteries of Lowell* (Boston: E. P. Williams, 1844), 21.

20. Bridenbaugh, *Cities in the Wilderness*, 155–58, 164–65, 169, 315–24; Bridenbaugh, *Cities in Revolt*, 31–33, 238–44; Greene, *Providence Plantations*, 114; Bacon, "American Municipal Police," 219, 254; "Communication," *New York Republican Watch-Tower*, July 4, 1806; Henry Bradshaw Fearon, *Sketches of America: A Narrative of a Journey of Five Thousand Miles through the Eastern and Western States of America* (London: Longman, Hurst, Rees, Orme and Brown, 1818), 134; S. Jones, *Pittsburgh in the Year Eighteen Hundred and Twenty-Six* (Pittsburgh: Johnston and Stockton, 1826), 41; B. Drake and E. D. Mansfield, *Cincinnati in 1826* (Cincinnati: Morgan, Lodge and Fisher, 1827), 52; City of Salem, Massachusetts, *City of Salem: An Ordinance to Prevent Unlawful and Injurious Practices in the Streets and Other Public Spaces of the City* (1836); Charles Haynes Haswell, *Reminiscences of New York by an Octogenarian (1816 to 1860)* (New York: Harper, 1896), 273, 299, 482, 542; Jacob Judd, "A City's Streets: A Case Study of Brooklyn, 1834–1855," *Journal of Long Island History* 9, no. 1 (1969): 32–43; "Accident," *NYTr*, Sept. 22, 1842; W. A. Kentish, "To Gas Consumers and the Public," *NYH*, Jan. 20, 1844; "Death by Drowning," *LVP*, Feb. 19, 1847; "Dark Nights! Dark Nights!!" (advertisement), *Daily Cincinnati Commercial*, Dec. 1 1847; "Crime, Poverty and the Streets of New York City: The Diary of William H. Bell, 1850–51," ed. Sean Wilentz, *History Workshop*, no. 7 (Spring 1979): 147.

21. Pomerantz, *New York*, 298; Bridenbaugh, *Cities in the Wilderness*, 169, 202, 382–84; Bridenbaugh, *Cities in Revolt*, quotation at 113, 299–303; "Brutal Outrage," *Baltimore Patriot*, May 23, 1818; Quincy, *Municipal History of the Town and City of Boston*, 102, 109.

22. Bridenbaugh, *Cities in the Wilderness*, 63–68; Bacon, "American Municipal Police," 22, 74, 327, 341, 349, 380, 505; Savage, *Boston Watch*, 12.

23. James F. Richardson, *The New York Police: Colonial Times to 1901* (New York: Oxford University Press, 1970), 11; Bacon, "American Municipal Police," 328, 381; Howard O. Sprogle, *The Philadelphia Police, Past and Present* (Philadelphia: Howard O. Sprogle, 1887), 48, 49; Savage, *Boston Watch*, 20.

24. Lawrence M. Friedman, *Crime and Punishment in American History* (New York: Basic Books, 1993), 224; Bacon, "American Municipal Police," 101, 656, 660, 663, 677, 688–89; Greene, *Providence Plantations*, 68–69; George Grafton Wilson, "The Political Development of the Towns," in Field, *State of Rhode Island and Providence Plantations*, 3:71; Henry C. Castellanos, *New Orleans as It Was: Episodes of Louisiana Life* (1895; Baton Rouge: Louisiana

State University Press and the American Revolution Bicentennial Commission, 1978), 205-6; James Mease, *The Picture of Philadelphia* (Philadelphia: B. and T. Kite, 1811), 166; Joyce D. Goodfriend, *Before the Melting Pot: Society and Culture in Colonial New York City, 1664-1730* (Princeton, NJ: Princeton University Press, 1992), 120-22; Bridenbaugh, *Cities in Revolt,* 299; Richard C. Wade, *Slavery in the Cities: The South, 1820-1860* (London: Oxford University Press, 1964), 98, 184.

25. Bacon, "American Municipal Police," 455-56; Castellanos, *New Orleans as It Was,* 164.

26. Tryon, *Mirror for Americans,* 85.

27. The foregoing is based in part on my earlier book, Peter C. Baldwin, *Domesticating the Street: The Reform of Public Space in Hartford, 1850-1930* (Columbus: Ohio State University Press, 1999), and on Harold Platt's *Shock Cities.* Similar observations appear in the introduction to John A. Jakle's *City Lights: Illuminating the American Night* (Baltimore: Johns Hopkins University Press, 2001).

CHAPTER TWO

1. George C. Foster, *New York by Gas-Light and Other Urban Sketches* (1850; Berkeley: University of California Press, 1990), 120-22; Edwin G. Burrows and Mike Wallace, *Gotham: A History of New York City to 1898* (New York: Oxford University Press, 1999), 474.

2. Tyler Anbinder, *Five Points: The 19th-Century New York City Neighborhood That Invented Tap Dance, Stole Elections, and Became the World's Most Notorious Slum* (New York: Free Press, 2001); Timothy J. Gilfoyle, *A Pickpocket's Tale: The Underworld of Nineteenth-Century New York* (New York: W. W. Norton, 2006), 18-22; Ned Buntline [Edward Z. C. Judson], *The Mysteries and Miseries of New York: A Story of Real Life* (New York: Berford, 1848), 2:92; 3:55.

3. Cornelius Mathews, "The Career of Puffer Hopkins" (1842) in *The Various Writings of Cornelius Mathews* (New York: Harper and Brothers, 1843), 198; see also B. Drake and E. D. Mansfield, *Cincinnati in 1826* (Cincinnati: Morgan, Lodge, and Fisher, 1827), 51-52; "Mayor's Message," *BS,* Nov. 21, 1850; "More Light," *NYT,* Jan. 1, 1859; Mark J. Bouman, "Luxury and Control: The Urbanity of Street Lighting in Nineteenth Century Cities," *Journal of Urban History* 14, no. 1 (Nov. 1987): 7-37.

4. Jack Larkin, *The Reshaping of Everyday Life, 1790-1840* (New York: Harper Perennial, Harper and Row, 1988), 136; Helen Brigham Hebard, *Early Lighting in New England, 1620-1861* (Rutland, VT: Charles E. Little, 1964), 41-42, 59; Herman Melville, *Moby-Dick* (1851; New York: Dell, 1959), 34.

5. David P. Erlick, "The Peales and Gas Lights in Baltimore," *Maryland Historical Magazine* 80, no. 1 (Spring 1985): 9-18; George T. Brown, "The Gas Light Company of Baltimore: A Study of a Natural Monopoly," *Johns Hopkins University Studies in Historical and Political Science* 54, no. 2 (1936): 15, 22-27; Consolidated Gas Electric and Power Company of Baltimore, "American Gas Centenary, 1816-1916," *Baltimore Gas and Electric News Centennial Number,* vol. 5, no. 6 (June 1916): 249. compared with George W. Boynton and Thomas G. Bradford, "Baltimore," in *Maryland* (Boston: Weeks, Jordan, 1838).

6. *AGLJ* 1, no. 1 (July 1, 1859): 2-3; Frederick L. Collins, *Consolidated Gas Company of New York: A History* (New York: Consolidated Gas Company of New York, 1934), 17-19, 48-59; Louis Bader, "Gas Illumination in New York City, 1823-1863" (PhD diss., New York Univer-

sity, 1970), 124, 171, 249, 262; David Biggs [J. W. Watson], "Gas and Gas-Making," *Harper's New Monthly Magazine* 26 (Dec. 1862): 14–28; John A. Jakle, *City Lights: Illuminating the American Night* (Baltimore: Johns Hopkins University Press, 2001), 24–32; "Philadelphia Gas," *Philadelphia North American*, Dec. 27, 1839; Allen Richmond, *The First Twenty Years of My Life* (Philadelphia: American Sunday School Union, 1859), 132; George Lippard, *New York: Its Upper Ten and Lower Million* (Cincinnati: H. M. Rulison, 1854), 137; George Lippard, *The Empire City, or New York by Night and Day: Its Aristocracy and Its Dollars* (1850; Philadelphia: T. B. Peterson, 1864), 134.

7. Bader, "Gas Illumination," 214; *Facts, in relation to the Introduction of Gas Light into the City of Philadelphia* (1824), 8; Jakle, *City Lights*, 30–31; "Important Communication concerning Gas," *AGLJ* 8, no. 22 (May 16, 1867): 338–39; "Lamps and Gas," *NYH*, Feb. 29, 1844; "In a Fog," *Atlantic Monthly*, June 1860, 650; "Gas," *NYT*, Feb. 20, 1866; "The Gas Monopoly," *BG*, Mar. 23, 1872; "Annual Report of the Superintendent of Lamps for 1876," in Boston Gas Company Records, box 14, folder 7, John J. Burns Library, Boston College. Early gas jets burned at twelve to sixteen "candlepower," but improvements in candle manufacturing meant that a candle could produce as much as two candlepower; Bader, "Gas Illumination," 214, 216, and 214 n. 3.

8. S. R. Seibert and J. M. Toner, *Outlines of Tracts of Land within Which the City of Washington Was Laid Out* (1870), GMLC; William Russell Smith, *As It Is* (Albany, NY: Munsell and Rowland, 1860), 38; *Report of the Committee of the Senate of the State of New York, Appointed to Investigate Several Departments of the Government of the City and County of New York* (Albany: State of New York, 1876), 24; "Gas Lamps for Churches," *AGLJ* 2, no. 28 (Feb. 15, 1861): 254; *Second Annual Report of the Board of Public Works to the Common Council of the City of Chicago* (Chicago: Tribune Book and Job Printing Office, 1863), 24; *Journal of the Proceedings of the Common Council of the City of Chicago*, vol. 5, May 15, 1865; *Ninth Annual Report of the Board of Public Works to the Common Council of the City of Chicago* (Chicago: Lakeside Press, 1870), 30–32.

9. Joel Tiffany, *Reports of Cases Argued and Determined in the Court of Appeal of the State of New York*, vol. 5 (Albany: Weare C. Little, 1865), 511; "The Gas Monopolists," *NYT*, Nov. 23, 1871; "The Gas Trouble," *NYT*, Nov. 25, 1871; "Gas Reform," *BG*, Oct. 21, 1872; Jeremy Loud, *Gabriel Vane: His Fortune and His Friends* (New York: Derby and Jackson, 1856), 250; Osgood Bradbury, *The Banker's Victim, or The Betrayed Seamstress* (New York: Robert M. DeWitt, 1857), 78; Ned Buntline, *Rose Seymour, or The Ballet Girl's Revenge* (New York: Hilton, 1865), 11; Fanny Warner, "'After Many Days,'" in *Beech Bluff: A Tale of the South* (Philadelphia: Peter F. Cunningham, [1873]), 305–6; Harriet Beecher Stowe, *We and Our Neighbors, or The Records of an Unfashionable Street* (New York: Fords, Howard and Hurlbert, [1875]), 151, 197; J. S. Buckingham, *America: Historical, Statistic, and Descriptive*, vol. 1 (London: Fisher, 1841), 221–22.

10. James D. McCabe, *Lights and Shadows of New York Life, or The Sights and Sensations of the Great City* (Philadelphia: National Publishing Company, 1872), 191, 295, 478, quotation at 133; James Rees, *Mysteries of City Life* (Philadelphia: J. W. Moore, 1849), 47; "The World of New York," *Putnam's Monthly Magazine*, Aug. 1856, 223; *AGLJ* 1, no. 1 (July 1, 1859): 2–3; *Sixteenth Annual Report of the Trustees of the Philadelphia Gas Works* (Philadelphia: L. R. Bailey, 1851), 36; Catherine M. Sedgwick, "Fanny McDermot," in *A New England Tale, and Miscellanies* (New York: George Putnam, 1852), 380.

11. Bader, "Gas Illumination," 345; "Important to Landlords and Gas Consumers," *AGLJ* 19, no. 3 (Aug. 2, 1873): 46; *Report of the Hearings before the Board of Mayor and Aldermen, upon the Remonstrances against the South Gasometer, and the Extension of the Works of the Gas Company at the North End* (Boston: Wright and Hasty, 1853); "Annual Report of the Superintendent of Lamps for 1876"; George Lippard, *Empire City*, 133; "Gas-Light Glare over Towns," *AGLJ* 17, no. 2 (July 16, 1872): 20.

12. Robert B. Taber, *The Advantages of Gas as an Illuminant, with a Few Hints on Gas Burners, and the Ease and Economy of Cooking and Heating by Gas* (Boston: Rockwell and Churchill, 1879), 4; "Annual Report of the Superintendent of Lamps for 1876"; Bader, "Gas Illumination," 219–26; Joel A. Tarr, "Transforming an Energy System: The Evolution of the Manufactured Gas Industry and the Transition to Natural Gas in the United States (1807–1954)," in *The Governance of Large Technical Systems*, ed. Olivier Cloutard (London: Routledge, 1999), 21, 22.

13. Seibert and Toner, *Outlines of Tracts of Land*; J. F. Cedney, *Exhibit Chart Showing Streets and Avenues of the Cities of Washington and Georgetown Improved under the Board of Public Works, D.C.* (Washington, DC: J. F. Cedney, 1873), GMLC; "Washington: Distribution of Population, 1880," New York Public Library; J. L. Lusk, *Statistical Map No. 7 Showing the Location of Street Lamps, City of Washington* (1891), GMLC; W. T. Rossell, *Statistical Map No. 1, Showing the Valuation of Real Property as Determined by the Assessment of 1887, City of Washington* (1891), GMLC; Constance McLaughlin Green, *Washington: Capital City, 1879–1950* (Princeton, NJ: Princeton University Press, 1963), 42–43, 47; James Borchert, *Alley Life in Washington: Family, Community, Religion, and Folklife in the City, 1850–1970* (Urbana: University of Illinois Press, 1980), 1–44.

14. Bouman, "Luxury and Control"; Kenneth T. Jackson, *Crabgrass Frontier: The Suburbanization of the United States* (New York: Oxford University Press, 1985), 20–44; Stuart M. Blumin, *The Emergence of the Middle Class: Social Experience in the American City, 1760–1900* (Cambridge: Cambridge University Press, 1999), 163–79, 358; Peter C. Baldwin, *Domesticating the Street: The Reform of Public Space in Hartford, 1850–1930* (Columbus: Ohio State University Press, 1999), 34–35, 39–45.

15. McCabe, *Lights and Shadows of New York Life*, 14, 398; Herman Melville, *Pierre, or The Ambiguities* (New York: Harper, 1852), 315, 322. For a different interpretation, see John F. Kasson, *Rudeness and Civility: Manners in Nineteenth-Century Urban America* (New York: Hill and Wang, 1990), 74–80.

16. Bader, "Gas Illumination," 171, 243; George Templeton Strong, *The Diary of George Templeton Strong: Young Man in New York, 1835–1849*, ed. Allan Nevins and Milton Halsey Thomas (New York: Macmillan, 1952), 150; J. H. Ingraham, "Glimpses at Gotham—No. II," *Ladies' Companion and Literary Expositor*, Feb. 1839, 177–78; Solon Robinson, *Hot Corn: Life Scenes in New York Illustrated* (New York: DeWitt and Davenport, 1854), 48; Edward Crapsey, *The Nether Side of New York, or The Vice, Crime and Poverty of the Great Metropolis* (1872; Montclair, NJ: Patterson Smith, 1969), 159; Joel H. Ross, *What I Saw in New York, or A Bird's Eye View of City Life* (Auburn, NY: Derby and Miller, 1851), 229.

17. "City Intelligence," *NYH*, Feb. 29, 1844; "More Light," *Brother Jonathan*, Sept. 16, 1843, 73; "Lighting the Streets of Brooklyn," *AGLJ* 7, no. 22 (May 6, 1866): 344; "Time Table for Lighting and Extinguishing the Gas Lamps for the Year 1861," *AGLJ* 2, no. 28 (Feb. 15, 1861):

254; Untitled article, *AGLJ*, vol. 2, no. 15 (Aug. 1, 1860): 37, col. 3; Kate Bolton, "The Great Awakening of the Night: Lighting America's Streets," *Landscape* 23, no. 3 (1979): 45; "Phases of Washington," *NYT,* Feb. 21, 1892; "Annual Report of the Superintendent of Lamps for 1876."

18. [William Richard], *The Gas-Consumer's Guide: A Handbook of Instruction on the Proper Management and Economical Use of Gas* (Boston: Alexander Moore, 1871), 25–26.

19. James F. Richardson, *The New York Police: Colonial Times to 1901* (New York: Oxford University Press, 1970), 7–11, 18; Douglas Greenberg, *Crime and Law Enforcement in Colonial New York, 1691–1776* (Ithaca, NY: Cornell University Press, 1974), 158; Selden D. Bacon, "The Early Development of American Municipal Police: A Study of the Evolution of Formal Controls in a Changing Society" (PhD diss. Yale University, 1939), 160, 327–28, 405–6, 501; "Assault and Battery—False Imprisonment," *New York City Hall Recorder,* Apr. 1819, 57; Augustine E. Costello, *Our Police Protectors: History of the New York Police from the Earliest Period to the Present Time* (New York: Augustine E. Costello, 1885), 8, 13, 22–24, 51; James F. Richardson, *Urban Police in the United States* (Port Washington, NY: Kennikat Press, National University Publications, 1974), 22–23; Edward H. Savage, *Police Records and Recollections, or Boston by Daylight and Gaslight for Two Hundred and Forty Years* (Boston: John P. Dale, 1873), 42; Charles Christian, *A Brief Treatise on the Police of the City of New York* (1812; New York: Arno Press and the New York Times, 1970), 7; Edward H. Savage, *A Chronological History of the Boston Watch and Police, from 1631 to 1865; Together with the Recollections of a Boston Police Office, or Boston by Daylight and Gaslight* (Boston: Edward H. Savage and J. E. Farwell, 1865), 49.

20. George W. Walling, *Recollections of a New York Chief of Police* (New York: Caxton Book Concern, 1887), 31–32; Howard O. Sprogle, *The Philadelphia Police, Past and Present* (Philadelphia: Howard O. Sprogle, 1887), 48; Bacon, "American Municipal Police," 74–75, 84, 328, 352, 355–56, 499; Welcome Arnold Greene, *The Providence Plantations for Two Hundred and Fifty Years* (Providence: J. A. and R. A. Reid, 1886), 114; Savage, *Boston Watch*, 20; "City Watchmen," *NYH,* Mar. 18, 1830.

21. Richardson, *New York Police,* 10; Savage, *Police Records and Recollections,* 42; Savage, *Boston Watch,* 31; Bacon, "American Municipal Police," 506–7; Costello, *Our Police Protectors,* 72–73; Walling, *Recollections,* 32; Greene, *Providence Plantations* 68, 114.

22. David R. Johnson, *Policing the Urban Underworld: The Impact of Crime on the Development of the American Police, 1800–1897* (Philadelphia: Temple University Press, 1979), 9, 15, 19, 35; *Letters from John Pintard to His Daughter, Eliza Noel Pintard Davidson, 1816–1833,* vol. 3, *1828–1831* (New York: New-York Historical Society, 1941), 209. Baltimore newspaper quotation is from Buckingham, *America,* 458.

23. Richardson, *Urban Police,* 19–22, 28; Allen Steinberg, *The Transformation of Criminal Justice: Philadelphia, 1800–1880* (Chapel Hill: University of North Carolina Press, 1989), 135–40; Costello, *Our Police Protectors,* 106, 127; Bacon, "American Municipal Police," 604, 655; "The World of Variety," *Quaker City,* May 18, 1850; Salomon de Rothschild, *A Casual View of America: The Home Letters of Salomon de Rothschild, 1859–1861* trans. and ed. Sigmund Diamond (Stanford, CA: Stanford University Press, 1961), 45.

24. Bacon, "American Municipal Police," 318; Samuel Walker, *A Critical History of Police Reform: The Emergence of Professionalism* (Lexington, MA: D. C. Heath, Lexington Books, 1977), 7; Costello, *Our Police Protectors,* 149, 294.

25. Walker, *Police Reform,* 23, 28–29; Lawrence M. Friedman, *Crime and Punishment in*

American History (New York: Basic Books, 1993), 70, 202; Theodore N. Ferdinand, "The Criminal Patterns of Boston since 1849," *American Journal of Sociology* 73, no. 1 (July 1967): 84–99; Roger Lane, "Crime and Criminal Statistics in Nineteenth-Century Massachusetts," *Journal of Social History* 2 (Winter 1968): 156–63.

26. Sprogle, *Philadelphia Police*, 142; Philadelphia Police Department, "About the Department," http://www.phillypolice.com/about; New York Police Department, "Frequently Asked Questions," http://www.nyc.gov/html/nypd/html/faq/faq.shtml; Wikipedia, "Boston Police Department," http://en.wikipedia.org/wiki/Boston_Police_Department; US Census Bureau, "American FactFinder," http://factfinder.census.gov. All websites accessed June 16, 2010.

27. "Quarterly report of the New-York Deputy Superintendent," in "Municipal," *NYT*, Feb. 13, 1858; A. E. Costello, *History of the Police Department of New Haven: From the Period of the Old Watch in Colonial Days to the Present Time* (New Haven, CT: Relief Books, 1892), 78, 112; George E. Waring Jr., US Department of the Interior Census Office, *Report on the Social Statistics of Cities, Part I, The New England and the Middle States* (1886; New York: Arno Press and the New York Times, 1970), 487.

28. Johnson, *Policing the Urban Underworld*, 103–9; Bacon, "Municipal Police," 319; *The Stranger's Guide in Philadelphia and Environs* (Philadelphia: Lindsay and Blakiston, 1854), 233–34; "Quarterly Report of the Superintendent of Police," *NYT*, Feb. 13, 1858; Walker, *Police Reform*, 20–21.

29. Howard Rock, "Independent Hours: Time and the Artisan in the New Republic," in *Worktime and Industrialization: An International History*, ed. Gary Cross (Philadelphia: Temple University Press, 1988); Paul E. Johnson, *A Shopkeeper's Millennium: Society and Revivals in Rochester, New York, 1815–1837* (New York: Hill and Wang, 1978), 55–60; Roy Rosenzweig, *Eight Hours for What We Will: Workers and Leisure in an Industrial City, 1870–1920* (Cambridge: Cambridge University Press, 1985).

30. John C. Schneider, "Public Order and the Geography of the City: Crime, Violence, and the Police in Detroit, 1845–1875," *Journal of Urban History* 4, no. 2 (Feb. 1978), 186–88; Walker, *Police Reform*, 15–16; Steinberg, *Criminal Justice*, 237; Johnson, *Policing the Urban Underworld*, 110–11, 128; Costello, *Our Police Protectors*, 116; "Ruffianism," *Philadelphia Evening Bulletin*, Nov. 30, 1857; "City Bulletin: Desperate Outrage," *Philadelphia Evening Bulletin*, Dec. 29, 1857.

31. Johnson, *Policing the Urban Underworld*, 5–6, 89, 152–61; McCabe, *Lights and Shadows of New York*, 480, 595, 603.

32. Edward Winslow Martin [James D. McCabe], *The Secrets of the Great City: A Work Descriptive of the Virtues and the Vices, the Mysteries, Miseries and Crimes of New York City* (Philadelphia: Jones Brothers, 1868), 46–47, 303–4; *Diary of George Templeton Strong*, 150, 217–18; Anbinder, *Five Points*, 213–14.

33. Robert M. Fogelson, *Downtown: Its Rise and Fall, 1880–1950* (New Haven, CT: Yale University Press, 2001), 14–32; Schneider, "Public Order and the Geography of the City," 185; James D. McCabe, *New York by Sunlight and Gaslight: A Work Descriptive of the Great American Metropolis* (Philadelphia: Douglass Brothers, 1882), 153, 379.

34. Schneider, "Public Order and the Geography of the City," 198; Johnson, *Policing the Urban Underworld*, 107; Ernest Ingersoll, "The Police of New York," *Scribner's Monthly*, July 1878, 353.

35. E. Porter Belden, *New-York: Past, Present, and Future* (New York: Prall and Lewis, 1850), 48; Ingersoll, "Police of New York," 353; John Bell Bouton, *Round the Block* (New York: Appleton, 1864), 288; "A New Boston Notion," *Wellman's Miscellany*, Mar. 1872, 99; Henry Morford, *Shoulder-Straps* (Philadelphia: T. B. Peterson, 1863), 28; Palmer Cox, *Squibs of California, or Every-day Life, Illustrated* (Hartford, CT: Mutual Publishing Company, 1874), 394; Allan Pinkerton, *Claude Melnotte as a Detective, and Other Stories* (Chicago: W. B. Keen, Cooke, 1875), 52; "Corner Loafing along Chestnut Street," *PPL*, July 19, 1859; David M. Henkin, *City Reading: Written Words and Public Spaces in Antebellum New York* (New York: Columbia University Press, 1998), 42–43; "Reform Backwards," *Philadelphia North American*, Sept. 20, 1855.

36. *The Tricks and Traps of New York City, Part 1* (New York: Dinsmore, 1858), 45–47; Johnson, *Policing the Urban Underworld*, 5, 78–79, 129–30; Costello, *Our Police Protectors*, 77; J. W. Buel, *Mysteries and Miseries of America's Great Cities: Embracing New York, Washington City, San Francisco, Salt Lake City, and New Orleans* (St. Louis, MO: Historical Publishing Company, 1883), 41; "The Prey of the 'Owls,'" *Pittsburg Times*, Nov. 29, 1886.

37. Johnson, *Policing the Urban Underworld*, 111–12; "More Police Protection Needed," *Pittsburg Times*, Apr. 28, 1886; John J. Flinn, *History of the Chicago Police, from the Settlement of the Community to the Present Time* (Chicago: Police Book Fund, 1887), 403; Sidney L. Harring, *Policing a Class Society: The Experience of American Cities* (New Brunswick, NJ: Rutgers University Press, 1983), 50–51.

38. Bacon, "Municipal Police," 299; Christian, *Brief Treatise on the Police*, 29; Sprogle, *Philadelphia Police*, 90–97; Johnson, *Policing the Urban Underworld*, 29; Peter Baldwin, "Becoming a City of Homes: The Suburbanization of Cranston, 1850–1910," *Rhode Island History* 51, no. 1 (Feb. 1993): 3–22.

39. "Lamps and Gas," *NYT*, Jan. 6, 1852; "Shantyboat Squatters," *Pittsburg Times*, Nov. 29, 1886; John Ernest McCann, "A Night with the River Police," *HW*, Jan. 23, 1892, 86; Walling, *Recollections*, 139–52; Crapsey, *Nether Side of New York*, 36–41; Collins, *Consolidated Gas Company*, 91; Johnson, *Policing the Urban Underworld*, 108–9; Sprogle, *Philadelphia Police*, 154–55, 283, 623, 627.

40. Strong, *Young Man in New York*, 320; "Garroters and Footpads," *Porter's Spirit of the Times*, Feb. 28, 1857; "Gas Reform," *Globe*, Oct. 21, 1872; "Cities as Gas-Makers," *Journal of Commerce*, Mar. 20, 1873; "'Dowsed Glims,'" *NYH*, Apr. 8, 1873; Brown, "Gas Light Company of Baltimore," 30; Bacon, "American Municipal Police," 461; Costello, *Our Police Protectors*, 105.

41. Tarr, "Transforming an Energy System"; Biggs, "Gas and Gas-Making"; "Gas," *Philadelphia Morning Post*, July 17, 1868; "The Gasmen's Strike," *NYT*, Apr. 6, 1873; "No Gas," *PPL*, July 18, 1868; *Report of the Commissioners Appointed to Investigate the Cause and Management of the Great Fire in Boston* (Boston, 1873), 308; "The Gas Famine," *New York World*, Apr. 7, 1873.

42. It was not the very first. A gasworks explosion in New York in 1835 interrupted service for at least a day. Also, an 1847 flood had shut Cincinnati's gasworks for over a week. "Explosion at the Gas Works," *Albany Evening Journal*, Dec. 26, 1835; "The Flood," *Daily Cincinnati Commercial*, Dec. 15, 1847; untitled item, *Daily Cincinnati Commercial*, Dec. 25, 1847.

43. "Serious Fire—Destruction of the New York Gas Works—the Lower Part of the City

in Darkness," *NYH*, Aug. 31, 1848; "Burning of the New York Gas Company's Establishment," *NYTr*, Aug. 31, 1848; "A Dark Night," *NYTr*, Aug. 31, 1848; "The New York Gas Light Works," *NYH*, Sept. 1, 1848.

44. "Stokers and Helpers at the Gas Works on a Strike," *PPL*, July 17, 1868; "The Strike at the Philadelphia Gas Works," *AGLJ* 10, no. 3 (Aug. 3, 1868): 36; "The Gas Strike," *PI*, July 18, 1868.

45. "Utter Darkness: The Quaker City in Eclipse," *Philadelphia Evening Star*, July 18, 1868; "The Troubles at the Gas Works," *Philadelphia Press*, July 18, 1868. For examples of how people greeted the first gas lighting of a city or town, see Wallace Rice, *75 Years of Gas Service in Chicago* (Chicago: People's Gas Light and Coke, 1925), 7; "A 'Quietly Happy Crowd,'" *AGLJ* 19, no. 9 (Nov. 3, 1873): 153. For an example of an initial response to electric arc lighting, see Harold L. Platt, *The Electric City: Energy and the Growth of the Chicago Area, 1880–1930* (Chicago: University of Chicago Press, 1991), 3. See also "Strike at the Philadelphia Gas Works"; "Philadelphia in Darkness," *PPL*, July 18, 1868; "No Gas," *PPL*, July 18, 1868; "A City in Darkness," *PI*, July 18, 1868; "This Morning's News," *PI*, July 20, 1868; "Gas Works Management," *PPL*, July 20 1868.

46. Jakle, *City Lights*, 40–52, 64–65; Platt, *Electric City*, 40, 46.

47. Philip M. Katz, *From Appomattox to Montmartre: Americans and the Paris Commune* (Cambridge, MA: Harvard University Press, 1998); Iver Bernstein, *The New York City Draft Riots: Their Significance for American Society and Politics in the Age of the Civil War* (New York: Oxford University Press, 1990), 17–40; Adrian Cook, *The Armies of the Streets: The New York City Draft Riots of 1863* (Lexington: University Press of Kentucky, 1974), 194; [William O. Stoddard], *The Volcano under the City* (New York: Fords, Howard and Hulbert, 1887), 200, 204, 207; Wolfgang Schivelbusch, *Disenchanted Night: The Industrialization of Light in the Nineteenth Century* (Berkeley: University of California Press, 1995), 97–114; untitled editorial, *AGLJ* 5, no. 87 (Aug. 1, 1863): 40.

48. "Communists in New York," *NYT*, June 7, 1871; "Yesterday's Disturbances," *NYT*, July 13, 1871; "'The Dangerous Classes,'" *NYT*, July 16, 1871. A third major blackout, in December 1871, stirred fears that looters would attack affluent neighborhoods in Midtown Manhattan; Peter C. Baldwin, "In the Heart of Darkness: Blackouts and the Social Geography of Lighting in the Gaslight Era," *Journal of Urban History* 30, no. 5 (July 2004): 749–68.

49. *Report of the Commissioners Appointed to Investigate*, passim; "The Great Calamity!" *BG*, Nov. 12, 1872; Christine Meisner Rosen, The *Limits of Power: Great Fires and the Process of City Growth in America* (Cambridge: Cambridge University Press, 1986), 177–79; [William Flint and Company], *History of the Great Conflagration, or Boston and Its Destruction* (Philadelphia: William Flint, 1872), 52; "Devastation!" *BG*, Nov. 11, 1872.

50. "Devastation!" *BG*, Nov. 11, 1872; Chandler and Company, *Full Account of the Great Fire in Boston! And the Ruins* (Boston: W. H. Chandler, 1872), 4; Russell H. Conwell, *History of the Great Fire in Boston, November 9 and 10, 1872* (Boston: B. B. Russell, 1873), 216–18; *Report of the Commissioners Appointed to Investigate*, 327.

51. "Devastation!" *BG*, Nov. 11 1872; Flint, *History of the Great Conflagration*, 46; Chandler, *Full Account of the Great Fire*, 21.

52. "Devastation!" *BG*, Nov. 11, 1872; "The Great Calamity," *Boston Journal*, Nov. 11, 1872; *Report of the Commissioners Appointed to Investigate*, xiv, 109–10, 308, 316, 329, 331, 332, 334;

Charles E. French diary, vol. 11, entry dated Nov. 11, 1872, MHS; "The Boston Fire," *NYT*, Nov. 12, 1872; "The Boston Fire," *HW*, Nov. 30, 1872, 934; "The Second Fire Extinguished," *NYTr*, Nov. 12, 1872; "After the Fire," *NYTr*, Nov. 13, 1872; untitled editorial, *Boston Evening Transcript*, Nov. 11, 1872, 2, col. 6; "The Roughs Abroad," *BG*, Nov. 12, 1872; "Miscellaneous Incidents," *Boston Evening Transcript*, Nov. 12, 1872; "No Gas," *Boston Evening Transcript*, Nov. 12, 1872; "Rain on the Ruins," *Boston Daily Advertiser*, Nov. 13, 1872; [Charles Carlton Coffin], *The Story of the Great Fire*, (Boston: Shepard and Gill, 1872), 29; William Gray Brooks diary, MHS, entry dated Nov. 13, 1872; "The Third Day," *BG*, Nov. 13, 1872; "The City Quiet," *BG*, Nov. 13, 1872.

53. "In Darkness," *NYT*, Apr. 7, 1873; "A Dark Deed," *NYH*, Apr. 7, 1873; "A City in the Shadow," *NYTr*, Apr. 7, 1873; "The Gas Famine," *New York World*, Apr. 7, 1873; "The Reign of Darkness," *NYT*, Apr. 7, 1873; "The Gasmen's Strike," *NYTr*, Apr. 8, 1873; "'Dowsed Glims,'" *NYH*, Apr. 8, 1873; "The Gas Strike," *NYT*, Apr. 8, 1873; "The Gasmen's Strike," *NYTr*, Apr. 9, 1873; "Politics," *Atlantic Monthly* 32 (July 1873): 127–28.

54. "The Gas-Men's Strike," *HW*, Apr. 26, 1873.

CHAPTER THREE

1. Descriptions of the open landscape appear in David R. Foster, *Thoreau's Country: Journey through a Transformed Landscape* (Cambridge, MA: Harvard University Press, 1999), 9, 16–22. Descriptions of the fall of darkness are based in part on Henry D. Thoreau, *The Journal of Henry D. Thoreau*, ed. Bradford Torrey and Francis F. Allen (New York: Dover, 1962), e.g., entries dated Aug. 18, 1841, Dec. 27, 1851, Jan. 7, Apr. 11, July 20, 1852. On evening chores and night work on farms, A. B. Cole, "Management of Milch Cows," *Western Farmer* 1, no. 10 (June 1840): 284; Thoreau, *Journal*, vol. 2 (Aug. 5, 1851), 371 (renumbered as 237); "Care of Tools," *Cultivator* 5, no. 8 (Aug. 1857): 237; Jack Larkin, *The Reshaping of Everyday Life, 1790–1840* (New York: Harper and Row, 1988), 21–22; Karen V. Hansen, *A Very Social Time: Crafting Community in Antebellum New England* (Berkeley: University of California Press, 1994), 83, 106–9; A. Roger Ekirch, *At Day's Close: Night in Times Past* (New York: W. W. Norton, 2005), 168–71.

2. The preceding two paragraphs are based on multiple sources. Michael O'Malley, *Keeping Watch: A History of American Time* (Washington, DC: Smithsonian Institution Press, 1990), 38–40; Henry David Thoreau, *Walden* (1854), in *The Portable Thoreau*, ed. Carl Bode (New York: Penguin Books, 1981), 364; Abigail [Abby D. Turner], "A Merrimack Reverie," *LO* 1, no. 2 (Dec. 1840): 29–30; Ella [Harriet Jane Farley], "A Weaver's Reverie," *LO*, n.s., 1, no. 6 (Aug. 1841): 188–90; [Elizabeth Emerson Turner], "Factory Girl's Reverie," *LO* 5, no. 6 (June 1845): 140; "A Week in the Mill," *LO* 5, no. 10 (Oct. 1845): 217–18; Lucy Larcom, *A New England Girlhood, Outlined from Memory* (1889; Williamstown, MA: Corner House, 1985), 153–54, 182; Josiah Curtis, *Brief Remarks on the Hygiene of Massachusetts* (Philadelphia: T. K. and P. G. Collins, 1849); J. L. B. [Josephine L. Baker], "A Second Peep at Factory Life," *LO* 5, no. 5 (May 1845): 97–100; William Scoresby, *American Factories and Their Female Operatives* (London: Longman, Brown, Green and Longmans, 1845); "Questions for Ministers, Physicians, Agents, Overseers, Operatives, Boarding-House Keepers, &c.," *Voice of Industry*, June 26, 1846; Mary, "A Wail from the Factory: By a Female Operative, Aged 15," *Voice of Indus-*

try, June 26, 1846; Herman Melville, "Bartleby, the Scrivener: A Story of Wall-Street" (1853), http://www.bartleby.com/129/; Thomas Augst, *The Clerk's Tale: Young Men and Moral Life in Nineteenth-Century America* (Chicago: University of Chicago Press, 2003), 83, 207–15; "Public Meeting to Promote the Seven o'Clock Closing of the Stores," *NYTr,* Nov. 22, 1850; T. DeWitt Talmage, "After Midnight," in *The Abominations of Modern Society* (New York: Adams, Victor, 1872), 61. Quotation from Susan [Harriet Jane Farley], "Letters from Susan: Letter Second," *LO* 4, no. 8 (June 1844): 171; in determining the authorship and proper citation information for *Lowell Offering* articles, I have relied on Judith Ranta, *The Lowell Offering Index,* Center for Lowell History, University of Massachusetts at Lowell, http://library.uml.edu/clh/Offering .htm (July 17, 2006).

3. Howard Rock, "Independent Hours: Time and the Artisan in the New Republic," in *Worktime and Industrialization: An International History,* ed. Gary Cross (Philadelphia: Temple University Press, 1988), 25–27; James Montgomery, *A Practical Detail of the Cotton Manufacture of the United States of America* (Glasgow: John Niven, 1840; repr., New York: Augustus M. Kelley, 1969), 175; "Mechanics Working Hours," *New England Artisan,* Feb. 2, 1832; "To the Merchants of the City of Boston," *New England Artisan,* June 21, 1832; "Almy and Brown's Acct. with Spinning Mill," Slater, Almy and Brown records, vol. 1, Slater Collection, Baker Library, Harvard Business School.

4. Theodore Steinberg, *Nature Incorporated: Industrialization and the Waters of New England* (Amherst: University of Massachusetts Press, 1991), 53; Thomas Dublin, *Women at Work: The Transformation of Work and Community in Lowell, Massachusetts, 1826–1860* (New York: Columbia University Press, 1979), 19–21; Thomas Dublin, *Transforming Women's Work: New England Lives in the Industrial Revolution* (Ithaca, NY: Cornell University Press, 1994), 83; *Statistics of Lowell Manufactures, January 1, 1835: Compiled from Authentic Sources* ([Lowell, 1835]); *Statistics of Lowell Manufactures, January, 1848: Compiled from Authentic Sources* (Lowell: James Atkinson, [1848]); *Statistics of Lowell Manufactures, January, 1850: Compiled from Authentic Sources* (Lowell: James Atkinson, [1850]); *Statistics of Lowell Manufactures, January, 1851: Compiled from Authentic Sources,* ([Lowell]: S. J. Varney, 1850); Charles Cowley, *Illustrated History of Lowell,* rev. ed. (Boston: Lee and Shepard, 1868), 140.

5. "Statement of Facts," *New England Artisan,* Mar. 22, 1832; Caroline F. Ware, *The Early New England Cotton Manufacture: A Study in Industrial Beginnings* (Boston: Houghton Mifflin, 1931), 249–50.

6. Testimony of Robert Kerr, May 12, 1837, in *Report of the Select Committee Appointed to Visit the Manufacturing Districts of the Commonwealth for the Purpose of Investigating the Subject of the Employment of Children in Manufactories* (Harrisburg, PA: Thompson and Clark, 1838), 18; "Statement of Facts," *New England Artisan,* Mar. 22, 1832; David R. Roediger and Philip S. Foner, *Our Own Time: A History of American Labor and the Working Day* (London: Verso, 1989), 51; S. G. B. [Sarah Bagley], in *Voice of Industry,* Jan. 16, 1846, reprinted in Philip Sheldon Foner, *The Factory Girls: A Collection of Writings on Life and Struggles in the New England Factories of the 1840s* (Urbana: University of Illinois Press, 1977), 225; Montgomery, *Practical Detail of the Cotton Manufacture,* 173–74; "Time Table of the Lowell Mills, to Take Effect on and after Oct. 21st, 1851," in O'Malley, *Keeping Watch,* 53; "Time Table of the Holyoke Mills, to Take Effect January 2d, 1854," AAS.

7. "Farmer's Evenings," *Genesee Farmer* 6, no. 44 (Oct. 29, 1836): 350; "Winter Evenings,"

New England Farmer 6, no. 1 (Jan. 1854): 38–39; quotation from Factory Operatives of Delaware County, Penna., *Address from the Factory Operatives of Delaware County, Pennsylvania, to Their Operative Brethren in the States of Massachusetts, Rhode Island, Connecticut, New Hampshire, Vermont, and Maine* (Delaware County, PA, 1847); Roediger and Foner, *Our Own Time*, 51; "Celebrating the 'Blow Out,'" *LVP*, Mar. 12, 1847; "Blowing Out Balls," *LVP*, Mar. 23, 1849; "Lighting Up," *LVP*, Sept. 21, 1849; "The Ten Hour System," *Voice of Industry*, Oct. 9, 1846; O'Malley, *Keeping Watch*, 3–7, 39–41, 82; "The Longitude of Manufacturing Villages," *New England Artisan*, Jan. 26, 1832; quotation from "New England Association of Farmers, Mechanics, and Other Working Men," *New England Artisan*, Feb. 23, 1832; An Overseer, "Children in Factories," *New England Artisan*, June 14, 1832; A Mechanic, "Facts for Operatives," *Voice of Industry*, Feb. 27, 1846; R., letter in *Voice of Industry*, Mar. 26, 1847, in Foner, *Factory Girls*, 88.

8. Josiah Curtis, *Brief Remarks on the Hygiene of Massachusetts but More Particularly of the Cities of Boston and Lowell* (Philadelphia: T. K. and P. G. Collins, 1849), 30–34; Eliza Hemmingway, Feb. 13, 1845, testimony, in John R. Commons, *Labor Movement, 1840–1860*, 2:135 (vol. 8 of John R. Commons, Ulrich B. Phillips, Eugene A. Gilmore, Helen L. Sumner, and John B. Andrews, *A Documentary History of American Industrial Society* [Cleveland: Arthur H. Clark, 1910]); "Growth of the Factory System," *Voice of Industry*, June 26, 1846; Ware, *Early New England Cotton Manufacture*, 69, 249–50; Teresa Murphy, "Work, Leisure, and Moral Reform: The Ten-Hour Movement in New England, 1830–1850," in Cross, *Worktime and Industrialization*.

9. Larcom, *New England Girlhood*, 199, 248; H. F. [Harriet Jane Farley], "Editorial: Two Suicides," *LO*, 4, no. 9 (July 1844): 214; Archibald Prentice, *A Tour in the United States* (London: Charles Gilpin, 1848), 125; Baker, "Second Peep at Factory Life," 99.

10. Elisha Bartlett, *A Vindication of the Character and Condition of the Females Employed in the Lowell Mills* (Lowell, MA: Leonard Huntress, 1841), 6–7; Henry A. Miles, *Lowell as It Was, and as It Is* (Lowell: Powers and Bagley and N. L. Dayton, 1845), 67–70, 128, 144–45, 187–91; Dublin, *Women at Work*, 77–79, 83–85, 143–44; An Observer, "Corporation Boarding-Houses," *Lowell Courier*, Aug. 25, 1842; Prentice, *Tour in the United* States, 122; Dublin, *Transforming Women's Work*, 89; *General Regulations to Be Observed by Persons Employed by the Lawrence Manufacturing Company, in Lowell* ([Lowell], 1833); Joseph Sturge, *A Visit to the United States in 1841* (London: Hamilton, Adams, 1842), 142; Capt. (Frederick) Marryat, *Second Series of a Diary in America* (Philadelphia: T. K. and P. G. Collins, 1840), 115.

11. Dublin, *Women at* Work, 80; Scoresby, *American Factories*, 25–26; Sturge, *Visit*, 142; A Lowell Factory Girl, letter in *Voice of Industry*, Apr. 17, 1846, in Foner, *Factory Girls*, 120; *Harbinger*, Nov. 14, 1846, in John R. Commons and Helen L. Sumner, *Labor Movement, 1820–1840*, 2:134–35 (vol. 6 of John R. Commons, Ulrich B. Phillips, Eugene A. Gilmore, Helen L. Sumner, and John B. Andrews, *A Documentary History of American Industrial Society* [Cleveland: Arthur H. Clark, 1910]); Lucy Larcom, *An Idyl of Work*, 2nd ed. (Boston: James R. Osgood, 1876), 95–96; Miles, *Lowell*, 67–69; Charles Richard Weld, *A Vacation Tour in the United States and Canada* (London: Longman, Brown, Green and Longmans, 1855), 52; [Harriet Jane Farley], "Editorial: Home in a Boarding-House," *LO* 3, no. 3, (Dec. 1842): 69–70; [Harriet Jane Farley], "Letters from Susan: Letter First," *LO*, May 1844, 146. A day in the mill left a worker

with "a moist, unpleasant body," but the boardinghouses lacked bathing facilities; [Harriet Jane Farley], "Editorial—Health," *LO* 3, no. 8 (May 1843), 190–92.

12. *Voice of Industry*, Jan. 16, 1846, quoted in Foner, *Factory Girls*, 224–25; Miles, *Lowell*, 71; F. G. A., "Susan Miller," *LO*, n.s. 1, no. 6 (Aug. 1841): 169; A Working Woman, letter in the *Boston Daily Evening Voice*, Feb. 23, 1867, quoted in Foner, *Factory Girls*, 343; "Questions for Ministers . . . &c.," *Voice of Industry*, June 26, 1846.

13. Lucinda [Harriet Jane Farley], "Evening before Pay-Day," *LO*, n.s., 1, no. 8 (Oct. 1841): 243–44; Scoresby, *American Factories*, 28–29; Josiah Curtis, *Brief Remarks*, 36; "Letters from Susan: Letter First"; Patrick Shirreff, *A Tour through North America* (Edinburgh: Oliver and Boyd, 1835), 45; Weld, *Vacation Tour*, 51, 53; "Ladies Look Out," *LVP*, May 4, 1849.

14. *Compendium of the Enumeration of the Inhabitants and Statistics of the United States as Obtained at the Department of State from the Returns of the Sixth Census* (Washington, DC: Thomas Allen, 1841), 8–10; *Statistics of Lowell Manufactures, January 1, 1840, Compiled from Authentic Sources* (Lowell: L. Huntress, [1840]); William Barry, *The Moral Exposure and Spiritual Wants of Manufacturing Cities* (Lowell, 1850), 4; G. W. Boynton, *Plan of the City of Lowell*, ca. 1845, in Miles, *Lowell*; John Coolidge, *Mill and Mansion: Architecture and Society in Lowell, Massachusetts, 1820–1865* (1942; Amherst: University of Massachusetts Press, 1993), 33–39; figs. 3, 89–90; "Hold!" *LVP*, June 15, 1849; "Keep to the Right!" *LVP*, Dec. 2, 1842.

15. "Soap Lock Rowdies," *LVP*, Jan. 15, 1842; untitled editorials, *LVP*, Jan. 15, 1842; "Mysteries of Lowell," *LVP*, Dec. 4, 1846; "Rowdyism," *LVP*, June 15, 1849; Argus, *Norton, or The Lights and Shades of a Factory Village* (Lowell: Vox Populi, 1849), 15–16, 18; Anglice, "Disturbers of Meetings," *LVP*, Oct. 8, 1842; Thomas L. Nichols, *Forty Years of American Life* (London: John Maxwell, 1864), 1:108–9; "A View of Lowell," *LVP*, Sept. 24, 1842; untitled item, *LVP*, Nov. 25, 1842; Quilp, letter in *LVP*, Sept. 8, 1843.

16. Almira, "The Spirit of Discontent," *LO* 1, no. 4 (July 1841): 112; L. T. H., "A Letter to Cousin Lucy, *LO* 5, no. 5 (May 1845): 111; Harriet H. Robinson, *Loom and Spindle, or Life among the Early Mill Girls* (1898; Kailua, HI: Press Pacifica, 1976), 25, 61; advertisements, *Lowell Courier*, July 7, 1842; "Editorial—Health," *LO* 3, no. 8 (May 1843): 192.

17. Larcom, *New England Girlhood*, 252; Miles, *Lowell*, 205; Teresa Anne Murphy, *Ten Hours' Labor: Religion, Reform, and Gender in Early New England* (Ithaca, NY: Cornell University Press, 1992), 60; Donald M. Scott, "The Popular Lecture and the Creation of the Public in Mid-Nineteenth-Century America," *Journal of American History* 66, no. 4 (Mar. 1980): 791–809; "Lectures and Lecturers," *Putnam's Monthly* 9 (Mar. 1857): 317–21; "Lowell Institute," *LVP*, Sept. 21, 1849; Testimony of Elizabeth Rowe, Massachusetts House Document No. 50 (Mar. 1845), in Robinson, *Loom and Spindle*, 139; "Letters from Susan: Letter Second"; Susan E. P. Brown Forbes diary, vol. 2 (1843), AAS, entry for Feb. 8, 1843; advertisements and articles in *Lowell Courier*, Sept. 6, 10, 17, Nov. 8, 26, Dec. 20, 1842; "Lowell Institute," *LVP*, Sept. 24, 1842; Poem by "Philo Musicus," *LVP*, Jan. 8, 1847.

18. "No Theatre in Lowell," *LO*, n.s., 1, no. 3 (Feb. 1841): 48; Cowley, *Illustrated History of Lowell*, 120–21; Fair Play, letter in *LVP*, Jan. 29, 1842; advertisement, *LVP*, Apr. 23, 1842; advertisement, *Lowell Courier*, Oct. 29, 1842; Harriet Farley, *Operatives' Reply to Hon. Jere. Clemens* (Lowell, MA: S. J. Varney, 1850), 15; "The Lowell Museum—Its Rise and Progress," *LVP*, Nov. 30, 1849; "Lowell Museum," *Lowell Advertiser*, Jan. 12, 1850; advertisement in *Low-*

ell Daily Journal and Courier, Nov. 1, 1848; William B. English, *Rosina Meadows, the Village Maid, or Temptations Unveiled* (Boston: Redding, 1843); Greenhorn, "The Absurdities of the Drama," *Life in Boston and New England Police Gazette*, May 18, 1850.

19. Forbes diary (1843); advertisement, *LVP*, Mar. 31, 1843.

20. Augusta, letter in *LVP*, Jan. 1, 1842; untitled items, *LVP*, Jan. 29 and Mar. 19, 1842.

21. Farley, *Operatives' Reply*, 15; "A Mechanic," letter in *Voice of Industry*, Apr. 3, 1846.

22. "The Moralist—No. 2," *LVP*, Apr. 30, 1847; Argus, *Norton*, 4, 46; see also Osgood Bradbury, *Mysteries of Lowell* (Boston: E. P. Williams, 1844); "Mysteries of Lowell," *LVP*, Dec. 4, 1846; "Nell Hathaway, or Moonlight Glimpses of the Darker Side of the 'Spindle City,'" *LVP*, Oct. 5, 19, 26, 1849.

23. Bartlett, *Vindication*; Scipio, letter in *Life in Boston, Sporting Chronicle, and Lights and Shadows of New England Morals*, Sept. 1, 1849; Harriet Farley, "Editorial: Conclusion of the Volume," *LO* 5, no. 12 (Dec. 1845): 282. On suspicions that the curfew was not properly enforced, see also An Observer, "Corporation Boarding-Houses," *Lowell Courier*, Aug. 25, 1842.

24. *Examination of Dr. William Graves, before the Lowell Police Court, from Sept. 25 to Sept. 29, 1837* ([Lowell, 1837]); "French Periodical Pills," *LVP*, Apr. 30, 1847; "Important to the Ladies," *Lowell Advertiser*, Jan. 12, 1850.

25. "Extract of a Letter from a Clergyman," *Voice of Industry*, Aug. 14, 1846.

26. Roediger and Foner, *Our Own Time*, 22–33; Murphy, *Ten Hours' Labor*, 63–65, 155, 164–90; "Hours of Labor," *New England Artisan*, Jan. 26, Feb. 2, 1832; *Boston Courier*, Sept. 27, 1836, in John R. Commons and Helen L. Sumner, *Labor Movement, 1820–1840*, 2:47 (vol. 6 of John R. Commons, Ulrich B. Phillips, Eugene A. Gilmore, Helen L. Sumner, and John B. Andrews, *A Documentary History of American Industrial Society* [Cleveland: Arthur H. Clark, 1910]); ten-hour circular from the *Man*, May 13, 1835, in ibid., 96–97.

27. Joseph F. Kett, *Rites of Passage: Adolescence in America, 1790 to the Present* (New York: Basic Books, 1977), 95; Bruce Dorsey, *Reforming Men and Women: Gender in the Antebellum City* (Ithaca, NY: Cornell University Press, 2002), 103–6; Allan Stanley Horlick, *Country Boys and Merchant Princes: The Social Control of Young Men in New York* (Lewisburg, PA: Bucknell University Press, 1975), 147–68; John Todd, *The Moral Influence, Dangers and Duties, Connected with Great Cities* (Northampton, MA, 1841); [Helen C. Knight], *Reuben Kent's First Winter in the City* (Philadelphia: American Sunday School Union, 1845); Patricia Cline Cohen, *The Murder of Helen Jewett: The Life and Death of a Prostitute in Nineteenth-Century New York* (New York: Vintage Books, 1999).

28. Paul Boyer, *Urban Masses and Moral Order in America, 1820–1920* (Cambridge, MA: Harvard University Press, 1978), 67; Campbell Gibson, "Population of the 100 Largest Cities and Other Urban Places in the United States: 1790 to 1990," Population Division Working Paper 27, US Bureau of the Census, June 1998, http://www.census.gov/population/www/documentation/twps0027.html; Richard B. Stott, *Workers in the Metropolis: Class, Ethnicity and Youth in Antebellum New York* (Ithaca, NY: Cornell University Press, 1990), 7–11; Edwin G. Burrows and Mike Wallace, *Gotham: A History of New York City to 1898* (New York: Oxford University Press, 1999), 446, 656–59; Kenneth T. Jackson, *Crabgrass Frontier: The Suburbanization of the United States* (New York: Oxford University Press, 1985), 25–33.

29. Burrows and Wallace, *Gotham*, 649, 656, 661–62, 873–75; Jacob Judd, "Brooklyn's

Changing Population in the Pre–Civil War Era," *Journal of Long Island History* 4, no. 2 (1964): 9–10; Ron Miller, Rita Seiden Miller and Stephen J. Karp, "The Fourth Largest City in America: A Sociological History of Brooklyn," in *Brooklyn USA: The Fourth Largest City in America*, ed. Rita Seiden Miller (New York: Columbia University Press, 1979), 17–19; Henry R. Stiles, *A History of the City of Brooklyn*, vol. 2 (Brooklyn, 1869), 292–94; Douglas V. Shaw, "The Making of an Immigrant City: Ethnic and Cultural Conflict in Jersey City, New Jersey, 1850–1877" (PhD diss., University of Rochester, 1972), 7, 11, 30–31; "Cities of Hoboken and Jersey City, Townships of West Hoboken and Weehawken and Town of Union," in *State Atlas of New Jersey*, ed. F. W. Beers (New York: Beers, Comstock and Cline, 1872); "Jersey City, State of New Jersey," *Ballou's Pictorial Drawing-Room Companion*, May 23, 1857; W. L. Alden, "From Pig to Pork," *Galaxy*, Dec. 15, 1866; "Table 10, Population of the 100 Largest Urban Places: 1870," in Gibson, "Population of the 100 Largest Cities."

30. Burrows and Wallace, *Gotham*, 726; Jacob Judd, "A City's Streets: A Case Study of Brooklyn, 1834–1855," *Journal of Long Island History* 9, no. 1 (1969): 32–33; Stiles, *History of the City of Brooklyn*, 497; "The Early Closing Movement," *BE*, Sept. 27, 1866; "Jersey City," *Ballou's Pictorial;* Shaw, "Jersey City," 33. Some of the storekeepers and clerks undoubtedly worked in New York.

31. *Compendium . . . from the Returns of the Sixth Census*, 123; Stott, *Workers in the Metropolis*, 92; Robert Ernst, *Immigrant Life in New York City, 1825–1863* (New York: King's Crown Press, 1949), 215.

32. Burrows and Wallace, *Gotham*, 656–59, 668, 943–45; Robert Greenhalgh Albion, *The Rise of New York Port, 1815–1860* (1939; New York: Charles Scribner's Sons, 1970), 260–80; Horlick, *Country Boys and Merchant Princes*, 31; Stott, *Workers in the Metropolis*, 210.

33. Walter Barrett [Joseph A. Scoville], *The Old Merchants of New York* (New York: Carleton, 1863), 57, 195; John Todd, *The Moral Influence, Dangers and Duties, Connected with Great Cities* (Northampton, MA, 1841), 126; "The Want of Employment," *NYTr*, Sept. 3, 1842; Horlick, *Country Boys and Merchant Princes*, 110–11, 168–69; Cohen, *Murder of Helen Jewett*, 11.

34. A Bachelor, "Boarding-Houses," *New-York Mirror*, Dec. 15, 1832; "The Perils of Pearl Street," *New-York Mirror*, Mar. 15, 1834; "Cheap Boarding and True Love," *New Mirror*, Oct. 21, 1843; "Dry-Goods Clerks—Early Closing," *New York Recorder*, Mar. 23, 1850; Helen Mar, "Boarding-Houses," *Herald of Health* 9, no. 4 (Apr. 1867): 158; *The Temptations of City Life: A Voice to Young Men Seeking a Home and Fortune in Large Towns and Cities* (New York: Edward H. Fletcher, 1849), 13; Thomas Butler Gunn, *The Physiology of New York Boarding-Houses* (New York: Mason Brothers, 1857); Stott, *Workers in the Metropolis*, 216; Stephen M. Griswold, *Sixty Years with Plymouth Church* (New York: Fleming H. Revell, 1907), 18–19.

35. *Temptations of City Life*, 31; Fisher, *Three Great Temptations of Young Men* (Cincinnati: Moore and Anderson, 1852), 11–12, 185, 194; Rev. Daniel C. Eddy, *The Young Man's Friend* (1855; Boston: Wentworth, Hewes, 1859), 157; Todd, *Moral Influence*, 227; "City Clerks," *Advocate of Moral Reform* 2, no. 5 (Mar. 1, 1836): 33.

36. Griswold, *Sixty Years*, 21, 179; Michael Floy Jr., *The Diary of Michael Floy Jr., Bowery Village, 1833–1837*, ed. Richard Albert Edward Brooks (New Haven, CT: Yale University Press, 1941), 193; Augst, *Clerk's Tale*, 62, 73–76.

37. Burrows and Wallace, *Gotham*, 777; Augst, *Clerk's Tale*, 158–75; Ralph Foster Weld, *Brooklyn Village, 1816–1834* (1938; New York: AMS Press, 1970), 194–99, 233–39, 253–54; Paul

Boyer, *Urban Masses and Moral Order in America, 1820–1920* (Cambridge, MA: Harvard University Press, 1978), 12–17; Stuart M. Blumin, *The Emergence of the Middle Class: Social Experience in the American City, 1760–1900* (Cambridge: Cambridge University Press, 1989), 192–215; Rev. Daniel C. Eddy, *The Young Man's Friend* (1855; Boston: Wentworth, Hewes, 1859), 88, 93–98; "A Learned Locality," *NYT*, Apr. 13, 1853.

38. Stott, *Workers in the Metropolis*, 206, 215–16; Warren Burton, *Moral Dangers of the City: To the Clergymen of Various Denominations throughout New England* (Boston: John Wilson, 1848); Fisher, *Three Great Temptations*, 92; quotation from *Temptations of City Life*, 14; "Adventures of a Bachelor," *New-York Mirror*, Dec. 29, 1832; Timothy J. Gilfoyle, *City of Eros: New York City, Prostitution, and the Commercialization of Sex, 1790–1920* (New York: W. W. Norton, 1992), 97.

39. Stott, *Workers in the Metropolis*, 74, 205–9; Cohen, *Murder of Helen Jewett*, 11.

40. Horlick, *Country Boys and Merchant Princes*, 54–55; Stott, *Workers in the Metropolis*, 206; Gilfoyle, *City of Eros*, 31–54; Burrows and Wallace, *Gotham*, 474; George G. Foster, *New York by Gas-Light and Other Urban Sketches*, ed. Stuart M. Blumin (1850; Berkeley: University of California Press, 1990), 69–76.

41. Cohen, *Murder of Helen Jewett*, 302, 328; Gilfoyle, *City of Eros*, 97–103, at 103.

42. "The Public Morals," *BE*, Feb. 19, 1842.

43. Fanny Trollope, *Domestic Manners of the Americans* (1832; London: Penguin Books, 1997), 273; "Hours of Store Keepers," *Working Man's Advocate*, Aug. 28, 1830; Augst, *Clerk's Tale*, 83–85, at 83.

44. "Hours of Store Keepers," *Working Man's Advocate*, Aug. 28, 1830; "Closing Stores," *BE*, Dec. 4, 1850; untitled item, *BE*, Sept. 10, 1842; "Early Closing," *BE*, Feb. 11, 1863; Marten Estey, "Early Closing: Employer-Organized Origin of the Retail Labor Movement," *Labor History* 13, no. 4 (Fall 1972): 560–70, quotation at 562; "Closing Stores at Night," *BE*, Sept. 9, 1843.

45. "Dry Goods Stores," *NYTr*, Nov. 27, 1849; "Public Meeting to Promote the Seven o'Clock Closing of the Stores," *NYTr*, Nov. 22, 1850; "Jersey City: Early Closing of Stores," *NYT*, Oct. 29, 1853; untitled items, *PPL*, May 25, June 1, 1850; "Early Closing," *Hartford Courant*, July 22, 1863; "Early Closing Movement," *BE*, Apr. 28, 1863; "The Early Closing Movement," *NYT*, June 17, 1864; "The Early-Closing Movement," *BE*, Oct. 26, 1850; "The Early Closing Movement," *BE*, Aug. 9, 1866; "The Early Closing Movement—Grand Torchlight Procession in Williamsburgh," *BE*, May 20, 1863.

46. Daniel T. Rodgers, *The Work Ethic in Industrial America, 1850–1920* (Chicago: University of Chicago Press, 1978), 94–99, at 96.

47. "Dry Goods Clerks—Early Closing," *New York Recorder*, Mar. 23, 1850; "Early Closing of Stores," *NYT*, Sept. 19, 1853; quotation from "The Early Closing Movement," *BE*, Oct. 26, 1850; "The Early Closing Movement," *BE*, Apr. 21, 1863; "Clerks' Congratulatory Meeting," *NYT*, Mar. 14, 1863.

48. A Fulton Avenue Merchant, "Wants Store Clerks Kept at Work till Bed-Time, to Keep them out of Mischief," *BE*, Aug. 5, 1869; untitled item, *BE*, Sept. 10, 1842; "Dry Goods Stores," *NYTr*, Nov. 27, 1849; quotation from "Closing Stores at Night," *BE*, Sept. 9, 1843.

49. "Glorious Demonstration of the Dry Goods Clerks!" *BE*, Nov. 22, 1850; "Early Closing Movement," *BE*, Apr. 21, 1863; "The Early Closing Movement," *NYT*, June 17, 1864.

50. "Closing Stores at Night," *BE,* Sept. 9, 1843; "The Early-Closing Movement," *BE,* Oct. 26, 1850; J. N., "The Early Closing Humbug," *BE,* Feb. 2, 1860.

51. J. N., "The Early Closing Humbug," *BE,* Feb. 2, 1860; Independent, "Early Closing," *BE,* Apr. 16, 1863; "Early Closing: The Movement Fizzling Out," *BE,* Oct. 27, 1863; "The Early Closing Movement," *BE,* Jan. 12, 1864; Mrs. Jones, "Close Them Up," *Journal of United Labor* 7, no. 1 (May 10, 1886): 2067; "Early Closing on Saturday," *NYT,* Aug. 2, 1878.

52. "Early Closing," *Journal of United Labor* 7, no. 8 (Aug. 25, 1886): 2152; "Grocery Clerks," *BE,* May 24, 1880; "Grocers Falling into Line," *Pittsburgh Commercial Gazette,* Apr. 14, 1886; "Laboring Men's Complaints," *NYT,* Nov. 9, 1885; "Early Closing," *Cincinnati Daily Gazette,* May 22, 1879; "Long Hours for Drug Clerks," *NYT,* Aug. 19, 1880; "The Carpet and Furniture Clerks," *BE,* July 16, 1869; "Seeking Early Hours," *NYT,* June 20, 1881; "The Boss Shavers," *Pittsburgh Commercial Gazette,* Mar. 26, 1886; "The Workingmen's Eight Hour Convention," *NYT,* June 23, 1867; "They Break the Windows," *NYT,* June 28, 1888; "Early Closing," *BE,* Jan. 23, 1887; "Early Closing on Saturday," *NYT,* Aug. 2, 1878; Susan Porter Benson, *Counter Cultures: Saleswomen, Managers, and Customers in American Department Stores, 1890–1940* (Urbana: University of Illinois Press, 1986), 25.

53. "Woman's Wrongs," *NYT,* July 29, 1870; "Saleswomen Worse off Than Salesmen," *BE,* June 6, 1872; "Consideration for Shop Women," *NYT,* Jan. 7, 1876; "Asking for Shorter Hours," *NYT,* July 28, 1886; "No Pity for Salesgirls," *NYT,* Mar. 24, 1895.

54. John F. Kasson, *Amusing the Million: Coney Island at the Turn of the Century* (New York: Hill and Wang, 1978), 4–8. See also chapter 9 below.

CHAPTER FOUR

1. Walt Whitman, "New York Dissected: IV. Broadway" (1856), in *New York Dissected: A Sheaf of Recently Discovered Newspaper Articles by the Author of "Leaves of Grass"* (New York: Rufus Rockwell Wilson, 1936), 119–22.

2. The development of middle-class consciousness is explored in Stuart M. Blumin, *The Emergence of the Middle Class: Social Experience in the American City, 1760–1900* (Cambridge: Cambridge University Press, 1989), and Karen Halttunen, *Confidence Men and Painted Women: A Study of Middle-Class Culture in America, 1830–1870* (New Haven, CT: Yale University Press, 1982). On deceptive appearances, see "Swells, the Incubi of Society," and "Broadway, New York, at 3 p.m.," *Day and Night,* Feb. 1869. Whitman's description of Broadway bore similarities to earlier ones by Nathaniel Parker Willis ("Broadway," *NYTr,* Apr. 5, 1843, reprinted from *National Intelligencer*) and Caroline M. Kirkland (*The Evening Book, or Fireside Talk on Morals and Manners, with Sketches of Western Life* [New York: Charles Scribner, 1851], 148–52). A description of workers walking home appears in "The Shop Girls of New York," *Broadway Belle and New-York Shanghai,* Sept. 3, 1855. The daily cycle of crime is described in "Dodgers' Exposition," *NPG,* May 23, 1857. Descriptions focusing on night schedules include "A Night Out," *NYT,* June 17, 1853; "Letters from the Railway Switchman, Number Four: The Night Side of Broadway," *NYT,* Nov. 8, 1852; Harrison Gray Buchanan, *Asmodeus, or Legends of New York; Being a Complete Exposé of the Mysteries, Vices and Doings, as Exhibited by the Fashionable Circles of New York* (New York: John D. Munson, 1848); Henry Morgan, *Ned Nevins, the Newsboy, or Street Life in Boston* (Boston: Lee and Shepard, 1866), 173–80; W. O.

Stoddard, "Bowery, Saturday Night," *Harper's New Monthly Magazine* 42 (Apr. 1871): 671–72; James Dabney McCabe, *Lights and Shadows of New York Life, or The Sights and Sensations of the Great City* (Philadelphia: National, 1872), 300; John H. Warren Jr., *Thirty Years' Battle with Crime, or The Crying Shame of New York* (1875; New York: Arno Press and the New York Times, 1970), 353–57; Julian Ralph, "The City of Brooklyn," *Harper's New Monthly Magazine*, Apr. 1893, 651–52; Stephen Crane, "In the Broadway Cars" (1902), in *The New York Sketches of Stephen Crane and Related Pieces*, ed. R. W. Stallman and E. R. Hagemann (New York: New York University Press, 1966), 185–89; and Stephen Graham, *With Poor Immigrants to America* (New York: Macmillan, 1914), 73.

3. George G. Foster, *New York by Gas-Light and Other Urban Sketches* (1850; Berkeley: University of California Press, 1990), 189; "Saturday Night," *NYTr*, Sept. 17, 1853.

4. Donald J. Bruggink, "Talmage, Thomas DeWitt," in *American National Biography*, vol. 21, ed. John A. Garraty and Mark C. Carnes (New York: Oxford University Press, 1999), 286–87; "Brother Talmage Goes to See the Elephant," *Puck* 4 (Oct. 2, 1878): 8–9; T. DeWitt Talmage, "After Midnight," in *The Abominations of Modern Society* (New York: Adams, Victor, 1872), 59–78, at 61–62.

5. Talmage, *Abominations*, 63, 73.

6. J. S. Buckingham, *America: Historical, Statistic and Descriptive* (London: Fisher, 1841) 2:72, 2:180–81, 2:328–29, 3:371–72, 3:485; J. S. Buckingham, *The Eastern and Western States of America* (London: Fisher, 1842), 1:103, 1:125–26, 3:26, 3:31; Robert H. Collyer, *Lights and Shadows of American Life* (1838; Boston: Brainard, 1843), 38; Emilie Marguerite Cowell, *The Cowells in America, Being the Diary of Mrs. Sam Cowell during Her Husband's Concert Tour in the Years 1860–1861*, ed. M. Willson Disher (London: Oxford University Press, 1934), 14, 15, 19, 155–58, 212.

7. John D. Vose, *Fresh Leaves from the Diary of a Broadway Dandy* (New York: Bunnell and Price, 1852), 70, 72; McCabe, *Lights and Shadows*, 133–34; Emily Thornwell, *The Lady's Guide to Perfect Gentility* (New York: Derby and Jackson, 1859), 78; Arthur Martine, *Martine's Hand-Book of Etiquette and Guide to True Politeness* (New York: Dick and Fitzgerald, 1866), 119, 131; S. A. Frost, *Frost's Laws and By-Laws of American Society* (New York: Dick and Fitzgerald, 1869), 94; "Brooklyn by Gas Light," *BE*, Nov. 19, 1850; "City News and Gossip," *BE*, June 12, 1867; Jane Briggs Smith Fiske diary, AAS, entries for Mar. 4, Apr. 1, May 5, June 23, Aug. 6, Sept. 8, 1871; George Ellington, *The Women of New York, or Social Life in the Great City* (New York: New York Book Company, 1870), 33.

8. "The World of New York," *Putnam's Monthly Magazine*, June 1856, 660; Fiske diary, e.g., Jan. 25, Feb. 2, 5, Apr. 23, May 5, and passim; Mary Bainard Smith diary for 1894, PHS, Jan. 7, 9–12, 14, 19, 21, 24, and passim; Frances Trollope, *Domestic Manners of the Americans*, ed. Donald Smalley (1832; Gloucester, MA: Peter Smith, 1974), 74, 110; Felix, "Night Meetings," *Boston Recorder and Telegraph* 11 (June 23, 1826): 100; "Ninth and Spruce," *Philadelphia Press*, Jan. 29, 1869; Michael Floy Jr., *The Diary of Michael Floy Jr., Bowery Village, 1833–1837*, ed. Richard Albert Edward Brooks (New Haven, CT: Yale University Press, 1941), e.g., 32; George Templeton Strong, *The Diary of George Templeton Strong: The Turbulent Fifties, 1850–1859*, ed. Allan Nevins and Milton Halsey Thomas (New York: Macmillan, 1952), 28, 82, 147, 197, 233; quotation from "Letters from the Railway Switchman, Number Four: The Night Side of Broadway," *NYT*, Nov. 8, 1852.

9. Paul E. Johnson, *A Shopkeeper's Millennium: Society and Revivals in Rochester, New York, 1815–1837* (New York: Hill and Wang, 1978); Mary P. Ryan, *Cradle of the Middle Class: The Family in Oneida County, New York, 1790–1865* (Cambridge: Cambridge University Press, 1981), 47–48; George Watson Cole diary for 1875, AAS, entries for Mar. 31, Apr. 10, 24, 25, May 1, 2, 8, 15, 16, 22, 23, 26; Helen R. Deese, ed., *Selected Journals of Caroline Healey Dall*, vol. 1, *1838–1855* (Boston: Massachusetts Historical Society, 2006), 27.

10. Blumin, *Emergence of the Middle Class*, 192–229; Bruce Dorsey, *Reforming Men and Women: Gender in the Antebellum City* (Ithaca, NY: Cornell University Press, 2002); Floy, *Diary*, e.g., 212–13.

11. Henry Morgan, *Ned Nevins*, 229–30; [William A. Alcott], *The Young Man's Guide* (Boston: Lilly, Wait, Colman and Holden, 1834), 225; A Citizen, "To Residents of Brooklyn," *NYTr*, Sept. 27, 1843; "Brooklyn Young Men's Christian Association," *NYTr*, Oct. 26, 1853; "New-York Young Men's Christian Association," *NYTr*, Feb. 25, 1863; "Lectures and Lecturers," *Putnam's Monthly Magazine* 9 (Mar. 1857): 317–21, quotation at 320; Donald M. Scott, "The Popular Lecture and the Creation of the Public in Mid-Nineteenth-Century America," *Journal of American History* 66, no. 4 (Mar. 1980): 791–809; R. Laurence Moore, *Selling God: American Religion in the Marketplace of Culture* (New York: Oxford University Press, 1994), 56–60; [Eliza Ware Rotch Farrar], *The Young Lady's Friend* (Boston: American Stationers' Company, 1836), 317; Florence Hartley, *The Ladies' Book of Etiquette and Manual of Politeness* (Boston: Locke and Bubier, 1875), 176; C. S., "Correspondence in New York," *Christian Observer*, Apr. 1, 1848, 14; Cumings journal, MHS, Feb. 26, 1833; George Templeton Strong, *The Diary of George Templeton Strong: Young Man in New York, 1835–1849*, ed. Allan Nevins and Milton Halsey Thomas (New York: Macmillan, 1952), 16; advertisement in *Albion*, Oct. 23, 1837; "A Learned Locality," *NYT*, Apr. 13, 1853; Robert Macoy, *Illustrated How to See New York and Its Environs* (New York: Robert Macoy, 1876), 60, 101.

12. The information in this paragraph is pieced together from the Nathan S. Beekley diary, AAS, and many other sources. Particularly relevant are the diary entries for Jan. 2–4, Feb. 6, 17, Apr. 8, June 8, 26, July 30 to Aug. 9, Oct. 22, Nov. 15, Dec. 21, 1849. Also helpful were the following: Edward P. Cody Sr., "Additions to the Genealogy—DeHaven Family: Data Showing Hookup of the Biddle-Beekley-Cody Families to the DeHaven Family" (1948), AAS; William J. Buck, *History of Montgomery County with the Schuylkill Valley* (Norristown, PA: E. L. Acker, 1859), 65, 83–86; J. Smith Futhey and Gilbert Cope, *History of Chester County, Pennsylvania, with Genealogical and Biographical Sketches* (Philadelphia: Louis H. Everts, 1881), 350, 708; Ellwood Roberts, ed., *Biographical Annals of Montgomery County, Pennsylvania* (New York: T. S. Benham, 1904), 1:385; Chester County Historical Society and County of Chester, "Poor School Children, 1810–1842" (2004), http://dsf.chesco.org/archives/site/default.asp; population schedules of the 1850, 1860, and 1870 US Census for Pennsylvania, via Ancestry.com; *McElroy's Philadelphia Directory for 1850* (Philadelphia: Edward C. and John Biddle, 1850), 344; *McElroy's Philadelphia Directory for 1855* (Philadelphia: Edward C. and John Biddle, 1855), 32; *Gopsill's Philadelphia City Directory for 1875* (Philadelphia: James Gopsill, 1875), 161; Death notice for Nathan S. Beekley, *PI*, Oct. 18, 1877; G. W. Colton, "Philadelphia," in *Colton's Atlas of the World* (New York: J. H. Colton, 1856).

13. Beekley diary, entries for Jan. 3, 13, Feb. 26, Apr. 18, May 10, 29, Aug. 14, 31, Oct. 2, Dec. 3, 1849; Theodore Hershberg et al., "The 'Journey to Work': An Empirical Investiga-

tion of Work, Residence, and Transportation, Philadelphia, 1850 and 1880," in *Philadelphia: Work, Space, Family and Group Experience in the Nineteenth Century,* ed. Theodore Hershberg (Oxford: Oxford University Press, 1981), 128–73.

14. Diary entries for Jan. 7, 16, Mar. 22, Nov. 27, Dec. 31, 1849. Information about friends, churches, and entertainment spots can be gleaned from the diary and from population schedules for the 1850 US Census for Philadelphia, via Ancestry.com; *McElroy's Philadelphia Directory for 1850* (Philadelphia: Edward C. and John Biddle, 1850), 351, 497–502, and the map of ward boundaries on the verso of the title page. The street addresses can be located using *Bywater's Philadelphia Business Directory and City Guide for the Year* 1850 (Philadelphia: Maurice Bywater, 1850), 171–74, together with *Colton's Atlas*, pl. 21.

15. Beekley diary, Nov. 2, 1849; John F. Kasson, *Rudeness and Civility: Manners in Nineteenth-Century Urban America* (New York: Hill and Wang, 1990), 173–76; Halttunen, *Confidence Men and Painted Women*, 102–8; *True Politeness, or Etiquette for Ladies and Gentlemen* (New York, 1848), 18; *The Illustrated Manners Book: A Manual of Good Behavior and Polite Accomplishments* (New York: Leland Clay, 1855), 124; Henry Lunettes [Margaret C. Conkling], *The American Gentleman's Guide to Politeness and Fashion* (New York: Derby and Jackson, 1858), 157–62; Farrar, *Young Lady's Friend*, 220–22; Sarah J. Hale, *Manners, or Happy Homes and Good Society All the Year Round* (1868; New York: Arno Press, 1972), 217–24; Frost, *Frost's Laws and By-Laws*, 30–37.

16. *The Handbook of the Man of Fashion* (Philadelphia: Lindsay and Blakiston, 1846), 198–99; Frost, *Frost's Laws and By-Laws*, 31; *Etiquette at Washington: Together with the Customs Adopted by Polite Society in the Other Cities of the United States*, 7th ed. (Baltimore: Murphy, 1860), 50; Cecil B. Hartley, *The Gentlemen's Book of Etiquette and Manual of Politeness* (Boston: G. W. Cottrell, 1860), 77–78, quotation at 78; Kirkland, *Evening Book*, 81.

17. Hartley, *Gentlemen's Book of Etiquette*, 222–26; *Etiquette at Washington*, 107; Alcott, *Young Man's Guide*, 58; Frost, *Frost's Laws and By-Laws*, 51–54, 61, 73; Cindy R. Lobel, "Consuming Classes: Changing Food Consumption Patterns in New York City, 1790–1860" (PhD diss., City University of New York, 2003), 145–46; Hale, *Manners*, 220–24; Beekley diary, Jan. 11, 22, Feb. 20, 21, Nov. 3, 12, Dec. 17, 26, 1849; Strong, *Turbulent Fifties*, 212, 333, 390; Anna Quincy Thaxter Cushing diary, AAS, vol. 4, Jan. 15, 25, Feb. 21, 1847; vol. 5, Oct. 26, Nov. 11, 1847; vol. 15, Dec. 14, 1857, Mar. 3, 1858; Halttunen, *Confidence Men and Painted Women*, 174–86. George W. Cole described semi-spontaneous parties at his boardinghouse: Cole diary, May 6, 19, 26, 1875.

18. "Places of Public Amusement: Theatres and Concert Rooms," *Putnam's Monthly Magazine,* Feb. 1854, 149; Mary C. Henderson, *The City and the Theatre: New York Playhouses from Bowling Green to Times Square* (Clifton, NJ: James T. White, 1973), 87; Louis Bader, "Gas Illumination in New York City, 1823–1863" (PhD diss., New York University, 1970), 145–47; J. H. Ingraham, "Glimpses at Gotham—No. II," *Ladies' Companion and Literary Expositor,* Feb. 1839, 178; Edward Winslow Martin [James Dabney McCabe], *The Secrets of the Great City* (Philadelphia: Jones Brothers, 1868), 255; McCabe, *Lights and Shadows*, 133; quotation from Foster, *New York by Gas-Light*, 71.

19. Richard Butsch, *The Making of American Audiences: From Stage to Television, 1750–1990* (Cambridge: Cambridge University Press, 2000), 25–33, 45; Charles Blake, *An Historical*

Account of the Providence Stage (Providence, RI: George H. Whitney, 1868), 25, 46; Buckingham, *America: Historical, Statistic and Descriptive*, 1:372; George G. Foster, *New York Naked* (New York: DeWitt and Davenport, [1855]), 143.

20. "Early American Imprints: Series I, Evans," playbills for performances in New York, Feb. 20, 1752, Nov. 12, 1753, Feb. 19, Mar. 19, 1787, Dec. 12, 1791, Oct. 4, 1797; playbills for performances in Baltimore, Oct. 1, 11, 15, 18, Nov. 15, Dec. 26, 1782, Jan. 17, 31, Feb. 18, Mar. 7, 28, 31, May 30, June 3, 9, 1783, Aug. 24, 1795; playbills for performances in Boston, Feb. 3, 1794, Mar. 20, May 27, 1799; playbills for performances in Providence, Sept. 16, 1795, and Hartford, Oct. 15, 1799; newspaper advertisements in *New-York Gazette and General Advertiser*, Jan. 23, Nov. 27, 1798, Dec. 6, 1799; *Baltimore Daily Intelligencer*, Jan. 4, Oct. 17, 1794; *Newport Mercury*, May 20, June 3, July 1, Aug. 26, 1794; Lynne Conner, *Pittsburgh in Stages: Two Hundred Years of Theater* (Pittsburgh: University of Pittsburgh Press, 2007), 8; Joseph N. Ireland, *Records of the New York Stage from 1750 to 1860* (1866; New York: Benjamin Blom, 1966), 1:3-56, 129, 174.

21. "Early American Imprints Series II (Shaw-Shoemaker)," playbill for Boston Theatre, Nov. 7, 1803; playbill for New Theatre, New York, Oct. 18, 1806; playbill for Chestnut Street Theatre, Philadelphia, Dec. 5, 1814, HTC; advertisements in *Franklin Gazette*, Philadelphia, Sept. 14, Dec. 15, 1819, Mar. 18, July 17, Oct. 18, Nov. 21, 1820; *Boston Daily Advertiser*, Jan. 6, 1820; *Baltimore Patriot*, Apr. 26, 1820, Nov. 30, 1824, Oct. 16, 1826, Jan. 12, June 6, 1829; *Providence Patriot*, Jan. 21, 24, 28, Feb. 14, Mar. 4, June 24, 29, 1829.

22. "Museum" theatres and garden theaters experimented in the 1840s with performances at 8:15 or 8:30, but soon dropped this innovation. Examples appear in *NYH* advertisements, July 31, 1840, May 19, June 11, 30, 1841, June 20, July 12, 1845; advertisements in *BS*, July 22, Aug. 16, 1848, July 14, 1852; playbills for Barnum's American Museum, June 26, 1843, June 7, 1847, HTC.

23. Playbills for Chestnut Street Theatre, Philadelphia, various dates 1814-24, 1840, 1849-50, HTC; playbills for Walnut Street Theatre, Philadelphia, various dates 1835-39, 1849, 1860, 1863-66, 1870-73, HTC; playbills for Park Theatre, New York, various dates 1841-44, HTC; playbills for Chatham Garden Theatre, New York, various dates 1825-26, 1844-46, HTC; playbills for National Theatre, New York, various dates 1837-39, HTC; playbills for National Varieties Theatre, New York, various dates, 1877-78, HTC; playbills for Arch Street Theatre, Philadelphia, various dates, 1872-78, HTC; advertisements in the *NYH*, Dec. 9, 22, 29, 1840, Jan. 4, 11, 1841; *NYT*, Jan. 20, Mar. 16, Apr. 24, June 22, Sept. 21, Dec. 13, 1860, Jan. 2, 5, 1875, June 17, July 10, 1876; *PI*, June 16, Dec. 29, 1874, Jan. 11, Dec. 17, 1875, Jan. 4, 7, 1876; *CIO*, Mar. 14, 25, 1874, Jan. 5, Feb. 12, 22, 1875, Oct. 14, 23, 1876, Jan. 30, 1877, Feb. 6, Mar. 17, 1877.

24. Mary C. Henderson, "Scenography, Stagecraft, and Architecture in the American Theatre, Beginnings to 1870," in *The Cambridge History of American Theatre*, vol. 1, *Beginnings to 1870*, ed. Don B. Wilmeth and Christopher Bigsby (Cambridge: Cambridge University Press, 1998), 393; playbills for Chestnut Street Theatre, Philadelphia, Dec. 5, 1814, Dec. 4, 1816, Dec. 10, 26, 1817, Dec. 30, 1818, Jan. 2, Feb. 3, 1819, Dec. 2, 1822, Jan. 10, 1823, HTC; Sol Smith, *Theatrical Management in the West and South for Thirty Years* (1868; New York: Benjamin Blom, 1968), 124; schedule of sunset from *Bywater's Philadelphia Business Directory and City Guide for the Year 1850*, 4-26. Sunset times were roughly consistent for cities across the same

234 *Notes to Pages 64–68*

latitude in this era before time zones and daylight saving time. Then as now, northern cities saw slightly greater seasonal variation in daylight than did southern cities, but the difference was negligible within the east-west urbanized belt of the Northeast and Midwest.

25. *BS*, May 19, 1852; *NYT*, Dec. 18, 1860.

26. Advertisements in *Spirit of the Times*, May 22, Nov. 6, 1841, Dec. 14, 1844, Sept. 20, Dec. 6, 14, 1845, Oct. 21, 1848; *Albion*, Mar. 4, Oct. 28, Nov. 25, Dec. 9, 1843, Feb. 17, Mar. 2, 23, Oct. 19, Dec. 21, 1844; *NYH*, Jan. 4, 1844, Jan. 16, 1845, Jan. 31, Feb. 5, 1848; *NYTr*, Nov. 29, 1849; *NYT*, Sept. 18, 20, Oct. 6, 11, Nov. 14, Dec. 1, 10, 27, 30, 1851, Jan. 27, Mar. 8, 16, Apr. 17, Sept. 21, Dec. 18, 1852, Jan. 16, 1861; *BS*, Oct. 15, Dec. 15, 1851, May 19, 1852, Mar. 17, 1855; *Hartford Courant*, Sept. 1, 1840, Dec. 15, 1860, Jan. 2, 1866; Delmer D. Rogers, "Public Music Performances in New York City from 1800 to 1850," *Anuario Interamericano de Investigación Musical* 6 (1970): 5–50; Robert C. Toll, *Blacking Up: The Minstrel Show in Nineteenth-Century America* (New York: Oxford University Press, 1974), 31–38; Karen Ahlquist, *Democracy at the Opera: Music, Theater, and Culture in New York City, 1815–1860* (Urbana: University of Illinois Press, 1997), xii, 42–55, 77–89, 143–44; Katherine K. Preston, *Opera on the Road: Traveling Opera Troupes in the United States, 1825–1860* (Urbana: University of Illinois Press, 1993), 13, 103, 121–22, 176, 198, 240–41, 255, 303, 308; playbills, Niblo's Garden (New York), Feb. 27, May 15 to Sept. 15, 1843, HTC; playbills, Palmo's Opera House (New York), Mar. 7, 13, 1844, HTC. (Even at the exclusive Italian Opera House at Astor Place, the curtain rose at 7:00 or 7:30 in the late 1840s and early 1850s; see advertisements in *Albion*, Feb. 14, 1852, and in the *NYH* and *NYT*, above.)

27. Hale, *Manners*, 222. On earlier schedules in rural America, see also "The Early Closing Movement," *BE*, Jan. 12, 1864; Buckingham, *Eastern and Western States of America*, 1:126; compare Buckingham, *America: Historical, Statistic and Descriptive*, 1:58.

28. Blake, *Historical Account of the Providence Stage*, 96–106; "Pleasures of the Theatre," *New York Evangelist*, Apr. 18, 1844; Samuel W. Fisher, *The Three Great Temptations of Young Men* (Cincinnati: Moore and Anderson, 1852), 138–39; George C. D. Odell, *Annals of the New York Stage* (New York: Columbia University Press, 1927–49), 4:246, 4:404, 4:412, 4:497, 4:634, 4:643; "Fashionable Amusements," *Christian Observer*, Nov. 7, 1845; Ahlquist, *Democracy at the Opera*, 18–19; Tracy Patch Cheever journal, MHS, Feb. 3, 1853.

29. Lawrence W. Levine, *Highbrow/Lowbrow: The Emergence of Cultural Hierarchy in America* (Cambridge, MA: Harvard University Press, 1988), 25–30; Butsch, *American Audiences*, 36–38; Blake, *Providence Stage*, 237; Claudia D. Johnson, "That Guilty Third Tier: Prostitution in Nineteenth-Century American Theaters," in *Victorian America*, ed. Daniel Walker Howe (Philadelphia: University of Pennsylvania Press, 1976), 111–20; "Bowery Theatre—the Third Tier—Prostitution Before and Behind the Curtain," *New York Sporting Whip*, Feb. 4, 1843; "The Chatham Theatre," *New York Sporting Whip*, Feb. 18, 1843.

30. Richard Stott, *Jolly Fellows: Male Milieus in Nineteenth Century America* (Baltimore: Johns Hopkins University Press, 2009), 90–92; Quincy excerpt in *The Articulate Sisters: Passages from Journals and Letters of the Daughters of President Josiah Quincy of Harvard University*, ed. M. A. DeWolfe Howe (Cambridge, MA: Harvard University Press, 1946), 218.

31. Buckingham, *America: Historical, Statistic and Descriptive*, 1:47–48, 2:72, 2:328–29; Buckingham, *Eastern and Western States of America*, 1:103; Capt. [John W.] Oldmixon, *Transatlantic Wanderings, or A Last Look at the United States* (1855; New York: Johnson Reprint

Corporation, 1970), 182–83; Bruce A. McConachie, *Melodramatic Formations: American Theatre and Society, 1820–1870* (Iowa City: University of Iowa Press, 1992), 157–60; Conner, *Pittsburgh in Stages*, 25; "Immorality of Theatres," *Advocate of Moral Reform*, Nov. 15, 1838; "The Third Tier," *Weekly Rake*, Sept. 24, 1842; "Correspondence of the Whip," *Whip and Satirist of New-York and Brooklyn*, May 7, 1842. Some theaters had separate entrances for the third tier seats, in a marginally successful effort to keep prostitutes apart from decent ladies; e.g., "Theatrical Notices," *Spirit of the Times*, Nov. 7, 1840; Johnson, "That Guilty Third Tier," 117.

32. Butsch, *American Audiences*, 44–52; Walt Whitman, "The Old Bowery: A Reminiscence of New York Plays and Acting Fifty Years Ago," in *Collected Writings: Prose Works 1892*, vol. 2, *Collect and Other Prose* (New York: New York University Press, 1964), 595; "Places of Public Amusement: Theatres and Concert Rooms," *Putnam's Monthly Magazine*, Feb. 1854, 150; quotation from Foster, *New York by Gas-Light*, 155–56.

33. *Pen and Sword: The Life and Journals of Randal W. McGavock*, ed. Herschel Gower and Jack Allen (Nashville: Tennessee Historical Commission, 1959), 140; Beekley diary, Oct. 6, 1849; Francis J. Grund, *The Americans in Their Moral, Social and Political Relations* (London: Longman, Rees, 1837), 1:120; *A Philadelphia Perspective: The Diary of Sidney George Fisher Covering the Years 1834–1871*, ed. Nicholas B. Wainwright (Philadelphia: Historical Society of Pennsylvania, 1967), 34; Cumings journal, vol. 2, Apr. 8, 1833; Matthew Hale Smith, *Counsels Addressed to Young Women, Young Men, Young Persons in Married Life and Young Parents* (Washington, DC: Blair and Rives, 1846), 32; Daniel C. Eddy, *The Young Man's Friend* (Boston: Wentworth, Hewes, 1859), 107–10; Fisher, *Three Great Temptations of Young Men*, 182; John Todd, *The Moral Influence, Dangers and Duties, Connected with Great Cities* (Northampton, MA, 1841), 204; T. DeWitt Talmage, *Sports That Kill* (New York: Harper, 1875), 21–26; "Pleasures of the Theatre," *New York Evangelist*, Apr. 18, 1844; Henry Ward Beecher, *Lectures to Young Men, on Various Important Subjects* (Boston: John P. Jewett, 1852), 235–36.

34. Ahlquist, *Democracy at the Opera*, 1–3; Charles Haynes Haswell, *Reminiscences of New York by an Octogenarian (1816 to 1860)* (New York: Harper, 1896), 50; Odell, *Annals of the New York Stage*, 4:246; playbills for Park Theatre, Jan. 1, 1844, and various dates 1844–45, HTC; playbills for Bowery Theatre, various dates, 1827–36, 1843–45, HTC.

35. "A Visit to a Theatre," *Spirit of the Times*, May 15, 1847; Conner, *Pittsburgh in Stages*, 1; "Reform in Theatres," *Spirit of the Times*, Aug. 21, 1852; Trollope, *Domestic Manners of the Americans*, 271; facsimile of an Oct. 12, 1841, playbill in William G. B. Carson, *Managers in Distress: The St. Louis Stage, 1840–1844* (St. Louis: St. Louis Historical Documents Foundation, 1949), opposite 172.

36. William B. Wood, *Personal Recollections of the Stage* (Philadelphia: Henry Carey Baird, 1855), 461; Hiram Mattison, *Popular Amusements: An Appeal to Methodists* (New York: Carlton and Porter, [1867]), 80; Talmage, *Sports That Kill*, 42–43; Joshua Bates, "Theatrical Amusements," *Happy Home and Parlor Magazine*, Oct. 1, 1856, 168; "The Devil's Sieve," *Christian Advocate and Journal*, Sept. 13, 1849, 146; "The Theatre," *Christian Review*, Sept. 1837, 401–2; Todd, *Moral Influence*, 238.

37. Wood, *Personal Recollections*, 461; "Theatrical Reform," *NYTr*, Sept. 11, 1843; "Theatre Going Boys," *New York Evangelist*, Jan. 27, 1859; Fisher, *Three Great Temptations of Young Men*, 182; H. Daniel, "Dangers of the Theatre," *Happy Home and Parlor Magazine*, Aug. 1, 1859, 134.

38. Kasson, *Rudeness and Civility*, 222–28; Ahlquist, *Democracy at the Opera*, 134–42; Foster, *New York Naked*, 36–44.

39. Butsch, *American Audiences*, 71–74; Neil Harris, *Humbug: The Art of P. T. Barnum* (Boston: Little, Brown, 105–8; Odell, *Annals of the New York Stage*, 5:56–60, 5:137–39; Bruce A. McConachie, *Melodramatic Formations: American Theatre and Society, 1820–1870* (Iowa City: University of Iowa Press, 1992), 157–76, 200–203; playbills for Barnum's American Museum, July 9, 1853, and various dates 1844–57, HTC; "Theatrical Progression," *Saturday Evening Post*, Dec. 23, 1848; "Niblo's Saloon and Garden," *New York Mirror*, July 6, 1839, 15; playbill for Niblo's Garden, June 5, 1844, HTC.

40. Rosemarie K. Bank, *Theatre Culture in America, 1825–1860* (Cambridge: Cambridge University Press, 1997), 50; Bruce McConachie, "American Theatre in Context, from the Beginnings to 1870," in *Cambridge History of American Theatre*, 1:147–64, 1:174; Johnson, "That Guilty Third Tier," 114–15; Olive Logan, *Before the Footlights and Behind the Scenes: A Book about "The Show Business" in All Its Branches* (Philadelphia: Parmelee, 1870), 308; John Vacha, *Showtime in Cleveland: The Rise of a Regional Theater Center* (Kent, OH: Kent State University Press, 2001), 11–12; Conner, *Pittsburgh in Stages*, 41; Dale Cockrell, *Demons of Disorder: Early Blackface Minstrels and Their World* (Cambridge: Cambridge University Press, 1997), 58; Foster, *New York Naked*, 146.

41. Barnum advertised performance times as late as 8:30 during the summer of 1843; playbills, Barnum's American Museum, June 26, 1843, Jan. 27, June 25, 1845, Sept. 17, 1849, Nov. 7, 1850, HTC; advertisement in *NYT*, Sept. 5, 1865; "The Amphitheatre," *NYH*, Jan. 29, 1841; [Solomon Southwick], *Five Lessons for Young Men, by a Man of Sixty* (Albany, NY: Alfred Southwick, 1837), 74; "Reform in Theatres," *Spirit of the Times*, Aug. 21, 1852, 321; "Theatres and Things Theatrical," *Wilkes' Spirit of the Times*, Jan. 14, 1865, 320; "Theatrical Reform," *NYTr*, Sept. 11, 1843; Wood, *Personal Recollections*, 437; Carson, *Managers in Distress*, 190; Levine, *Highbrow/Lowbrow*, 33.

42. *NYH*, Apr. 8, July 6, 1856; *NYT*, Dec. 18, 1860, Sept. 4, 1865, Jan. 1, June 8, July 31, 1866, Jan. 5, 10, 13, Dec. 19, 1875, Jan. 19, July 6, 1876; *CIO*, Aug. 31, 1874, Oct. 9, 1875, Mar. 12, 1877; *BS*, Dec. 21, 1874; *Hartford Courant*, Oct. 25, 1875; *Columbus Daily Inquirer*, Sept. 26, 1876; *PI*, Dec. 17, 1875, Jan. 7, Sept. 30, 1876; *Philadelphia Evening Bulletin*, Nov. 1 1876; quoted example from *NYT*, May 21, 1875.

43. "Letters from the Railway Switchman, Number Four: The Night Side of Broadway," *NYT*, Nov. 8, 1852; playbill for the Walnut Street Theatre, Philadelphia, Jan. 12, 1835, HTC; Cushing diary, vol. 4, Nov. 14, 1846; advertisement for Front Street Theatre, *BS*, Oct. 15, 1851; advertisement for American Museum, *PI*, Apr. 13, 1871; advertisements for Roberts Opera House, *Hartford Courant*, Mar. 6, Dec. 14, 1875; playbills for Niblo's Garden, New York, June 19, 1837, June 4, 1841. By 1848, Niblo's Garden had ceased its omnibus service to City Hall Park, instead reminding its audience of the omnibuses and horsecars that "constantly" passed along Broadway; playbill, Aug. 26, 1848, HTC.

44. "Hissing," *Philadelphia Press*, Mar. 11, 1869; "Music and Drama at Home and Abroad," *Wilkes' Spirit of the Times*, Dec. 10, 1864, 235; playbill, Walnut Street Theatre, Philadelphia, June 29, 1866, HTC; Logan, *Before the Footlights*, 308; Levine, *Highbrow/Lowbrow*, 193.

45. Butsch, *American Audiences*, 37; Nicholas E. Tawa, *High-Minded and Low-Down: Music in the Lives of Americans, 1800–1861* (Boston: Northeastern University Press, 2000),

35–45, 57–60, 66, 124–28; Ahlquist, *Democracy at the Opera*, 21; T. S. Arthur, *Advice to Young Men on Their Duties and Conduct in Life* (Philadelphia: J. W. Bradley, 1860), 81, 94–95; Hale, *Manners*, 49–51, 177; Scott Gac, "The Hutchinson Family Singers and the Culture of Reform in Antebellum America" (PhD diss., City University of New York, 2003); Rogers, "Public Music Performances in New York City," 5–50; Vera Brodsky Lawrence, *Strong on Music: The New York Music Scene in the Days of George Templeton Strong, 1836–1875* (New York: Oxford University Press, 1988), 1:22, 1:30, 1:84n41, 2:73; *The Diary of Philip Hone, 1828–1851* (New York: Dodd, Mead, 1936), 183; W. E. Channing, "Popular Amusements," *Western Messenger*, Jan. 1838, 343–49, quotation at 344; Eddy, *Young Man's Friend*, 88, 94; Talmage, *Sports That Kill*, 103; Buckingham, *America*, 2:72; Cushing diary, vol. 3: Sept. 29, Oct. 11, Nov. 2, 14, 1846, Feb. 4, 1847.

46. Advertisement, *NYH*, Jan. 23, 1841; advertisement, *NYT*, Oct. 27, 1851; Cushing diary, vol. 4: Dec. 5, 1846, Jan. 30, 1847; Benjamin W. Crowninshield, *A Private Journal, 1856–1858* (Cambridge, MA: Riverside Press, 1941), 88; Cole diary, Mar. 28–31, 1875; Lawrence, *Strong on Music*, 1:19, 1:42, 1:110, 1:198; 1:271–72, 1:299, 1:338–39, 1:613, 2:286, 2:499, Hone quotation at 1:203, Strong quotation at 1:447; Levine, *Highbrow/Lowbrow*, 134–40, 186–94.

47. Walt Whitman, "The Opera," *Life Illustrated*, Nov. 10, 1855, reprinted in *New York Dissected*, 19; William G. B. Carson, *The Theatre on the Frontier: The Early Years of the St. Louis Stage* (Chicago: University of Chicago Press, 1932), 31, 46, 224, 232, 270, 305; Smith, *Theatrical Management*, 146–47; playbill, Niblo's Garden, Aug. 8, 1838, HTC; Rogers, "Public Music Performances in New York City," 26; Peter G. Buckley, "Paratheatricals and Popular Stage Entertainment," in *Cambridge History of American Theatre*, 1:472; "Commencement of the Theatrical Season," *NYH*, Aug. 18, 1840; "Music and the Drama: New York Amusements," *Porter's Spirit of the Times*, July 18, 1857, 320; "Music and the Drama," *Wilkes' Spirit of the Times*, Aug. 18, 1860. The historian John Henry Hepp IV develops this theme of bourgeois security in *The Middle-Class City: Transforming Space and Time in Philadelphia, 1876–1926* (Philadelphia: University of Pennsylvania Press, 2003).

48. Butsch, *American Audiences*, 75–76; Foster, *New York Naked*, 50–51; "Swells at the Opera," *Sporting Times*, Apr. 24, 1869; Frost, *Frost's Laws and By-Laws*, 106; Hartley, *Ladies' Book of Etiquette*, 173–74.

49. "Original Sketches: Lithographs, by an Amateur," *New-York Mirror*, Dec. 7, 1833, 180; Robinson quoted in Patricia Cline Cohen, *The Murder of Helen Jewett: The Life and Death of a Prostitute in Nineteenth-Century New York* (New York: Random House, 1998), 324.

50. J. M. Buckley, *Christians and the Theater* (New York: Nelson and Phillips, 1876), 47.

CHAPTER FIVE

1. Horace Bushnell, "Of Night and Sleep," in *Moral Uses of Dark Things* (1867; New York: Charles Scribner's Sons, 1903), 12–13; Ned Buntline [Edward Zane C. Judson], *The Mysteries and Miseries of New York: A Story of Real Life* (New York: Berford, 1848), 3:55.

2. Richard L. Bushman, *The Refinement of America: Persons, Houses, Cities* (New York: Vintage Books, 1993), 409–15; Frederic Cople Jaher, *The Urban Establishment: Upper Strata in Boston, New York, Charleston, Chicago, and Los Angeles* (Urbana: University of Illinois Press, 1982), 173–208, 245–50; Edwin G. Burrows and Mike Wallace, *Gotham: A History of New York*

238   Notes to Pages 76–80

City to 1898 (New York: Oxford University Press, 1999), 712–14; Sidney George Fisher, *A Philadelphia Perspective: The Diary of Sidney George Fisher, Covering the Years 1834–1871*, ed. Nicholas B. Wainwright (Philadelphia: Historical Society of Philadelphia, 1967), 246–47.

3. Sven Beckert, *The Monied Metropolis: New York City and the Consolidation of the American Bourgeoisie, 1850–1896* (Cambridge: Cambridge University Press, 2001), 35–38; "Swells, the Incubi of Society," and "Broadway, New York, at 3 p.m.," *Day and Night*, Feb. 1869; C. Astor Bristed, *The Upper Ten Thousand* (New York: Stringer and Townsend, 1852), 9, 18; Buntline, *Mysteries and Miseries*, 2:89; John D. Vose, *Fresh Leaves from the Diary of a Broadway Dandy* (New York: Bunnell and Price, 1852), 13; John F. Kasson, *Rudeness and Civility: Manners in Nineteenth-Century Urban America* (New York: Hill and Wang, 1990), 118–19, 123, 127; Karen Halttunen, *Confidence Men and Painted Women: A Study of Middle-Class Culture in America, 1830–1870* (New Haven, CT: Yale University Press, 1982), 67, 154–56; George G. Foster, *New York Naked* (New York: DeWitt and Davenport, [1855]), 48–49; Osgood Bradbury, *Ellen Grant, or Fashionable Life in New York* (New York: Dick and Fitzgerald, n.d.), 13–14; Patricia Cline Cohen, *The Murder of Helen Jewett: The Life and Death of a Prostitute in Nineteenth-Century New York* (New York: Vintage, 1999), 309–10; George Thompson, *The Gay Girls of New York, or Life on Broadway* (New York, 1853), 63–81.

4. "The Post Office," *Baltimore Monument*, June 9, 1838; Abba Goold Woolson, *Woman in American Society* (Boston: Roberts Brothers, 1873), 125–29, quotation at 129.

5. Foster, *New York Naked*, 147; Thomas L. Nichols, *Forty Years of American Life* (London: John Maxwell, 1864), 1:279–80.

6. *Albion*, Mar. 14, 1829; George G. Foster, *New York by Gas-Light and Other Urban Sketches* (1850; repr., with an introduction by Stuart M. Blumin, Berkeley: University of California Press, 1990), 164; J. Frank Kernan, *Reminiscences of the Old Fire Laddies and the Volunteer Fire Departments of New York and Brooklyn* (New York: M. Crane, 1885), 167.

7. *The Art of Good Behaviour* (1846), quoted in Halttunen, *Confidence Men and Painted Women*, 95; Fisher, *Philadelphia Perspective*, 29.

8. Fisher, *Philadelphia Perspective*, 48; Hiram Mattison, *Popular Amusements: An Appeal to Methodists, in regard to the Evils of Card-Playing, Billiards, Dancing, Theatre-Going, etc.* (New York: Carlton and Porter, 1867), 11–12.

9. Elizabeth Aldrich, *From the Ballroom to Hell: Grace and Folly in Nineteenth-Century Dance* (Evanston, IL: Northwestern University Press, 1991), 11–13, 22, 25, 45–46; "The Fireman's Ball at the Park Theatre," *New York Sporting Whip*, Feb. 4, 1843; "Balls and Society in New York," *NYH*, Jan. 6, 1849; Foster, *New York by Gas-Light*, 164.

10. Edward Ferrero, *The Art of Dancing, Historically Illustrated* (New York, 1859), 74, quoted in Aldrich, *From the Ballroom to Hell*, 119; "Row at a Ball—Stabbing and Arrest," *NYTr*, Oct. 10, 1850; Anna Quincy Thaxter Cushing diary, AAS, vol. 4, Jan. 25, 1847; Cushing diary, vol. 5, Oct. 26, Nov. 11, 1847; George Templeton Strong, *The Diary of George Templeton Strong: The Turbulent Fifties, 1850–1859*, ed. Allan Nevins and Milton Halsey Thomas (New York: Macmillan, 1952), 209; J. S. Buckingham, *America: Historical, Statistic and Descriptive* (London: Fisher, 1841), 1:56–57; Thomas Hillgrove, *A Complete Practical Guide to the Art of Dancing* (New York: Dick and Fitzgerald, 1863), 237; Beckert, *Monied Metropolis*, 154–57; Burrows and Wallace, *Gotham*, 963.

11. Aldrich, *From the Ballroom to Hell*, 20, 154–56; Matthew Hale Smith, *Counsels Addressed*

to Young Women, Young Men, Young Persons in Married Life, and Young Parents (Washington, DC: Blair and Rives, 1846), 31–32; Harvey Newcomb, *How to Be a Man: A Book for Boys* (Boston: Gould, Kendall, and Lincoln, 1847), 186; "Balls and Society in New York," *NYH*, Jan. 6, 1849; James Dabney McCabe, *Lights and Shadows of New York Life, or The Sights and Sensations of the Great City* (Philadelphia: National Publishing, 1872), 164–66; Thomas L. Nichols, *The Lady in Black: A Story of New York Life, Morals, and Manners* (New York, 1844), 14; Samuel W. Fisher, *The Three Great Temptations of Young Men* (Cincinnati: Moore and Anderson, 1852), 125–26.

12. C. M. Kirkland, *The Evening Book, or Fireside Talk on Morals and Manners, with Sketches of Western Life* (New York: Charles Scribner, 1851), 78; Frances Trollope, *Domestic Manners of the Americans*, ed. Donald Smalley (Gloucester, MA: Peter Smith, 1974), 299; "The Fancy Ball: An Account for 'Outsiders,'" *Spirit of the Times*, Jan. 27, 1849; Woolson, *Woman in American Society*, 128; Aldrich, *From the Ballroom to Hell*, 109–16; *The Laws of Etiquette, or Short Rules and Reflections for Conduct in Society, by a Gentleman* (1836; Philadelphia: Carey, Lea, and Blanchard, 1836), 180; Strong, *Young Man in New York*, 270.

13. John H. Griscom, *The Uses and Abuses of Air: Showing Its Influence in Sustaining Life and Producing Disease, with Remarks on the Ventilation of Houses* (New York: Redfield, 1854), 54, 62; [William A. Alcott], *The Young Man's Guide* (Boston: Lilly, Wait, Colman, and Holden, 1834), 177; M. M. Noah, "Fashionable Parties and Late Hours," *New York Mirror*, Feb. 21, 1835; George Ellington, *The Women of New York, or Social Life in the Great City* (New York: New York Book Company, 1870), 59; Sarah Josepha Hale, *Manners, or Happy Homes and Good Society All the Year Round* (Boston: J. E. Tilton, 1868), 287; [Robert Tomes], *The Bazar Book of Decorum: The Care of the Person, Manners, Etiquette, and Ceremonials* (New York: Harper, 1870), 229; James K. Paulding, *The New Mirror for Travellers; and Guide to the Springs, by an Amateur* (New York: G. and C. Carvill, 1828), 35, 48; T. DeWitt Talmage, *The Abominations of Modern Society* (New York: Adams, Victor, 1872), 88.

14. Tomes, *The Bazar Book of Decorum*, 225; *Etiquette at Washington: Together with the Customs Adopted by Polite Society in the Other Cities of the United States*, 7th ed. (Baltimore: Murphy, 1860), 72; Buckingham, *America*, 1:58; Arthur Martine, *Martine's Hand-Book of Etiquette and Guide to True Politeness* (New York: Dick and Fitzgerald, 1866), 104; *The American Book of Genteel Behavior* (New York: Hurst, 1875), 42; Fisher, *Philadelphia Perspective*, 15, 74; Paulding, *New Mirror for Travellers*, 28–37, quotation at 36–37. On earlier hours in Cincinnati and Chicago, see Richard Champion Rawlins, *An American Journal, 1839–40*, ed. John L. Tearle (Madison, NJ: Farleigh Dickinson University Press, 2002), 73; Jaher, *Urban Establishment*, 470.

15. S. A. Frost, *Frost's Laws and By-Laws of American Society* (New York: Dick and Fitzgerald, 1869), 69; Tomes, *Bazar Book of Decorum*, 228; "Gossip with Readers and Correspondents," *Knickerbocker*, Oct. 1846, 372.

16. George D. Budd, "The German Cotillion," *Galaxy*, June 1867, 145–51; Aldrich, *From the Ballroom to Hell*, 17–18, 181–86; quotations at 17 and 185; "Our Best Society," *Putnam's Monthly Magazine*, Feb. 1853, 176; Ellington, *Women of New York*, 58.

17. McCabe, *Lights and Shadows*, 166; Benjamin W. Crowninshield, *A Private Journal, 1856–1858* (Cambridge, MA: Riverside Press, 1941), 108; Frost, *Frost's Laws and By-Laws*, 67; Mary Lydig Daly, *Diary of a Union Lady, 1861–1865*, ed. Harold Early Hammond (1962; Lin-

coln: University of Nebraska Press, 2000), 218; "The Fireman's Ball at the Park Theatre," *New York Sporting Whip*, Feb. 4, 1843; Talmage, *Abominations*, 88.

18. Hartley, *Gentlemen's Book of Etiquette*, 100; Halttunen, *Confidence Men and Painted Women*, 58–60; Aldrich, *From the Ballroom to Hell*, 5–9, 109–16; Woolson, *Woman in American Society*, 129.

19. Aldrich, *From the Ballroom to Hell*, 22–30, 113–14; Hillgrove, *Complete Practical Guide to the Art of Dancing*, 26–39.

20. Sidney Irving Pomerantz, *New York, an American City, 1783–1803: A Study of Urban Life* (New York: Columbia University Press, 1938), 464–65; Peter Thompson, *Rum Punch and Revolution: Taverngoing and Public Life in Eighteenth-Century Philadelphia* (Philadelphia: University of Pennsylvania Press, 1999), 1–11, 75–110; Richard B. Stott, *Workers in the Metropolis: Class, Ethnicity, and Youth in Antebellum New York City* (Ithaca, NY: Cornell University Press, 1990), 217–22; Richard Stott, *Jolly Fellows: Male Milieus in Nineteenth-Century America* (Baltimore: Johns Hopkins University Press, 2009), 8–26.

21. W. J. Rorabaugh, *The Alcoholic Republic: An American Tradition* (New York: Oxford University Press, 1979), 5–21; Paul E. Johnson, *A Shopkeeper's Millennium: Society and Revivals in Rochester, New York, 1815–1837* (New York: Hill and Wang, 1978), 55–61; Bruce Dorsey, *Reforming Men and Women: Gender in the Antebellum City* (Ithaca, NY: Cornell University Press, 2002), 90–102, 120–24; Mary P. Ryan, *Cradle of the Middle Class: The Family in Oneida County, New York, 1790–1865* (Cambridge: Cambridge University Press, 1981) 132–41; Stuart M. Blumin, *The Emergence of the Middle Class: Social Experience in the American City, 1760–1900* (Cambridge: Cambridge University Press, 1989), 195–204; Paul Boyer, *Urban Masses and Moral Order in America, 1820–1920* (Cambridge, MA: Harvard University Press, 1978), 77–80.

22. Stott, *Workers in the Metropolis*, 217–22, Englishman's quotation at 218; *Sanitary Condition of the City: Report of the Council of Hygiene and Public Health of the Citizens' Association of New York* (1866; New York: Arno Press and New York Times, 1970), 214, 137.

23. *Sanitary Condition of the City*, cxxi, 27, 59; Charles Loring Brace, *The Dangerous Classes of New York, and Twenty Years' Work among Them* (New York: Wynkoop and Hallenbeck, 1880), 64–65; E. R. Pulling, "Sanitary and Social Chart of the Fourth Ward of the City of New-York," in *Sanitary Condition of the City*.

24. Junius Henri Browne, *The Great Metropolis: A Mirror of New York* (Hartford, CT: American Publishing Company, 1869), 159–66, quotation at 162; "The Pewter Mug," *Weekly Rake*, Oct. 22, 1842.

25. "New-York Bars," *Appletons' Journal*, Apr. 1, 1876; McCabe, *Lights and Shadows*, 310–12; Carolyn Brucken, "In the Public Eye: Women and the American Luxury Hotel," *Winterthur Portfolio* 31, no. 4 (Winter 1996): 203–20; Cindy R. Lobel, "Consuming Classes: Changing Food Consumption Patterns in New York City, 1790–1860" (PhD diss., City University of New York, 2003), 104–6.

26. *The Diary of Philip Hone, 1828–1851* (New York: Dodd, Mead, 1936), 151, 209, 230, 348, 385, 395–96; Edward Pessen, "Philip Hone's Set: The Social World of the New York City Elite in the 'Age of Egalitarianism,'" *New-York Historical Society Quarterly* 56, no. 4 (Oct. 1872): 285–308; Crowninshield, *Private Journal*, 24, 100, 21.

27. Timothy J. Gilfoyle, *City of Eros: New York City, Prostitution, and the Commercialization of Sex* (New York: W. W. Norton, 1992), 81; "The Drinking Clubs of the City," *NYH*, Mar. 3, 1844; Amy S. Greenberg, *Cause for Alarm: The Volunteer Fire Department in the Nineteenth-Century City* (Princeton, NJ: Princeton University Press, 1998), 52–70.

28. "Clubs—Club Life—Some New York Clubs," *Galaxy*, Aug. 1876; *The Diary of George Templeton Strong: Young Man in New York, 1835–1849*, ed. Allan Nevins and Milton Halsey Thomas (New York: Macmillan, 1952), 260; Jaher, *Urban Establishment*, 67, 109–10, 277, 501; Beckert, *Monied Metropolis*, 58, 263–64; John S. Gilkeson Jr., *Middle-Class Providence, 1820–1940* (Princeton, NJ: Princeton University Press, 1986), 64; Sarah H. Killikelly, *The History of Pittsburgh: Its Rise and Progress* (Pittsburgh: B. C. and Gordon Montgomery, 1906), 538; Browne, *Great Metropolis*, 442; Talmage, *Abominations*, 190–97.

29. Nichols, *Forty Years of American Life*, 1:269–70; Crowninshield, *Private Journal*, 86, 120; Charles MacKay, *Life and Liberty in America, or Sketches of a Tour in the United States and Canada, in 1857–8* (London: Smith, Elder, 1859), 1:25–27; Charles Haynes Haswell, *Reminiscences of New York by an Octogenarian (1816 to 1860)* (New York: Harper, 1896), 61–62; Bristed, *Upper Ten Thousand*, 243.

30. Lobel, "Consuming Classes," 128–30; Foster, *New York by Gas-Light*, 73, 108; Capt. [John W.] Oldmixon, *Transatlantic Wanderings, or A Last Look at the United States* (1855; New York: Johnson Reprint Corporation, 1970), 28; "Vice Underground," *HW*, Jan. 24, 1857; "Late Suppers and Licentiousness," *Broadway Belle and Mirror of the Times*, Jan. 22, 1855; "The Ellsler Saloon," *Flash*, Dec. 11, 1841; "The Invisible Spy—No. 2," *Weekly Rake*, Oct. 1, 1842; advertisement for "Eating House, No. 88 Cedar Street," *Flash*, Dec. 18, 1841; "Oyster Suppers," *Weekly Rake*, Nov. 5, 1842; Harrison Gray Buchanan, *Asmodeus, or Legends of New York* (New York: John D. Munson, 1848), 56.

31. Eric Homberger, *Scenes from the Life of a City: Corruption and Conscience in Old New York* (New Haven, CT: Yale University Press, 1994), 30–36; George Thompson, *The Gay Girls of New York, or Life on Broadway* (New York, 1853), 30; see also [George Thompson], *Mysteries and Miseries of Philadelphia* (New York: Williams, n.d.), 80.

32. "Vice Underground," *Harper's Weekly*, Jan. 24, 1857; Thompson, *Gay Girls of New York*, 30.

33. Gilfoyle, *City of Eros*, 99–116; Buntline quotations in ibid., 104; Elliott J. Gorn, *The Manly Art: Bare-Knuckle Prize Fighting in America* (Ithaca, NY: Cornell University Press, 1986), 27–32, at 28; Stott, *Jolly Fellows*, 119–20, 215–16; Cohen, *Murder of Helen Jewett*, 302, 309–10, 328. As in England, the words "sporting" and "fancy" both carried a whiff of sexuality: a "sporting woman" was a prostitute, and pimps were sometimes called "fancy men." The word "fancy," derived from "fantasy," also suggested extravagance and pretense, the qualities of the dandy. *Oxford English Dictionary*, 2nd ed., s.v. "dandy," "fancy," "fancy man," and "sporting."

34. Brooks McNamara, *The New York Concert Saloon: The Devil's Own Nights* (Cambridge: Cambridge University Press, 2002), 2–10; advertisements, *BS*, Jan. 24, Aug. 8, 1856; advertisement, *NYH*, May 10, 1857; "The Latest Nuisance on the Board," *Broadway Omnibus*, Nov. 1, 1858; "A Story about Concert Saloons," *PI*, May 3, 1862; "The Police Commissioners' Report," *NYT*, Jan. 5, 1866; Robert C. Allen, *Horrible Prettiness: Burlesque and American Culture*

(Chapel Hill: University of North Carolina Press, 1991), 73–76; "The Elephant in Brooklyn," *BE*, Dec. 9, 1861; *The Diary of George Templeton Strong: The Civil War, 1860–1865*, ed. Allan Nevins and Milton Halsey Thomas (1952; New York: Octagon Books, 1974), 73–74.

35. Allen, *Horrible Prettiness*, 74–76; "The Concert Saloon Reform," *New York Times*, Apr. 25, 1862; "The Police among the Concert Saloons," *New York Times*, May 6, 1862.

36. McNamara, *New York Concert Saloon*, 17–24, 31, 61; [Henry William Herbert], *The Tricks and Traps of Chicago*, part 1 (New York: Dinsmore, 1859), 67; quotation from "A Nuisance to Be Squelched," *New York Times*, Dec. 21, 1861; Gilfoyle, *City of Eros*, 129–30; John H. Warren Jr., *Thirty Years' Battle with Crime, or The Crying Shame of New York* (1875; New York: Arno Press and New York Times, 1970), 117–18; "Morals and Vices in New York," *New York Observer and Chronicle*, Mar. 22, 1866; McCabe, *Lights and Shadows*, 594–96; "Concert Saloon Keepers Defiant," *NYT*, May 7, 1876; "Down in the Depths," *Detroit Evening News*, July 30, 1875; "A Nice Place," *Detroit Evening News*, Aug. 9, 1875; "Crusading the Cyprians," *National Police Gazette*, Dec. 6, 1879; "Robbed in a Concert Saloon," *Philadelphia Inquirer*, Dec. 9, 1867; M. Edward Winslow Martin [James Dabney McCabe], *The Secrets of the Great City* (Philadelphia: Jones Brothers, 1868), 312–13; Edward Crapsey, *The Nether Side of New York, or The Vice, Crime and Poverty of the Great Metropolis* (1872; Montclair, NJ: Patterson Smith, 1969), 161–63.

37. "Cigar Shops," *Weekly Rake*, Oct. 22, 1842; Joel Best, *Controlling Vice: Regulating Brothel Prostitution in St. Paul* (Columbus: Ohio State University Press, 1998), 22; "Abominations in Our City," *Advocate of Moral Reform*, Feb. 1, 1836; Strong, *Turbulent Fifties*, 57; Buchanan, *Asmodeus*, 55–58; Crapsey, *Nether Side*, 122, 143–45; Foster, *New-York by Gas-Light*, 134–35; "A Nice Little Party in an Engine House," *BE*, Apr. 4, 1857; "Expose of the Social Evil," *BE*, Sept. 4, 1871; "A Clerk's Amusement," *New York Sporting Whip*, Feb. 4, 1843; "Philadelphia," *Whip*, July 30, 1842.

38. Walt Whitman, "On Vice," Brooklyn *Daily Times*, June 20, 1857; in *The Uncollected Poetry and Prose of Walt Whitman*, ed. Emory Holloway (Garden City, NY: Doubleday, Page, 1921) 2:5–8; Gilfoyle, *City of Eros*, 99, 103; "Cincinnati Correspondence," *NPG*, Jan. 14, 1860; Warren Burton, *Moral Dangers of the City: To the Clergymen of Various Denominations throughout New England* (Boston: John Wilson, 1848), 1; [George Thompson], *New-York Life, or The Mysteries of Upper-Tendom Revealed* (New York: Charles S. Atwood, n.d.), 56, 57; "Providence Correspondence," *NPG*, Apr. 21, 1860.

39. *Pen and Sword: The Life and Journals of Randal W. McGavock*, ed. Herschel Gower and Jack Allen (Nashville: Tennessee Historical Commission, 1959), 159, 165.

40. Gilfoyle, *City of Eros*, 58–59; Buckingham, *America*, 1:463; "Pittsburg Correspondence," *NPG*, Apr. 28, 1860; *New Orleans as It Is* (Utica, NY: Dewitt C. Grove, 1849), 3, 34–45; "Chicago: The Snares and Pitfalls of the City," *NPG*, Nov. 11, 1866; *Life in Chicago, or Day and Night in the World's Wickedest City* (Chicago: Society Publishing, 1876), 5; William W. Sanger, *The History of Prostitution: Its Extent, Causes and Effects throughout the World* (1858; New York: Eugenics Publishing Company, 1937), 613.

41. McCabe, *Secrets of the Great City*, 46–47, 255, quotation at 303; James Dabney McCabe, *Lights and Shadows*, 133–34; Buntline, *Mysteries and Miseries*, 1:9, 1:43; Foster, *New-York by Gas-Light*, 71; "Street Walkers," *Whip and Satirist of New York*, May 7, 1842; James D. McCabe, *New York by Sunlight and Gaslight: A Work Descriptive of the Great American Metropolis* (Phila-

delphia: Douglass Brothers, 1882), 254, 643; Crapsey, *Nether Side*, 138–39, 157; "Picture of the City," *Monthly Cosmopolite*, Oct. 1, 1849.

42. Letter by Lothario, *Ely's Hawk and Buzzard*, Aug. 31, 1833; "The Battery," *New York Mirror*, May 30, 1840; "Evening Amusements in the Parade Ground," *Sunday Flash*, Sept. 14, 1841; Ellington, *Women of New York*, 303–5; "The Whip Wants to Know," *Whip and Satirist of New York*, Apr. 16, 1842; "Diary of a Rake—No. 15," *Whip and Satirist of New York*, Apr. 30, 1842; "Scenes in the Park," *Flash*, June 23, 1842; "Fallow Deer," *Flash*, Sept. 18, 1842; "The Sodomites," *Whip and Satirist of New York*, Jan. 29, 1842; Thompson, *City Crimes*, 246; "Brooklyn," *New York Sporting Whip*, Jan. 28, 1843; "Municipal Reform," *NYT*, Feb. 22, 1860; "Expose of the Social Evil," *BE*, Sept. 4, 1871. On Boston, "Wants to Know," *Flash*, Sept. 18, 1842; "Advertisement," *Life in Boston and New England Police Gazette*, Jan. 4, 1851; "A Moonlight Colloquy," *Life in Boston and New York*, May 13, 1854. On Philadelphia, Marcia Carlisle, "Disorderly City, Disorderly Women: Prostitution in Ante-Bellum Philadelphia," *Pennsylvania Magazine of History and Biography* 110, no. 4 (Oct. 1986): 557. On Worcester, "The Whip Wants to Know," *Whip and Satirist of New-York and Brooklyn*, June 4, 1842; "Wants to Know," *Flash*, July 31, 1842. On New London, "Wants to Know," *Flash*, Sept. 11, 1842. On Newark, "Success of the Flash," *Flash*, Sept. 25, 1842; "Wants to Know," *Flash*, Nov. 12, 1842. On Brooklyn, "The Whip Wants to Know," *Whip*, July 9, 1842. On Providence, "The Whip Wants to Know," *Whip*, Sept. 10, 1842. On New Haven, "The Whip Wants to Know," *Whip and Satirist of New-York and Brooklyn*, Apr. 16, 1842. On "mistaken identity" see "Albany Soaplocks Exposed," *Whip*, Aug. 6, 1842; "Boys after Night Fall," *Whip*, Nov. 26, 1842; "Ninth and Chestnut after Dark—No. II," *Philadelphia Press*, Jan. 25, 1869; "Crusading the Cyprians," *NPG*, Dec. 6, 1879.

43. Jeffrey S. Adler, "Streetwalkers, Degraded Outcasts, and Good-for-Nothing Huzzies: Women and the Dangerous Class in Antebellum St. Louis," *Journal of Social History* 25, no. 4 (Summer 1992): 737–55; John C. Schneider, *Detroit and the Problem of Order, 1830–1880* (Lincoln: University of Nebraska Press, 1980), 32, 41–42, 66, 104; "Down in the Depths," *Detroit Evening News*, July 30, 1875.

44. Adler, "Streetwalkers," 741–43; Best, *Controlling Vice*, 5, 23; Henry Morgan, *Ned Nevins, the Newsboy, or Street Life in Boston* (Boston: Lee and Shepard, 1866), 176, 261; Neil Larry Shumsky, "Tacit Acceptance: Respectable Americans and Segregated Prostitution, 1870–1910," *Journal of Social History* 19, no. 4 (Summer 1986): 665–79; "Philadelphia Correspondence of the Whip," *Whip*, Sept. 10, 1842; Marcia Carlisle, "Disorderly City, Disorderly Women: Prostitution in Ante-Bellum Philadelphia," *Pennsylvania Magazine of History and Biography* 110, no. 4 (Oct. 1986): 549–68; Beekley diary, Mar. 10, 1849; G. W. Colton, "Philadelphia," in *Colton's Atlas of the World* (New York: J. H. Colton, 1856); "Detroit at Night," *Detroit Evening News*, Aug. 3, 1875; "Down in the Depths," *Detroit Evening News*, July 30, 1875; Eugene Robinson, "Map of the City of Detroit," in *Atlas of the State of Michigan*, ed. H. F. Walling (Claremont, NH: Claremont Manufacturing Company, 1873); David R. Johnson, *Policing the Urban Underworld: The Impact of Crime on the Development of the American Police, 1800–1887* (Philadelphia: Temple University Press, 1979), 152–56; Gilfoyle, *City of Eros*, 29–54.

45. Warren, *Thirty Years' Battle*, 121; Morgan, *Ned Nevins*, 176, 261, quotation at 264; George G. Foster, "Philadelphia in Slices" (1848–49), ed. George Rogers Taylor, *Pennsylvania Magazine of History and Biography* 93, no. 1 (Jan. 1969): 38–41; Foster, *New-York by Gas-Light*,

143-44; Whitman, "On Vice," 6; Crapsey, *Nether Side,* 142-43, 158-63; Nathaniel Parker Willis, "Daguerreotype Sketches of New York," *New Mirror,* May 20, 1843; "The City Expositor," *NPG,* July 20, 1861; Bayard Taylor, "A Descent into the Depths," *Nation* 2 (Mar. 15, 1866): 332-33; Buntline, *Mysteries and Miseries,* 2:15-16, 2:79-85.

46. "Detroit at Night," *Detroit Evening News,* Aug. 3, 1875; Johnson, *Policing the Urban Underworld,* 156; Richard C. Lindberg, *To Serve and Collect: Chicago Politics and Police Corruption from the Lager Beer Riots to the Summerdale Scandal, 1855-1960* (1991; Carbondale: Southern Illinois University Press, 1998), 8-9; "The City on Sunday," *NYH,* Sept. 4, 1848; "The Demi-Monde of New York," *NPG,* Dec. 28, 1867; Gilfoyle, *City of Eros,* 39-40, 51, 170; Tyler Anbinder, *Five Points: The 19th-Century New York City Neighborhood That Invented Tap Dance, Stole Elections, and Became the World's Most Notorious Slum* (New York: Free Press, 2001), 208-14, 220-21; J. R. McDowall, *Magdalen Facts* (New York, 1832), 13-14, 25, 78, 83.

47. Buchanan, *Asmodeus,* 11-14, 27-30; Buntline, *Mysteries and Miseries,* 1:46-47; McCabe, *Lights and Shadows,* 580-82; "The Actual State of Public Morals in New York," *New York Arena,* May 27, 1842; "City Life—a Contrast," *Broadway Belle and Mirror of the Times,* Jan. 8, 1855; "The Demi-Monde of New York," *NPG,* Dec. 28, 1867; Warren, *Thirty Years' Battle,* 16-20; Asmodeus [pseud.], *Sharps and Flats, or The Perils of City Life* (Boston: William Berry, 1850), 39; Sanger, *History of Prostitution,* 549-51; "Picture of the City," *Monthly Cosmopolite,* Oct. 1, 1849; Cohen, *Murder of Helen Jewett,* 112-13, 128-51; Halttunen, *Confidence Men and Painted Women,* 92.

48. Gilfoyle, *City of Eros,* 130, 170, 232-36; Foster, *New-York by Gas-Light,* 164-67; "A Grand Bal Given by Julia Brown," *Flash,* Jan. 14, 1843; "Buffalo Correspondence," *NPG,* Apr. 3, 1858; "A Carnival of Vice," *NPG,* Mar. 2, 1867. On masquerades, Haswell, *Reminiscences of New York by an Octogenarian,* 238, 406; "The Military and Fancy Ball at Winthrop Hall, Boston," *Flash,* Dec. 3, 1842; "The Masquerade," *NYH,* Feb. 11, 1844; *Tricks and Traps of Chicago,* 68-69; McCabe, *Lights and Shadows,* 604-11; Warren, *Thirty Years' Battle,* 125-31; "Masked Ball at Philadelphia," *NYT,* Jan. 26, 1866; "Notes," *Philadelphia Press,* Jan. 11, 1869; "Masked Balls," *NYT,* Jan. 2, 1876.

49. *First Annual Report of the Executive Committee of the New York Magdalen Society* (New York, 1831), 5; Crapsey, *Nether Side,* 143-45; Sanger, *History of Prostitution,* 566-68; McCabe, *Lights and Shadows,* 587-89, at 588; Solon Robinson, *Hot Corn: Life Scenes in New York Illustrated* (New York: DeWitt and Davenport, 1854), 353-54; Ellington, *Women of New York,* 275-77; "A Masher Mashed," *NPG,* July 26, 1879; "Midnight Pictures," *NPG,* Dec. 6, 1879.

50. Henry Mann, ed., *Our Police: History of the Pittsburgh Police Force, under the Town and City* (Pittsburgh, 1889), 275; Thompson, *City Crimes,* 282; McCabe, *Sunlight and Gaslight,* 256, 595; "Murder in the Seventeenth Ward," *Philadelphia Daily Evening Bulletin,* Feb. 16, 1858; "Midnight Lawlessness," *NYT,* Apr. 15, 1870; Crapsey, *Nether Side,* 163; McDowall, *Magdalen Facts,* 78, 83, 88; Buntline, *Mysteries and Miseries,* 44; Cohen, *Murder of Helen Jewett,* 6; "Dance Houses," *NYTr,* Oct. 22, 1850; S. Leavitt, "An Idyl of the Night," *Appletons' Journal,* May 11, 1872; "Midnight Pictures," *NPG,* Dec. 6, 1879.

51. Gilfoyle, *City of Eros,* 79-81; Anbinder, *Five Points,* 225; Johnson, *Policing the Urban Underworld,* 78; "Watchmen," *New-York Mirror,* Aug. 25, 1832; Crapsey, *Nether Side,* 28-29; Henry Morgan, *Boston Inside Out! Sins of a Great City! A Story of Real Life* (Boston: Shawmut Publishing, 1883) 104; Augustine E. Costello, *Our Police Protectors: History of the New York*

Police from the Earliest Period to the Present Time (New York, 1885), 120; "Late Hours," *BS,* Sept. 24, 1839; Buntline, *Mysteries and Miseries,* 1:10.

52. Henry Leander Foote, *A Sketch of the Life of Henry Leander Foote* (New Haven, CT: Thomas J. Stafford, 1850), 6.

53. Foote, *Sketch of the Life,* 6–8. On "model artist" exhibitions, see Foster, *New-York by Gas-Light,* 77–83.

54. Ann Fabian, *Card Sharps and Bucket Shops: Gambling in Nineteenth-Century America* (New York: Routledge, 1999), 40–41; "Desecrating the Sabbath," *BS,* Mar. 31, 1840; "A Scouring Out," *BS,* July 21, 1840; "Brooklyn," *Whip,* Nov. 26, 1842; "Gambling on Sunday," *BS,* Dec. 23, 1851; William J. Snelling, *Exposé of the Vice of Gaming: As It Lately Existed in Massachusetts* (Boston, 1833), 10; Foster, "Philadelphia in Slices," 48; Necromancer, letter in *Life in Boston and New York,* Oct. 1, 1853.

55. "Cock Fighting," *Sporting Times,* Apr. 25, 1868; "Badger Bait," *Whip,* Aug. 6, 1842; "A Fight between a Dog and a Boar," *Sporting Times,* Sept. 19, 1868; "Kit Burns' Rat-Killing Bear, 'Bruin,'" *Sporting Times,* July 4, 1868; "A Night at Kit Burns'," *Sporting Times,* Sept. 19, 1868; Edward H. Savage, *A Chronological History of the Boston Watch and Police, from 1631 to 1865; Together with the Recollection of a Boston Police Officer, or Boston by Daylight and Gaslight* (Boston: Edward H. Savage and J. E. Farwell, 1865), 160–63. *Wilkes' Spirit of the Times,* a New York sporting paper, ran a regular column called "Canine Sports" and even published box scores of dogs' prowess in killing rats (e.g., Jan. 5, 1861); McCabe, *Secrets of the Great City,* 389; Snelling, *Exposé of the Vice of Gaming,* 5–8; "Sports in the Bowery," *Sunday Flash,* Oct. 3, 1841; Crowninshield, *Private Journal,* 15.

56. Fisher, *Three Great Temptations,* 11–12, quotation at 92; "Gambling Houses," *Philadelphia Inquirer,* Nov. 23, 1860; "The Gambling Houses and the Grog Shops," *NYH,* Apr. 14, 1844; "Gambling and Gaming Houses," *NYH,* Dec. 11, 1856; Foster, "Philadelphia in Slices," 47; "The Faro Bank and the Gamblers," *NYTr,* Feb. 4, 1858; Tony Pastor [Harlan Page Halsey], *Night Scenes in New York; in Darkness and by Gaslight* (New York: American News Company, 1876), 26–27; Asmodeus, *Sharps and Flats,* 39; "Peep into a Washington Gambling-House," *BS,* Jan. 12, 1856; J. H. Green, *A Report on Gambling in New York* (New York: J. H. Green, 1851), 58; Warren, *Thirty Years' Battle,* 287; Mason Long, *The Life of Mason Long, the Converted Gambler* (Chicago: Donnelley, Loyd, 1878), 135.

57. Ellington, *Women of New York,* 391; McCabe, *Sunlight and Gaslight,* 546; "Gambling," *BS,* Mar. 20, 1838; "Gambling and Late Hours," *BS,* Aug. 1, 1844; "Peep into a Washington Gambling-House," *BS,* Jan. 12, 1856; Buntline, *Mysteries and Miseries,* 3:32; T. DeWitt Talmage, *The Masque Torn Off* (Chicago: J. Fairbanks, 1879), 57; "Descent on a Gambling House," *BS,* Feb. 27, 1850; "Descent upon an Alleged Gambling House," *NYH,* Mar. 21, 1858; "Descents on Gambling Houses," *BS,* Apr. 26, 1858; "Gambling on Sunday," *BS,* Jan. 17, 1859; Long, *Life of Mason Long,* 138–39.

58. Henry Ward Beecher, *Seven Lectures to Young Men on Various Important Subjects* (Indianapolis: Thomas B. Cutler, 1844), 116; Buntline, *Mysteries and Miseries,* 4:4; Buchanan, *Asmodeus,* 34; Burton, *Moral Dangers of the City,* 2; "Chicago: The Snares and Pitfalls of the City," *NPG,* Nov. 3, 1866; Green, *Report on Gambling in New York,* 47–48.

59. Talmage, *Masque Torn Off,* 64; Fisher, *Three Great Temptations,* 182; John Todd, *The Moral Influence, Dangers and Duties, Connected with Great Cities* (Northampton, MA, 1841),

238; Beecher, *Seven Lectures*, 116; Stephen Nissenbaum, *Sex, Diet and Debility in Jacksonian America: Sylvester Graham and Health Reform* (Westport, CT: Greenwood Press, 1980), 109; Daniel C. Eddy, *The Young Man's Friend* (Boston: Wentworth, Hewes, 1859), 160; *Life in Chicago*, 12; *Sanitary Condition of the City*, 10.

60. Junius Henry Browne, *The Great Metropolis: A Mirror of New York* (Hartford, CT: American Publishing Company, 1869), 245; "Peep into a Washington Gambling-House," *BS*, Jan. 12, 1856; "The Faro Bank and the Gamblers," *NYTr*, Feb. 4, 1856; "Gambling and Gaming Houses, *NYH*, Dec. 11, 1856; McCabe, *Lights and Shadows*, 715–25; "Gambling," *NYH*, Aug. 26, 1866.

61. Warren, *Thirty Years' Battle*, 287, 295; Beecher, *Seven Lectures*, 118–21; "Gambling," *Providence Patriot*, Dec. 13, 1834; "Gambling," *BS*, Mar. 12, 1838; Jackson Lears, *Something for Nothing: Luck in America* (New York: Viking, 2003), 94, 129–31; J. H. Green, "Gambling in Connecticut," *NYT*, Nov. 13, 1855; "Execution of Foote and McCaffrey," *Middletown (CT) Constitution*, Oct. 9, 1850.

62. Long, *Life of Mason Long*, 127–28; Beecher, *Seven Lectures*, 124; Asmodeus, *Sharps and Flats*, 39–43; Warren, *Thirty Years' Battle*, 282–86, 294–97.

63. Lears, *Something for Nothing*, 99–102; "The Gambling Vice," *NPG*, Dec. 8, 1866; Warren, *Thirty Years' Battle*, 298; Fabian, *Card Sharps and Bucket Shops*, 56–57, 65.

64. Nicholas E. Tawa, *High-Minded and Low-Down: Music in the Lives of Americans, 1800–1861* (Boston: Northeastern University Press, 2000), 246–52; James Wickes Taylor, *"A Choice Nook of Memory": The Diary of a Cincinnati Law Clerk*, ed. James Taylor Dunn (Columbus: Ohio State Archaeological and Historical Society, 1950), 26; "Rather Unpleasant," *Cummings' Evening Telegraphic Bulletin*, July 9, 1847; "Brooklyn," *Whip*, Dec. 17, 1842.

CHAPTER SIX

1. US Bureau of the Census, *Special Reports: Occupations at the Twelfth Census* (Washington, DC: Government Printing Office, 1904), xxxvii, 457, 460, 473; "Toilers after Dark," *Milwaukee Sentinel*, Mar. 14, 1897; *The Revised Charter and Ordinances of the City of Chicago* (Chicago: Chicago Common Council, 1851), chap. 35, sec. 8, p. 180; William Mack, ed., *Cyclopedia of Law and Procedure*, vol. 32 (New York: American Law Book Company, 1909), 731–36; Howard B. Woolston, *Prostitution in the United States*, vol. 1, *Prior to the Entrance of the United States into the World War* (New York: Century Company, 1921), 25–26.

2. The sociologist Murray Melbin describes night work in terms of three distinct schedules: "early phased," "late phased," and "incessant." I use "countercyclical" to describe jobs that were deliberately scheduled to have minimal overlap with normal waking hours. Murray Melbin, *Night as Frontier: Colonizing the World after Dark* (New York: Free Press, 1987), 82–85.

3. Martin V. Melosi, *The Sanitary City: Urban Infrastructure from Colonial Times to the Present* (Baltimore: Johns Hopkins University Press, 2000), 41, 91, 93; "The German Night Scavengers," *NYT*, May 25, 1869; Jon C. Teaford, *The Unheralded Triumph: City Government in America, 1870–1900* (Baltimore: Johns Hopkins University Press, 1984), 219–22.

4. Joel A. Tarr, *The Search for the Ultimate Sink: Urban Pollution in Historical Perspective* (Akron, OH: University of Akron Press, 1996), 115–16, 295–99; George E. Waring Jr., *Report on the Social Statistics of Cities, Part I: The New England and the Middle States* (Washington,

DC: Government Printing Office, 1886), 36, 174; George E. Waring Jr., *Report on the Social Statistics of Cities, Part II: The Southern and the Western States* (1887; New York: Arno Press and the *New York Times*, 1970), 22, 289, 424, 454, 558, 701, 748, 771, 811; Charles V. Chapin, *Municipal Sanitation in the United States* (Providence, RI: Snow and Farnham, 1900), 742–52; New York Bureau of Municipal Research, *The City of Pittsburgh, Pennsylvania: Report on a Survey* . . . (Pittsburgh: Pittsburgh City Council, 1913), 34; B. A. Segur, "Privy-Vaults and Cesspools," American Public Health Association, *Public Health Reports and Papers* 3 (1875–76): 185; Terry F. Yosie, "Retrospective Analysis of Water Supply and Wastewater Policies in Pittsburgh, 1800–1959" (PhD diss., Carnegie Mellon University, 1981), 104–5; S. J. Kleinberg, *The Shadow of the Mills: Working-Class Families in Pittsburgh, 1870–1907* (Pittsburgh: University of Pittsburgh Press, 1989), 91–92.

5. Tarr, *Ultimate Sink*, 115, 296; Yosie, "Retrospective Analysis," 67; Stuart Galishoff, *Newark, the Nation's Unhealthiest City, 1832–1895* (New Brunswick, NJ: Rutgers University Press, 1988), 100; US Bureau of the Census, *Special Reports, Statistics of Cities Having a Population of Over 30,000: 1905* (Washington, DC: Government Printing Office, 1907), 337.

6. Edward March Blunt, *Blunt's Stranger's Guide to the City of New York* (New York: Seymour for Blunt, 1817), 245; *Ordinances of the Mayor, Aldermen and Commonalty of the City of New York* (New York: Chas. W. Baker, 1859), 318; *Laws and Ordinances Governing the City of Chicago, January 1, 1866* ([Chicago]: E. B. Myers and Chandler, 1866), 322; *The Revised Charter and Ordinances of the City of Detroit* (Detroit: William Storey, 1855), 124; Waring, *Social Statistics, Part I*, 712, 811; Waring, *Social Statistics, Part II*, 590, 675; Tarr, *Ultimate Sink*, 113, 181; "A New Monopoly," *NYT*, Oct. 16, 1872.

7. John Duffy, *A History of Public Health in New York City, 1625–1866* (New York: Russell Sage Foundation, 1968), 211, 378, 408–9, 412–13; "Violating City Ordinances," *BS*, July 12, 1855; E. H., letter in *Washington Daily National Intelligencer*, Aug. 8, 1857; "City Nuisances and Waste of Manures," *NYTr*, Sept. 9, 1853; "Presentments of the Grand Jury," *NYT*, Aug. 9, 1853; *Ordinances of . . . the City of New York* (1859), 320; J. Smith Homans Jr., ed., *A Cyclopedia of Commerce and Commercial Navigation* (New York: Harper, 1859), 1434.

8. "The City Government," *NYT*, Dec. 23, 1863; "The City Government," *NYT*, Feb. 24, 1865; "Board of Health," *NYT*, Jan. 3, 1867; G. W. B., "Something for the Health Commissioners to Look After," *NYT*, July 29, 1866; *Report of the Council of Hygiene and Public Health of the Citizens' Association of New York upon the Sanitary Condition of the City*, 2nd ed. (New York: D. Appleton, 1866), 309.

9. "A New Monopoly" *NYT*, Oct. 16, 1872; Waring, *Social Statistics, Part I*, 129, 484, 829–30. "The Proceedings," *NYT*, Nov. 13, 1873; W. W. Thomson, *A Digest of the Acts of Assembly relating to, and the General Ordinances of the City of Pittsburgh, from 1804 to Jan. 1, 1897*, 2nd ed. (Pittsburgh: W. T. Nicholson Sons, 1897), 367; Henry I. Bowditch, *Public Hygiene in America* (Boston: Little, Brown, 1877), 229, 276; quotation from J. W. Magruder, "Exchanging 70,000 Earth Closets for a $20,000,000 Sewer System—Baltimore," *Survey* 26 (Sept. 2, 1911): 811.

10. Charles V. Chapin, *Municipal Sanitation in the United States* (Providence, RI: Snow and Farnham, 1900), 742–52; "City and County: The Nuisance Cases before the Board of Health," *Cincinnati Commercial*, Nov. 23, 1881; Yosie, "Retrospective Analysis," 109.

11. "Street Cleaning," *NYTr*, Oct. 25, 1850; "Street-Cleaning in New York," *HW*, Nov. 24,

1877; "The City Refuse," *HW,* Sept. 27, 1884; Henry Smith Williams, "How New York Is Kept Partially Clean," *HW,* Oct. 13, 1894; George E. Waring Jr., *Street-Cleaning and the Disposal of a City's Wastes: Methods and Results and the Effect upon Public Health, Public Morality, and Municipal Prosperity* (New York: Doubleday and McClure, 1898): 37, 43–44, 91–109; Rudolph Hering and Samuel A. Greeley, *Collection and Disposal of Municipal Refuse* (New York: McGraw-Hill, 1921), 602; G. A. Soper, *Modern Methods of Street Cleaning* (New York: Engineering News Publishing Company, 1909), 22–23, 45–46, 170, 177.

12. "Disposal of Refuse in American Cities," *Scientific American,* Aug. 29, 1891; Waring, *Street-Cleaning,* 44; Soper, *Modern Methods of Street Cleaning,* 175; Hering and Greeley, *Collection and Disposal of Municipal Refuse,* 94, 146, 158–59; William F. Morse, *The Collection and Disposal of Municipal Waste* (New York: Municipal Journal and Engineer, 1908), 3.

13. William G. Panschar, *Baking in America: Economic Development* (Evanston, IL: Northwestern University Press, 1956), 1–49; Hazel Kyrk and Joseph Stancliffe Davis, *The American Baking Industry, 1849–1923: As Shown in Census Reports* (Stanford, CA: Stanford University Press, 1925), 9, 23, 34, 53–55; "A Large Bakery," *NYTr,* June 16, 1862.

14. "Trade Strikes," *NYT,* June 9, 1868; "Points Gained by Bakers," *NYT,* May 3, 1881; R. Ovens Bakery, "Letters to the Editor: They Are with the Bakers," *Buffalo Express,* Mar. 29, 1886.

15. E. Melanie DuPuis, *Nature's Perfect Food: How Milk Became America's Favorite Drink* (New York: New York University Press, 2002), 18–31; John T. Schlebecker, *Whereby We Thrive: A History of American Farming, 1607–1972* (Ames: Iowa State University Press, 1975), 95.

16. Chester Linwood Roadhouse and James Lloyd Henderson, *The Market-Milk Industry* (New York: McGraw-Hill, 1941), 4–5, 149; "The Railroad Inquiry," *NYT,* June 20, 1879; "Events in the Metropolis," *NYT,* July 8, 1883; "Riding with the Milkman," *NYT,* Oct. 21, 1888.

17. Thomas F. DeVoe, *The Market Book: Containing a Historical Account of the Public Markets in the Cities of New York, Boston, Philadelphia and Brooklyn* (New York: Thomas F. DeVoe, 1862), 1:345; James D. McCabe, *New York by Sunlight and Gaslight: A Work Descriptive of the Great American Metropolis* (Philadelphia: Douglass Brothers, 1882), 665; Clyde Lyndon King, "Municipal Markets," *Annals of the American Academy of Political and Social Science* 50 (Nov. 1913): 112, 116; James Mease, *The Picture of Philadelphia* (Philadelphia: B. and T. Kite, 1811), 118; Waring, *Social Statistics, Part I,* 147, 711, 738, 759, 893; Waring, *Social Statistics, Part II,* 43–44, 283, 611–12, 700.

18. John R. Stilgoe, *Borderland: Origins of the American Suburb, 1820–1939* (New Haven, CT: Yale University Press, 1988), 75; Schlebecker, *Whereby We Thrive,* 95–96.

19. Robert Shackleton, "A Night Encampment on Manhattan Island," *HW,* Nov. 20, 1909, 11–12.

20. Simeon Strunsky, "When the City Wakes," *Harper's New Monthly Magazine* 129 (Oct. 1914): 700; Joseph Nelson Pardee, "With a Boston Market Man," *New England Magazine* 27 (Dec. 1902): 450–63.

21. Waring, *Social Statistics, Part I,* 146–47; Achsah Lippincott, "Municipal Markets in Philadelphia, *Annals of the American Academy of Political and Social Science* 50 (Nov. 1913): 135; James M. Mayo, *The American Grocery Store: The Business Evolution of an Architectural Space* (Westport, CT: Greenwood Press, 1993), 39–40, 48–49; Julian Ralph, "Ancient New

York Market Rights," *HW,* Apr. 11, 1896, 351; Donald L. Miller, *City of the Century: The Epic of Chicago and the Making of America* (New York: Simon and Schuster, 1996), 206-9; Harvey Levenstein, *Revolution at the Table: The Transformation of the American Diet* (New York: Oxford University Press), 31-32; Nellie M. Reed, "Protecting the Food of the City," *Outlook* 124, no. 3 (Jan. 21, 1920): 114-17; quotation from G. T. Ferris, "How a Great City Is Fed," *HW,* Mar. 22, 1890, 229.

22. Edwin Griswold Nourse, *The Chicago Produce Market* (Boston: Houghton Mifflin, 1918), 24; "Peas, Beans and Cabbage," *NYT,* July 7, 1872; "An All-Night Truck Market," *NYT,* June 18, 1893; Edward Hungerford, "Food for a Cityful," *HW,* June 3, 1911, 11-12; Reed, "Protecting the Food," 115.

23. "Huckster Methods Must Be Improved," *Hartford Times,* July 26, 1912; Waring, *Social Statistics, Part II,* 43, 612; McCabe, *New York by Sunlight and Gaslight,* 268; Scott Thompson, "A Saturday Night Street Market," *HW,* Mar. 29, 1890, 242.

24. "The Manufacture of Newspapers," *Emerson's United States Magazine,* July 1857, 28; Franklin Matthews, "A Night in a Metropolitan Newspaper Office," *Chautauquan* 23, no. 3 (June 1896): 332-36; "How a Newspaper Is Produced," *Scientific American,* Nov. 14, 1903, 338-39; Vincent Richard DiGirolamo, "Crying the News: Children, Street Work, and the American Press, 1830s-1920s" (PhD diss., Princeton University, 1997), 165-68; George G. Foster, *New York by Gas-Light and Other Urban Sketches* (1850; repr., with an introduction by Stuart M. Blumin, Berkeley: University of California Press, 1990), 113-14; McCabe, *New York by Sunlight and Gaslight,* 595; Jacob A. Riis, *The Making of an American* (New York: Macmillan, 1901), 245, 331; "City News," *Chicago Times-Herald,* Nov. 19, 1896; Junius Henry Browne, *The Great Metropolis: A Mirror of New York* (Hartford, CT: American Publishing Company, 1869), 109. On night streetcars, see chapter 8.

25. "20,000 Toilers in Greater New York Earn Their Bread at Night," *BE,* July 7, 1901; "The Manufacture of Newspapers"; George P. Rowell and Company, *Centennial Newspaper Exhibition, 1876* (New York: George P. Rowell, 1876), 196. By the time of the *Brooklyn Eagle* article, New York included Brooklyn, Queens, Staten Island, and the Bronx as well as Manhattan.

26. "The Manufacture of Newspapers," 22; Edward Winslow Martin [James Dabney McCabe], *The Secrets of the Great City: A Work Descriptive of the Virtues and Vices, the Mysteries, Miseries and Crimes of New York City* (Philadelphia: Jones Brothers, 1868), 136-37; Rowell, *Centennial Newspaper Exhibition,* 197-202, 212-13; Matthews, "Night in a Metropolitan Newspaper Office"; "How a Newspaper Is Produced."

27. Theodore Dreiser, "Out of My Newspaper Days, II: St. Louis," *Bookman,* Jan. 1922, 427-33; Matthews, "Night in a Metropolitan Newspaper Office," 234; Riis, *Making of an American,* 245.

28. In addition to the articles published in newspapers, mid-nineteenth-century journalists also produced much of the nonfiction and fiction purporting to expose the "mysteries and miseries" of the city. On this genre, see John F. Kasson, *Rudeness and Civility: Manners in Nineteenth Century America* (New York: Hill and Wang, 1990), 74-8, and David S. Reynolds, *Beneath the American Renaissance: The Subversive Imagination in the Age of Emerson and Melville* (Cambridge, MA: Harvard University Press, 1989), 169-210. These works described the nocturnal city as a place of deep mystery and lurking horror, where daylight rules and rationality did not apply. Journalists continued to describe night like this through the turn of the cen-

tury; e.g., James B. Carrington, "New York at Night," *Scribner's Magazine*, Mar. 1900, 326–36; Richardson Wright "On Walking Home at Night," *Lippincott's Magazine*, Dec. 1914, 78–82. This convention persists in popular histories based on these sources, such as Herbert Asbury's *The Gangs of New York: An Informal History of the Underworld* (1927; New York: Thunder's Mouth Press, 2001); Luc Sante's *Low Life: Lures and Snares of Old New York* (New York: Vintage, 1992); and Mark Caldwell's *New York Night: The Mystique and Its History* (New York: Scribner, 2005). Some academic writing on the night has also emphasized mystery, irrationality, and deviance. "Fear and fascination meet us as we approach the nocturnal city and accompany us along our way," declares Joachim Schlör at the opening of *Nights in the Big City*. "We enter a world that is both familiar and strange, a landscape full of light and rich with shadows, the receptacle of our desires and of our hidden, unspeakable fears" (*Nights in the Big City: Paris, Berlin, London, 1840–1930* [London: Reaktion Books, 1998], 9).

29. Ernest Poole, *Child Labor: The Street* (New York: National Child Labor Committee, 1903), 23, 27; David Nasaw, *Children of the City: At Work and at Play* (New York: Oxford University Press, 1985), 48; George Thompson, *Venus in Boston: A Romance of City Life* (New York, 1849), 13; William F. Howe and A. H. Hummel, *In Danger, or Life in New York: A True History of a Great City's Wiles and Temptations* (New York: J. S. Ogilvie, 1888), 20–24; Jacob A. Riis, *The Children of the Poor* (1892; New York: Arno Press and the New York Times, 1971), 272–73; Edward N. Clopper, *Child Labor in City Streets* (New York: Macmillan, 1912), 52.

30. DiGirolamo, "Crying the News," 50, 124–25, 159, 177, 219, 335; Nasaw, *Children of the City*, 63, 72–77; Poole, *Child Labor*, 1–5; Scott Nearing, "The Newsboy at Night in Philadelphia," *Charities and the Commons*, Feb. 2, 1907, 778–84; *The Diary of George Templeton Strong: The Turbulent Fifties, 1850–1859*, ed. Allan Nevins and Milton Halsey Thomas (New York: Macmillan, 1952), 38; Maria Lydig Daly, *Diary of a Union Lady, 1861–1865*, ed. Harold Earl Hammond (1962; Lincoln: University of Nebraska Press, 2000), 351.

31. DiGirolamo, "Crying the News," 125–30, 172; Nasaw, *Children of the City*, 80–86, 115–20; "Out among the Newsboys," *NYT*, Aug. 17, 1879; William Hard, "'De Kid Wot Works at Night," *Everybody's Magazine*, Jan. 1908, 25–37; Riis, *Children of the Poor*, 260–61.

32. Poole, *Child Labor*, 2, 17–18; "What of the Newsboy of the Second Cities?" *Charities*, Apr. 11, 1903, 371; "Child Labor Reform in New York," *Charities*, Jan. 10, 1903, 56; "Herald Examiner News Alley, Monday, December 3, 1923," folder 84, JPAP; Florence Kelley, "The Street Trader under Illinois Law," in *The Child in the City: A Series of Papers Presented at the Conferences Held during the Chicago Child Welfare Exhibit* (Chicago: Chicago School of Civics and Philanthropy, 1912), 296; Nearing, "Newsboy at Night."

33. "20,000 Toilers in Greater New York Earn Their Bread at Night," *BE*, July 7, 1901.

CHAPTER SEVEN

1. Willard Glazier, *Peculiarities of American Cities* (Philadelphia: Hubbard Brothers, 1886), 332. See also Frank G. Carpenter, *Carpenter's Geographical Reader: North America* (New York: American Book Company, 1898), 218; J. M. Kelly, *J. M. Kelly's Handbook of Greater Pittsburg, 1895* (Pittsburgh: J. M. Kelly Company, [1895]), 83–84; James Parton, "Pittsburg," *Atlantic Monthly* 21 (Jan. 1868): 17–36; Charles Henry White, "Pittsburg," *Harper's Monthly Magazine* 117 (Apr. 1908): 902.

2. Samuel Young, *The Smoky City: A Tale of Crime* (Pittsburgh: A. A. Anderson, 1845), 134-35.

3. Alfred Mathews, "Pittsburgh. II: An Outline of the City's Industrial and Commercial Development," *Magazine of Western History* 2, no. 2 (June 1885): 175-91; Waldon Fawcett, "The Center of the World of Steel," *Century Magazine*, 62, no. 2 (June 1901); George H. Thurston, *Pittsburgh's Progress, Industries and Resources, 1886* (Pittsburgh: A. A. Anderson, 1886), 193; Frank C. Harper, *Pittsburgh of Today: Its Resources and People* (New York: American Historical Society, 1931), 2:628-33; US Bureau of the Census, *Occupations at the Twelfth Census* (Washington, DC: Government Printing Office, 1904), 461, 463, 628.

4. Charles B. Barnes, *The Longshoremen* (New York: Russell Sage Foundation, 1915), 57-59; Ernest Poole, "The Men on the Docks," *Outlook* 86, no. 4 (May 25, 1907): 142-44. Philadelphia longshoremen also endured very long shifts; Lisa McGirr, "Black and White Longshoremen in the IWW: A History of the Philadelphia Marine Transport Workers Union Local 8," *Labor History* 36, no. 3 (Summer 1995): 382.

5. Barnes, *Longshoremen*, 76-81, 122, 169-70, 184-85; Poole, "Men on the Docks"; "Labor Topics," *NYT*, Nov. 11, 1874; "Longshoremen Tell of Their Hard Life," *NYT*, June 9, 1914; "Longshoremen End 3 Weeks' Strike," *PI*, May 30, 1913; Michael Coleman Connolly, "The Irish Longshoremen of Portland, Maine, 1880-1923" (PhD diss., Boston College, 1988), 168.

6. John R. Stilgoe, *Metropolitan Corridor: Railroads and the American Scene* (New Haven, CT: Yale University Press, 1983), 4, 27, 30, 44; quotation from "A Little Rough on Pittsburgh," *Pittsburgh Commercial Gazette,* Mar. 20, 1886; Brad S. Lomazzi, *Railroad Timetables, Travel Brochures and Posters: A History and Guide for Collectors* (Spencertown, NY: Golden Hill Press, 1995), 6-7, 44, 46; *Everybody's Guide, January 1865* (Boston: A. F. Pollock, 1865); *Travelers' Official Railway Guide for the United States and Canada* (New York: J. W. Pratt, June 1868); Sam Bass Warner, *The Urban Wilderness: A History of the American City* (1972; Berkeley: University of California Press, 1995), 104-5; Walter Licht, *Working for the Railroad: The Organization of Work in the Nineteenth Century* (Princeton, NJ: Princeton University Press, 1983), 34, 174-77; "3,500 Railroad Men Join Fight for All Night Cars," *Harrisburg Patriot,* Jan. 17, 1910; *Occupations at the Twelfth Census*, 436; John R. Commons and William M. Leiserson, "Wage Earners of Pittsburgh," in *Wage Earning Pittsburgh*, ed. Paul Underwood Kellogg, *Pittsburgh Survey*, vol. 6 (1914; New York: Arno Press, 1974), 124-26; Ralph D. Fleming, *Railroad and Street Transportation* (Cleveland: Survey Committee of the Cleveland Foundation, 1916), 16, 35.

7. "New York and Harlem Railroad: The First Street Railroad in the World: Particulars of Its Early History," *SRJ* 9, no. 1 (Jan. 1893): 11; Joel A. Tarr, *Transportation Innovation and Changing Spatial Patterns in Pittsburgh, 1850-1934* (Chicago: Public Works Historical Society, 1978), 5; untitled item, *Hartford Courant*, Apr. 22, 1863; "Sixth-Avenue Railroad Company," *NYT*, Mar. 3, 1856; *The Boston Guide for Railroad, Steamboat and Horse Car Travel* (Boston, Railway Steam Printing Works), no. 1 (Dec. 1871): n.p.; "The Cars to Run All Night," *BE*, Mar. 10, 1859; "The Progress of the City Passenger Railway," *Philadelphia Daily Evening Bulletin,* Nov. 27, 1857; *Boston Railway Guide*, Jan. 1879, 38; New England Railway Publishing Company, *ABC Pathfinder: Once a Week,* Jan. 1-7, 1883, 6; "The Mighty Aldermen," *Chicago Times,* Dec. 28, 1880; "Street Railway Construction and Management," *SRJ* 2, no. 7 (May 1886): 247-48; "The News in Brooklyn," *NYT*, Nov. 9, 1890; "People's Street R.R.

Time Table," *Scranton Republican,* Oct. 10, 1888; "The Owl Car Booming," *St. Louis Globe-Democrat,* Sept. 19, 1879; "May Agree on Fares," *Milwaukee Sentinel,* Sept. 12, 1896; F. W. Doolittle, *Studies in the Cost of Urban Transportation Service* (New York: American Electric Railway Association, 1916), 280.

8. US Bureau of the Census, Department of Commerce and Labor, *Street and Electric Railways, 1902* (Washington, DC: Government Printing Office, 1905), 9, 14; George W. Hilton, *The Cable Car in America,* rev. ed. (Stanford, CA: Stanford University Press, 1997), 149, 156; Tarr, *Transportation Innovation,* 16–20; Doolittle, *Studies in the Cost of Urban Transportation,* 266.

9. "Facts and Figures relating to Street Railway Practice," *SRJ* 9, no. 12 (Dec. 1893): 811; "To Run All-Night Cars," *NYT,* May 21, 1894; Frederic W. Speirs, *The Street Railway System of Philadelphia: Its History and Present Condition* (Baltimore: Johns Hopkins Press, 1897), 121; W. W. Wheatly, "The Philadelphia Rapid Transit System—II," *SRJ* 21, no. 14 (Apr. 4, 1903): 503; "Improved 'Owl' Car Service in Philadelphia," *ERJ* 40, no. 22 (Dec. 7, 1912): 1177; "'Owl' Service in Louisville," *ERJ* 40, no. 23 (Dec. 14, 1912): 1215; "20,000 Toilers in Greater New York Earn Their Bread at Night," *BE,* July 7, 1901; Doolittle, *Studies in the Cost of Urban Transportation,* 392–93; Henry W. Blake and Walter Jackson, *Electric Railway Transportation* (New York: McGraw-Hill, 1917), 203–4; Alan R. Lind, *Chicago Surface Lines: An Illustrated History* (Park Forest, IL: Transport History Press, 1974), 201.

10. Speirs, *Street Railway System,* 100, 103, 105, 113, quotation at 101; "Life on a Street-Car," *Buffalo Express,* Mar. 14, 1886; untitled editorial, *SRJ* 2, no. 6 (Apr. 1886): 202; "The Swing System," *SRJ* 2, no. 9 (July 1886): 334; Blake and Jackson, *Electric Railway Transportation,* 142; "Street Railway Policy," *SRJ* 2, no. 9 (July 1886): 338.

11. Clay McShane and Joel A. Tarr, *The Horse in the City: Living Machines in the Nineteenth Century* (Baltimore: Johns Hopkins University Press, 2007), 109–10, 119–25; "The Horse Cars," *CIO,* July 30, 1881; "Street-Car Labor," *BS,* Feb. 23, 1886; "The Lot of the Car Horse," *NYT,* Mar. 18, 1883; "New York Third Avenue Street Railway." *Street Railway Journal* 2, no. 5 (March 1886): 191–92.

12. J. W. Watson, "Gas and Gas-Making," *Harper's New Monthly Magazine* 26 (Dec. 1862): 14–28; Kenneth Warren, *The American Steel Industry, 1850–1970: A Geographic Interpretation* (Oxford: Clarendon Press, 1973), 11, 15–16; Robert B. Gordon, *A Landscape Transformed: The Ironmaking District of Salisbury, Connecticut* (New York: Oxford University Press, 2001); Peter Temin, *Iron and Steel in Nineteenth Century America: An Economic Inquiry* (Cambridge: MIT Press, 1964), 82–88, 98; US Bureau of Labor, *Report on Conditions of Employment in the Iron and Steel Industry of the United States,* vol. 3, *Working Conditions and the Relations of Employers and Employees* (Washington, DC: Government Printing Office, 1913), 164; Charles Rumford Walker, *Steel: The Diary of a Furnace Worker* (Boston: Atlantic Monthly Press, 1922), 136.

13. Untitled editorial, *AGLJ* 4, no. 75 (Feb. 2, 1863): 232; "Local Intelligence: Progress of Our Iron-Clads," *NYT,* Sept. 16, 1862; "City Intelligence," *Hartford Courant,* Jan. 12, 1863; "Another Powder Mill Explosion," *NYT,* July 14, 1863; Edward K. Spann, *Gotham at War: New York City, 1860–1865* (Wilmington, DE: Scholarly Resources, 2002), 52; Emerson David Fite, *Social and Industrial Conditions in the North during the Civil War* (1910; Williamstown, MA: Corner House, 1976), 84.

14. National Industrial Conference Board, *Night Work in Industry* (New York: National Industrial Conference Board, 1927), 1–8; Andrew Carnegie, "Results of the Labor Struggle," *Forum*, Aug. 1886, 544–45; Judith A. McGaw, *Most Wonderful Machine: Mechanization and Social Change in Berkshire Paper Making, 1801–1885* (Princeton, NJ: Princeton University Press, 1987), 231, 305–6; Joseph D. Weeks, *Report on the Statistics of Wages in Manufacturing Industries* (Washington, DC: Government Printing Office, 1886), 74, 265–88; US Bureau of Labor, *Conditions of Employment in the Iron and Steel Industry*, 3:164–65; George E. Barnett, *Chapters on Machinery and Labor* (Cambridge, MA: Harvard University Press, 1926), 67, 95; Warren C. Scoville, *Revolution in Glassmaking: Entrepreneurship and Technological Change in the American Industry, 1880–1920* (Cambridge, MA: Harvard University Press, 1948), 38, 76; *Proceedings of the Sixteenth Annual Convention of the American Flint Glass Workers' Union* (Pittsburgh: American Flint Glass Workers' Union, 1893), 47–48; *Proceedings of the Forty-Seventh Convention of the American Flint Glass Workers' Union* (1923), 164–65; "The Colburn Machine Glass Co.," *National Glass Budget*, Sept. 8, 1906; Constance McLaughlin Green, *Holyoke Massachusetts: A Case History of the Industrial Revolution in America* (New Haven, CT: Yale University Press, 1939), 107; Martha Shiells and Gavin Wright, "Night Work as a Labor Market Phenomenon: Southern Textiles in the Interwar Period," *Explorations in Economic History* 20, no. 4 (Oct. 1983): 331–50. There appear to be no comprehensive data on the numbers of shift workers before 1950; Martha Ellen Koopman Shiells, "Hours of Work and Shiftwork in the Early Industrial Labor Markets of Great Britain, the United States and Japan" (PhD diss., University of Michigan, 1985), 26.

15. Scoville, *Revolution in Glassmaking*, 14–18, 27, 38, 74–76; Editorial, *American Flint* 17, no. 4 (Feb. 1926): 22–24; US Department of Labor, *Productivity of Labor in the Glass Industry* (Washington, DC: Government Printing Office, 1927), 18, 22–23; Harper, *Pittsburgh of Today*, 1:605–9; description of working conditions in Louis Cadley, letter in *American Flint* 12, no. 9 (July 1921): 35; Weeks, *Report on the Statistics of Wages in Manufacturing Industries*, 78–80, 85–86, quotation at 85.

16. Scoville, *Revolution in Glassmaking*, 14–18, 27–28, 38, 81, 158–59, 190, 211; John A. Voll, "Collective Bargaining in the Glass Bottle Industry," *American Flint* 12, no. 11 (Sept. 1921): 7; Editorial, *National Glass Budget*, Aug. 29, 1903; "New Machine Marks the Passing of the Glassmakers' Unions," *NYH*, May 3, 1908; *Proceedings of the Sixteenth Annual Convention of the American Flint Glass Workers' Union* (Pittsburgh: American Flint Glass Workers' Union, 1893), 47–48; *Forty-Seventh Convention of the American Flint Glass Workers' Union*, 164–65; "The Colburn Machine Glass Co.," *National Glass Budget*, Sept. 8, 1906.

17. Scoville, *Revolution in Glassmaking*, 77–78, 81, 202–8, 321; "Why the Summer Stop Should Be Abolished," *American Flint* 11, no. 10 (Aug. 1920): 1–11; Voll, "Collective Bargaining in the Glass Bottle Industry," 8–10; *Proceedings of the Forty-Sixth Convention of the American Flint Glass Workers' Union* (1922), 47–49, 170; *Proceedings of the Forty-Ninth Convention of the American Flint Glass Workers' Union* (1925), 33, 131–39; *Proceedings of the Fiftieth Convention of the American Flint Glass Workers' Union* (1926), 71, 88–89, 111; Patrick J. Fallon, letter in *American Flint* 17, no. 4 (Feb. 1926): 39.

18. Paul Krause, *The Battle for Homestead, 1880–1892: Politics, Culture, and Steel* (Pittsburgh: University of Pittsburgh Press, 1992), 47–56; Temin, *Iron and Steel*, 155–59, 221; Fawcett, "Center of the World of Steel," 189–203; John A. Fitch, *The Steel Workers* (New York:

Russell Sage Foundation, 1910), 166; US Bureau of Labor, *Conditions of Employment in the Iron and Steel Industry*, 3:164–65; US Bureau of Labor Statistics, *Wages and Hours of Labor in the Iron and Steel Industry: 1907 to 1926* (Washington, DC: Government Printing Office, 1927), 9–10, 27, 64.

19. Quoted in S. J. Kleinberg, *The Shadow of the Mills: Working-Class Families in Pittsburgh, 1870–1907* (Pittsburgh: University of Pittsburgh Press, 1989), 9; John N. Ingham, *Making Iron and Steel* (Columbus: Ohio State University Press, 1991), 130–31; Andrew Carnegie, "Results of the Labor Struggle," *Forum*, Aug. 1886, 544; David R. Roediger and Philip S. Foner, *Our Own Time: A History of American Labor and the Working Day* (New York: Greenwood Press, 1989), 186; US Bureau of Labor Statistics, *Wages and Hours of Labor in the Iron and Steel Industry*, 5–6; David Brody, *Steelworkers in America: The Nonunion Era* (1960; New York: Harper and Row, 1969), 36, 273–75.

20. US Bureau of Labor, *Report on Conditions of Employment in the Iron and Steel Industry in the United States*, vol. 1, *Wages and Hours of Labor* (Washington, DC: Government Printing Office, 1911), 11; "Perils That Menace Pittsburg Workers," *NYT*, Jan. 3, 1909; US Bureau of Labor, *Conditions of Employment in the Iron and Steel Industry*, 3:159–60, 202.

21. McGaw, *Most Wonderful Machine*, 306; Weeks, *Report on the Statistics of Wages in Manufacturing Industries*, 269, 281, 284, 288; Commons and Leiserson, "Wage Earners of Pittsburgh," 126; Kleinberg, *Shadow of the Mills*, 9–10; Fitch, *Steel Workers*, 63; "Yardmen Seeking Relief," *Pittsburg Times*, Apr. 20, 1886; "Strike at West End Plant of the A.T. & S. Company," *Bridgeport Post*, July 16, 1907. Some restaurants also had their waitresses alternate between night and day shifts; Consumers' League of New York City, *Behind the Scenes in a Restaurant* (1916), reprinted in *The Consumers' League of New York: Behind the Scenes of Women's Work*, ed. David J. Rothman and Sheila M. Rothman (New York: Garland, 1987), 17.

22. Walker, *Steel*, v, 147, 150, 154; Walker's assessment is supported by a Labor Department report, which found that managers believed men on the night shift were so sleep deprived that they were not "fit for service immediately after coming on duty"; US Bureau of Labor, *Conditions of Employment in the Iron and Steel Industry*, 3:383.

23. Thomas Bell, *Out of This Furnace* (1941; Pittsburgh: University of Pittsburgh Press, 1976), 47, 167; Walker, *Steel*, 66, 72, 151; Margaret F. Byington, *Homestead: The Households of a Mill Town* (New York: Russell Sage Foundation, 1910), 127n; "Danger of Night Work," *Journal of United Labor* 5, no. 2 (May 25, 1884): 702; Hamlin Garland, "Homestead and Its Perilous Trades," *McClure's* 3, no. 1 (June 1894): 3–20; "Perils That Menace Pittsburg Workers," *NYT*, Jan. 3, 1909; Mary D. Hopkins, *The Employment of Women at Night*, Bulletin of the Women's Bureau 64 (Washington, DC: Government Printing Office, 1928), 46–48.

24. "20,000 Toilers in Greater New York Earn Their Bread at Night," *BE*, July 7, 1901; Licht, *Working for the Railroad*, 233–34; Barnes, *Longshoremen*, 58; "River Sharks," *NYH*, July 25, 1870.

25. Walker, *Steel*, 58, 63, 154; Bell, *Out of This Furnace*, 47; Mary Heaton Vorse, *Men and Steel* (London: Labour Publishing Company, 1922), 29; Byington, *Homestead*, 36; *Report on the Statistics of Wages*, 284.

26. "Perils That Menace Pittsburg Workers," *NYT*, Jan. 3, 1909; US Bureau of Labor, *Conditions of Employment in the Iron and Steel Industry*, 3:159–60; "'Double Shift' Wife Vanishes after Day Mate Compares Notes with Night Spouse," *Pittsburgh Post*, May 27, 1922; Blake and

Jackson, *Electric Railway Transportation*, 401; Bell, *Out of This Furnace*, 135; "Courtship on the 'L,'" *Buffalo Express*, Dec. 10, 1885.

27. Byington, *Homestead*, 37, 148, 172–73; Perry R. Duis, *Challenging Chicago: Coping with Everyday Life, 1837–1920* (Urbana: University of Illinois Press, 1998), 113; Peter Roberts, "Immigrant Wage Earners," in Kellogg, *Wage Earning Pittsburgh*, 47.

28. G. M. Hopkins Company, *Map of Greater Pittsburgh, Pa.* (Philadelphia: G. M. Hopkins Company, 1910); US Department of the Interior, US Geological Survey, "Pittsburgh East Quadrangle," 7.5 minute series, 1997; "Jones and Laughlin's Coke Plant," *Mines and Minerals* 29, no. 5 (Dec. 1908): 195–99; Citizens Committee on City Plan of Pittsburgh, *Railroads of the Pittsburgh District: A Part of the Pittsburgh Plan*, report no. 5 (Pittsburgh: Municipal Planning Association, 1924), 28; US Bureau of the Census, Department of Commerce and Labor, "Thirteenth Census of the United States: 1910. Manuscript Population Schedule. Pittsburgh, Allegheny County, Pennsylvania," enumeration districts 469–79, Ward 15; *Parks: A Part of the Pittsburgh Plan*, report no. 4 (Pittsburgh: Citizens Committee on City Plan of Pittsburgh, 1923); photograph captioned "Homes with Privies Above the Railroad Tracks, Sylvan Avenue, Hazelwood, February 1907," in Kleinberg, *Shadow of the Mills*, following 196.

29. Pittsburgh Bureau of Police Records, No. 10 Station (Hazelwood), ser. 18, vol. 1, Complaint Book, May 1 to Dec. 31, 1922, AIS; entries dated May 8, 14, 18, 19, 22, 23, 24, 29, 30, 31, June 9, 12, 23, July 15, Sept. 4, 10, 11, Oct. 4, 6, 1922; Sanborn Map Company, *Insurance Maps of Pittsburgh Pennsylvania*, vol. 4 (New York: Sanborn Map Company, 1925), pl. 477; location of police station from G. M. Hopkins Company, digitized map with heading "1923 Volume 2—East End (South): Wards 7 and 14–15," pl. 31B, Historic Pittsburgh Maps Collection, University of Pittsburgh Library, http://digital.library.pitt.edu/maps/; 1920 US Census Population Schedules, Pittsburgh, 15th Ward, enumeration district 565, 10B and 11B. On peddlers, Pittsburgh Police, Complaint Book, entries dated May 31, Sept. 8, 21, 26, 1922); "Mayor Repels Night Workers," *CT*, June 23, 1911; see also Duis, *Challenging Chicago*, 279.

30. US Bureau of Labor, *Conditions of Employment in the Iron and Steel Industry*, 3:96; Kleinberg, *Shadow of the Mills*, 236–40; Walker, *Steel*, 28, 151–52; Barnes, *Longshoremen*, 22–24, 129–30; McGirr, "Black and White Longshoremen in the IWW," 382.

31. Licht, *Working for the Railroad*, 44, 149–53, 214–21; Fleming, *Railroad and Street Transportation*, 14, 67; Thomas Conway Jr., "Street Railway Employment in the United States," *ERJ*, Sept. 29, 1917, 570–73; "Picked up in the Streets," *NYT*, Jan. 16, 1882.

32. "My Experiences as a Street-Car Conductor," *Lippincott's Monthly Magazine*, June 1886, 633–34; Licht, *Working for the Railroad*, 73–78, 167–70; Conway, "Street Railway Employment," 572; Kleinberg, *Shadow of the Mills*, 27–34; Barnes, *Longshoremen*, 57–58, 129–40; Poole, "Men on the Docks"; Testimony of M. E. Griffin, "Public Hearing before the Special Commission on Social Insurance . . . ," Oct. 25, 1916, in American Association for Labor Legislation pamphlets, No. 5001-P, box 4, folder 1, KCCU.

33. Owen R. Lovejoy, "Child Labor in the Glass Industry," *Annals of the American Academy of Political and Social Science* 27, no. 2 (Mar. 1906): 42–53; Scoville, *Revolution in* Glassmaking, 321; Boris Stern, *Productivity of Labor in the Glass Industry*, Bulletin of the United States Bureau of Labor Statistics 441 (Washington, DC: Government Printing Office, 1927), 24–25.

34. Kleinberg, *Shadow of the Mills*, 25; Alice Kessler-Harris, *Out to Work: A History of Wage-Earning Women in America* (New York: Oxford University Press, 1982), 128, 179, 185, 203–4;

US Bureau of the Census, *Fourteenth Census of the United States Taken in the Year 1920*, vol. 4, *Population, 1920, Occupations* (Washington, DC: Government Printing Office, 1923), 208.

35. Koopman Shiells, "Hours of Work and Shiftwork," 176–78; Shiells and Wright, "Night Work as a Labor Market Phenomenon"; "Another Horror," *PI*, Oct. 13, 1881; "The Fifty-Five Hour Law," *Trenton Times*, Aug. 11, 1892; "Manufacturing News," *Boston Daily Advertiser*, Oct. 18, 1893; "Day Spinners Strike," *Boston Morning Journal*, Oct. 23, 1898; "Mills Will Go on Double Time," *PI*, Aug. 31, 1899; "Object to Night Work," *Pawtucket Evening Times*, Oct. 5, 1899; "Question of Night Work in Mills," *Springfield Republican*, Jan. 19, 1900; "Little Girls Toil at Night in Big Woolen Factory," *Trenton Times*, June 3, 1902; "Night Work by Girls Stopped," *Pawtucket Evening Times*, Jan. 15, 1903; J. P. Munroe, "What Are Operatives to Do?" *Charlotte Observer*, Feb. 5, 1893; "The Argument Continues," *Charlotte Observer*, Feb. 7, 1893; Jacquelyn Dowd Hall, James Leloudis, Robert Korstad, Mary Murphy, Lu Ann Jones, and Christopher B. Daly, *Like a Family: The Making of a Southern Cotton Mill World* (1987; New York: W. W. Norton, 1989), 200; Hopkins, *Employment of Women at Night*, 1, 4–7; Agnes De Lima, "Night Working Mothers in Textile Mills of Passaic, N.J. and Vicinity," *American Flint* 18, no. 3 (Jan. 1927): 4–9, quotation at 4.

36. Junius Henri Browne, *The Great Metropolis: A Mirror of New York* (Hartford, CT: American Publishing Company, 1869), 102–5; Stephen Crane, "In the Broadway Cars" (1902), in *The New York City Sketches of Stephen Crane and Related Pieces*, ed. R. W. Stallman and E. R. Hagemann (New York: New York University Press, 1966), 185.

37. H. C. Bunner, *The Suburban Sage: Stray Notes and Comments on His Simple Life* (New York: Keppler and Schwarzmann, 1896), quotations at 92, 94–95, 97.

CHAPTER EIGHT

1. This summary is based mainly on court testimony, as paraphrased in "The Opening and Testimony for the Prosecution," *NYT*, May 23, 1871, and "The Foster Trial," *NYTr*, May 23, 1871, as well as on J. Edwards Remault, *The "Car-Hook" Tragedy: The Life, Trial, Conviction and Execution of William Foster for the Murder of Avery D. Putnam* (Philadelphia: Barclay, 1873). Additional information comes from 1870 US Census schedules, New York, NY, ward 15, enumeration district 9, sheet 11, and enumeration district 11, sheet 65; F. W. Beers, "Map of the Central Portions of the Cities of New York and Brooklyn," in *State Atlas of New Jersey* (New York: Beers, Comstock and Cline, 1872); "Opening of the Gilsey House," *NYT*, Apr. 16, 1871; "Shocking Murder," *NYT*, Apr. 28, 1871; "The Car Murder," *NYT*, May 2, 1871; "Foster's Conviction," *NYT*, May 26, 1871; and "Foster to Be Hanged," *NYT*, Jan. 22, 1873. Jeanne Duval usually went by "Anna"; I use "Mrs." here to distinguish her from her daughters.

2. "Ruffianism in Street-Cars," *NYT*, Apr. 28, 1871; "Mr. Putnam Dead," *NYT*, Apr. 30, 1871; "The Duty of Self-Defense," *NYT*, May 7, 1871; "Shall Murder Be Punished?" *NYTr*, May 16, 1871.

3. Remault, *"Car-Hook" Tragedy*, 29; *Some of the Papers Laid before the Governor in Support of the Petition for the Commutation of the Sentence of William Foster to Imprisonment for Life* (New York: Evening Post, 1873); "Foster's Autobiography," *NYTr*, May 8, 1871; "Foster to Be Hanged," *NYT*, Jan. 22, 1873; "Twain on Foster," *Denver Daily Rocky Mountain News*, Mar. 19, 1873.

4. "The Prisoner Indicted," *NYT*, May 3, 1871; "Foster's Conviction," *NYT*, May 26, 1871; "The Foster-Putnam Murder Trial," *San Francisco Daily Evening Bulletin*, May 27, 1871; untitled editorial, *Cleveland Morning Daily Herald*, Jan. 24, 1873; "The Governor in the Case of Foster," *New York Observer and Chronicle*, Mar. 20, 1873. Foster had spent the first two days after the attack in the friendlier confines of a police station, courtesy of police court magistrate George W. Plunkitt (a career politician), rather than committed to the Tombs prison as was customary in homicide cases; "The Broadway Car Outrage," *NYTr*, May 1, 1871.

5. Iver Bernstein, *The New York City Draft Riots: Their Significance for American Society and Politics in the Age of the Civil War* (New York: Oxford University Press, 1990), 228–35; Eric Homberger, *Scenes from the Life of a City: Corruption and Conscience in Old New York* (New Haven, CT: Yale University Press, 1994), 193, 199–200; quotations from "A Revival of Justice," *NYTr*, May 27, 1871. See also chapter 2 above. Reports of the Commune's bloody defeat in May appeared alongside news of the Foster trial in New York newspapers.

6. "Summer Travelling," *Philadelphia Evening Bulletin*, Aug. 19, 1857; "Outrages on Ladies in Cars and Stages," *NYT*, Oct. 17, 1869; "A Good Lesson to Young Loafers," *BE*, Aug. 28, 1880; "A 'Masher' in Merited Misery," *BE*, Mar. 11, 1881; "Ruffianism in Public Vehicles," *BE*, Mar. 24, 1885. For a thoughtful examination of the issue, see Amy G. Richter, *Home on the Rails: Women, the Railroad, and the Rise of Public Domesticity* (Chapel Hill: University of North Carolina Press, 2005), and Patricia Cline Cohen, "Safety and Danger: Women on American Public Transport, 1750–1850," in *Gendered Domains: Rethinking Public and Private in Women's History*, ed. Dorothy O. Helly and Susan B. Reverby (Ithaca, NY: Cornell University Press, 1992). For a different interpretation, see John Henry Hepp IV, *The Middle-Class City: Transforming Space and Time in Philadelphia, 1876–1926* (Philadelphia: University of Pennsylvania Press, 2003).

7. On scrutiny by fellow passengers, see "Outrages on Ladies in Street Cars—Testimony of a Young Lady," *NYT*, Oct. 23, 1869; "The Passing Hour," *Buffalo Express*, Dec. 28, 1885. On working-class ridership, *The Diary of George Templeton Strong: The Turbulent Fifties, 1850–1859*, ed. Allan Nevins and Milton Halsey Thomas (New York: Macmillan, 1952), 97; Edward Winslow Martin [James Dabney McCabe], *The Secrets of the Great City: A Work Descriptive of the Virtues and Vices, the Mysteries, Miseries and Crimes of New York City* (Philadelphia: Jones Brothers, 1868), 117–18; John Noble Company, *Facts respecting Street Railways: The Substance of a Series of Official Reports from the Cities of New York, Brooklyn, Boston, Philadelphia, Baltimore, Providence, Newark, Chicago, Quebec, Montreal, and Toronto* (London: P. S. King, 1866), 26–27. On lack of police, "The Mayor and the Car Companies," *NYTr*, May 31, 1871.

8. Richter, *Home on the Rails*, 23–24; "Table-Talk," *Appleton's Journal*, May 27, 1871; "Ruffianism on Rails," *NYTr*, May 1, 1871; "Avery D. Putnam's Death," *NYT*, Feb. 10, 1874.

9. George Thompson, *The Gay Girls of New-York, or Life on Broadway* (New York, 1853), 14–15.

10. Does the City Need Reformation?" *Advocate of Moral Reform* 2, no. 12 (July 1, 1836): 94; "A Monstrous Outrage by Ruffians on a Girl," *NPG*, July 3, 1847; "Brutal Assault," *The Monthly Cosmopolite*, July 1, 1850; "Outrageous Conduct," *BE*, July 20, 1853; "Pittsburg Correspondence," *NPG*, May 12, 1860; "Police Intelligence: A Brutal Outrage," *NYH*, Sept. 27, 1866; "Assaulting Unprotected Women," *NYT*, Dec. 3, 1888; Christine Stansell, *City of Women: Sex and Class in New York, 1789–1860* (1986; Urbana: University of Illinois Press, 1987), 26–27.

11. "Rowdies," *Daily Cincinnati Commercial*, Dec. 29, 1847; "Insulting Females," *BE*, Aug. 14, 1852; "A Masher Captured," *Milwaukee Daily Sentinel*, May 25, 1881; "Where Are the Police?" *Toledo Blade*, Dec. 27, 1890.

12. Stansell, *City of Women*, 27–28; "A Few Facts," *Advocate of Moral Reform* 1, no. 4 (Apr. 1835): 20; Eliza Potter, *A Hairdresser's Experience in High Life* (1859; New York: Oxford University Press, 1991), 225.

13. "Street Rowdies," *NYTr*, Sept. 7, 1843; "Corner Loafers," *BE*, Mar. 17, 1864; "Summer Annoyances," *BE*, May 17, 1865; "Genteel Corner Loafers and the Other Sort," *BE*, June 13, 1865; "Our Correspondence Column: Corner Loafers," *BE*, Apr. 9, 1869; "Running the Gauntlet," *Frank Leslie's Illustrated Newspaper*, May 16, 1874; "Kill the Masher," *Milwaukee Daily Sentinel*, May 18, 1881; "How to Help Toledo," *Toledo Blade*, Dec. 31, 1890; "Monthly Report for Month of March, 1913," in Philadelphia Police, Letter Book, Feb.–Apr. 1913, Philadelphia City Archives, box A-999.

14. Stansell, *City of Women*, 86, 97–98; Timothy J. Gilfoyle, *City of Eros: New York City, Prostitution and the Commercialization of Sex, 1790–1920* (New York: W. W. Norton, 1992); "Fulton Street Flirtations," *BE*, May 21, 1855; "Saturday Night," *BE*, Sept. 4, 1871; Captain's report for 1882, in Fairmount Park Commission, bound volume titled "Statistics," Philadelphia City Archives, FPC-64.

15. J. R. McDowall, *Magdalen Facts* (New York, 1832), 87–88; "Street Walkers," *Whip and Satirist of New York*, May 7, 1842; "Insulting Females," *BE*, May 6, 1851; "Trials of Policemen," *NYH*, Sept. 27, 1866; "Broadway Statues," *NPG*, Dec. 8, 1866; "The Treatment of Ladies in Stages and Cars," *NYT*, Oct. 31, 1869; "The Foster Trial," *NYTr*, May 23, 1871; "Foster's Conviction," *NYT*, May 26, 1871; "Crusading the Cyprians," *NPG*, Dec. 6, 1879; "Washington Notes," *NPG*, May 8, 1880. See also *Diary of George Templeton Strong*, 218.

16. "Improper Style of Dress for the Street," *Advocate of Moral Reform* 3, no. 1 (Jan. 1, 1837): 188; "Attire of the Harlot," *Advocate of Moral Reform* 3, no. 24 (Dec. 15, 1837): 372.

17. Daphne Dale [Mrs. Charles F. Beezley], *Our Manners and Social Customs: A Practical Guide to Deportment, Easy Manners, and Social Etiquette* (Chicago: Elliott and Beezley, 1892), 139; Florence Hartley, *The Ladies' Book of Etiquette, and Manual of Politeness* (Boston: Locke and Bubier, 1875), 112–13; S. A. Frost [Sarah Annie (Frost) Shields], *Frost's Laws and By-Laws of American Society* (New York: Dick and Fitzgerald, 1869), 90, 161; Emily Thornwell, *The Lady's Guide to Perfect Gentility* (New York: Derby and Jackson, 1859), 78; Censor [Oliver Bunce], *Don't: A Manual of Mistakes and Improprieties More or Less Prevalent in Conduct and Speech* (New York: D. Appleton, 1883), 48–49; Pro Bono Publico, "The Weaker Sex," *Philadelphia Evening Bulletin*, June 2, 1911. See also Diane Shaw, *City Building on the Eastern Frontier: Sorting the New Nineteenth-Century City* (Baltimore: Johns Hopkins University Press, 2004), 90, 99.

18. "The Treatment of Ladies in Stages and Cars," *NYT*, Oct. 31, 1869; "A Real Reform Inaugurated by Judge Walsh," *BE*, Feb. 5, 1874; "Metropolitan Mashers," *BE*, July 21, 1884; quotation from "Insulting Ladies in the Public Vehicles," *BE*, Dec. 16, 1870.

19. Brian J. Cudahy, *Cash, Tokens, and Transfers: A History of Urban Mass Transit in North America* (New York: Fordham University Press, 1990), 7, 13; Hartley, *Ladies' Book of Etiquette*, 34–38, 115; Frost, *Frost's Laws and By-Laws*, 93–94; Annie R. White, *Polite Society at Home and Abroad* (Chicago: L. P. Miller, 1891), 36; "Prowling Wolves," *BE*, Aug. 18, 1869; "Outrages

on Ladies in Street Cars—Testimony of a Young Lady," *NYT*, Oct. 23, 1869; "Life in New York City," *BE*, Apr. 1, 1883; Stephen Crane, "A Lovely Jag in a Crowded Car" (1895), in *The New York City Sketches of Stephen Crane and Related Pieces*, ed. R. W. Stallman and E. R. Hagemann (New York: New York University Press, 1966), 125–28.

20. Noble and Company, *Facts respecting Street Railways*, 26–27; Stephen Crane, "In the Broadway Cars" (1902), in *New York Sketches of Stephen Crane*, 185; "Insulting Females," *BE*, Aug. 14, 1852; Delta, "A Trip in the Street Cars," *Godey's Lady's Book* 70 (June 1865): 509–11; "Insulting Females in Public Vehicles—Letter from a Lady," *NYT*, Oct. 9, 1869; "A Pair of Puppies," *BE*, June 10, 1870; "Insulting Ladies in the Public Vehicles," *BE*, Dec. 16, 1870; "The Wall Street Fops Who Insult Ladies," *BE*, May 22, 1871; "The Street Cars as a School of Immorality," *BE*, Mar. 20, 1876; "Correspondence: In Reference to Over-Crowded Cars," *PPL*, Mar. 22, 1890; "The Metropolitan Street Railway System," *Street Railway Review* 11, no. 9 (Sept. 15, 1901): 536; House Committee on the District of Columbia, *Report of Hearings of January 31, February 6 and 13, 1907* (Washington, DC: Government Printing Office, 1907), 90; "Report on Traffic Conditions in Pittsburgh," *ERJ* 34, no. 6 (Aug. 7, 1909): 212; Virginia Scharff, *Taking the Wheel: Women and the Coming of the Motor Age* (New York: Free Press, 1991), 6–7; quotation from "Street-Car Salad, *HW*, Mar. 23, 1867, 189. On eye contact, "Street-Car Sketches," *St. Louis Globe-Democrat*, June 20, 1880; untitled editorial, *BE*, Sept. 9, 1868; "A Good Lesson to Young Loafers," *BE*, Aug. 28, 1880; "Life in New York City," *BE*, Oct. 12, 1884.

21. Thornwell, *Lady's Guide*, 78; Hartley, *Ladies' Book of Etiquette*, 115; Beezley, *Our Manners*, 139–40; *Etiquette for Americans, by a Woman of Fashion* (Chicago: Herbert S. Stone, 1898), 242.

22. "New Scheme to Fight the Masher," *Milwaukee Sentinel*, Jan. 2, 1898; Mrs. Tucker C. Laughlin Pocket Diary, 1888–91, Historical Society of Pennsylvania; 1900 US Census schedule for Philadelphia, ward 28, enumeration district 695, sheet 2; "A Lady Insulted in a Street-Car," *NYT*, May 10, 1871; "An Impudent Street-Car Thief," *NYT*, Aug. 6, 1878; "Robbed in a Street Car," *NYT*, Dec. 3, 1887; "Thieves in Street Cars," *NYT*, July 16, 1889.

23. "A Railroad Ride through the City by Moonlight," *Philadelphia Daily Evening Bulletin*, Jan. 26, 1858; "Street-Car Sketches," *St. Louis Globe-Democrat*, June 20, 1880; House Committee, *Report of Hearings*, 90; "Owl Service Not Justified in La Crosse," *ERJ* 39, no. 13 (Mar. 30, 1912): 520–21. See also *Report of Transit Commissioner to the Honorable Mayor and the City Council of the City of Pittsburgh* (Pittsburgh, 1917), table A.

24. House Committee, *Report of Hearings*, 90–91; "The Bridge Transportation System between New York and Brooklyn," *SRJ* 13, no. 2 (Feb. 1897): 70–71; "System of the Brooklyn Rapid Transit Co.," *Street Railway Review* 11, no. 9 (Sept. 15, 1901): 548; "Daily Traffic on Lower Broadway," *Street Railway Review* 11, no. 9 (Sept. 15, 1901): 536; George H. Davis, "The Adjustment of American Street Railway Rates to the Expansion of City Areas," *ERJ* 37, no. 5 (Feb. 4, 1911): 212; "Regulation of Transportation Service by the Metropolitan Street Railway, New York," *ERJ* 36, no. 1 (July 10, 1910): 7; Ralph D. Fleming, *Railroad and Street Transportation* (Cleveland: Survey Committee of Cleveland Foundation, 1916), 73; "Traffic Curves on the New Orleans and Carrollton Railroad," *Street Railway Review* 11, no. 9 (Sept. 15, 1901): 565; Barclay Parsons and Klapp, *Report on a Rapid Transit System for the City of Detroit* (1918), 23; George H. Davis, "The Adjustment of American Street Railway Rates to the Expansion of City Areas," *ERJ* 37, no. 5 (Feb. 4, 1911): 212.

25. "Comments of Our Kicker," *SRJ* 1, no. 3 (Jan. 1885): 56; untitled editorial, *NYT*, Dec. 6, 1886; "Is Your Car Lighting Good?" *Street Railway Review* 6, no. 4 (Apr. 15, 1896): 215; US Bureau of the Census, Department of Commerce and Labor, *Street and Electric Railways, 1902* (Washington, DC: Government Printing Office, 1905), 206-7; "Straw in Street Cars," *PPL*, Dec. 25, 1889; Writers' Program of the Works Progress Administration, *Story of Old Allegheny City* (Pittsburgh: Allegheny Centennial Committee, 1941), 41.

26. "Street Railway Construction and Management," *SRJ* 2, no. 7 (May 1886): 247-48; D. H. Williams, "The Cable Railroad," *Pittsburg Times*, Dec. 6, 1886; *Boston Nickel Pathfinder* 15, no. 2 (Feb. 1913): 178-81; "Stay Out as Late as You Wish! 'Owl Cars' Are Running," advertisement in *Toledo Blade*, Apr. 27, 1910.

27. "Public Knowledge of City Schedules," *ERJ* 37, no. 17 (Apr. 29, 1911): 735; "Night Cars Must Run," *Chicago Times-Herald*, Nov. 17, 1896; Paul Barrett, *The Automobile and Urban Transit: The Formation of Public Policy in Chicago, 1900-1930* (Philadelphia: Temple University Press, 1983), 113; *Geer's Hartford City Directory for 1874-75* (Hartford, CT: Elihu Geer, 1874), 268; "Notes and Items," *SRJ* 2, no. 8 (June 1886): 294; untitled item, *SRJ* 2, no. 11 (Sept. 1886): 441; *Boston Pocket Manual: Street Railway Guide*, vol. 2, no. 20 (July 17, 1899); "Owl Service Campaign in Toledo," *ERJ* 35, no. 18 (Apr. 30, 1910): 801; Stephen Crane, "When Every One Is Panic Stricken" (1894), in *New York Sketches of Stephen Crane*, 97; "Frequency of Cars," *SRJ* 4, no. 1 (Jan. 1888): 16; Henry W. Blake and Walter Jackson, *Electric Railway Transportation* (New York: McGraw-Hill, 1917), 70; untitled editorial, *NYT*, Dec. 19, 1884.

28. *The Tricks and Traps of New York City*, part 1 (New York: Dinsmore, 1858), 46-47; "Rampant Robbers," *St. Louis Globe-Democrat*, June 23, 1881; "The Robbery Record," *St. Louis Globe-Democrat*, July 19, 1886; "Fiends in Human Guise," *Pittsburg Times*, Nov. 27, 1886; "Accused of Street Hold Up," *Philadelphia Evening Bulletin*, June 2, 1911; Frost, *Frost's Laws and By-Laws*, 94; "A Curious Story of Attempted Highway Robbery," *BE*, Jan. 18, 1866.

29. "Outrage by a Ruffian Gang—Assault and Robbery of a Stage Conductor," *BE*, Jan. 30, 1855; "The Metropolitan Police Commissioners," *BE*, July 28, 1857; "Row on an Owl Car," *St. Louis Globe-Democrat*, July 25, 1881; "Bound to Die," *St. Louis Globe-Democrat*, Aug. 25, 1881; "An Affray on an Owl Car," *St. Louis Globe-Democrat*, Mar. 4, 1882; "A Car-Driver's Philosophy," *NYT*, Dec. 25, 1882; "An Owl-Car Tragedy," *St. Louis Globe-Democrat*, Oct. 24, 1887; "Street Railway Employes Assaulted," *PPL*, June 5, 1890; George J. Manson, "Work Indoors and Out: A Conductor on a City Railway Car," *Independent* 48 (Oct. 8, 1896): 1378; "Fight in Trolley Car," *Pittsburgh Post*, July 27, 1907. On robbers and pickpockets, "Street-Robbers," *NYT*, Aug. 15, 1878; "Thieves in Street Cars," *NYT*, July 16, 1889; "Robbing Street Railway Passengers in Chicago," *SRJ* 9, no. 1 (Jan. 1893): 21.

30. "The Night Cars on the City Railroads," *NYT*, July 9, 1865; "The City Cars at Night," *NYT*, Jan. 6, 1866; "Profanity on the Cars," *Philadelphia Press*, Feb. 19, 1869; "Car Conduct: A Correspondent Relates an Incident of Too Frequent Occurrence," *BE*, Aug. 28, 1873; "Owl-Car Passengers," *St. Louis Globe-Democrat*, July 31, 1881; "Street Railway News," *SRJ* 4, no. 1 (Jan. 1888): 24; Crane, "In the Broadway Cars," 188-89; *Forty-First Annual Report of the [Massachusetts] Board of Railroad Commissioners, Jan. 1910* (Boston: Wright and Potter, 1910), 90.

31. "Owl-Car Passengers," *St. Louis Globe-Democrat*, July 31, 1881; James D. McCabe, *New York by Sunlight and Gaslight: A Work Descriptive of the Great American Metropolis* (Philadelphia: Douglass Brothers, 1882), 190; "The Night Car's Freight," *NYT*, Mar. 11, 1883; "Owl

Cars," *Globe-Democrat,* Mar. 23, 1880. On night workers' need for owl cars, "Common Council," *BE,* Mar. 31, 1857; "Will the Cars Run All Night?" *Philadelphia Daily Evening Bulletin,* Jan. 23, 1858; "Owl Cars for Night Workers," *Milwaukee Sentinel,* Nov. 3, 1895; Barclay Parsons and Klapp, *Report on Detroit Street Railway Traffic and Proposed Subway* (1915), 45. On Harrisburg, "Councils Ask for Night Cars," *Harrisburg Patriot,* Jan. 11, 1910; "Just and Reasonable," *Harrisburg Patriot,* Jan. 12, 1910; "3,500 Railroad Men Join Fight for All Night Cars," *Harrisburg Patriot,* Jan. 17, 1910; "The Railroad Men's Reasonable Request, *Harrisburg Patriot,* Jan. 18, 1910; "Night Car Service Much Too Costly," *Harrisburg Patriot,* Mar. 15, 1910.

32. McCabe, *Secrets of the Great City,* 121; "A Scene on the New York Horse Cars," *Sporting Times,* Mar. 13, 1869; "Twenty Chorus Maidens," *NPG,* Jan. 3, 1903; "A Night Car," *NYT,* Mar. 24, 1872; "New Yorkers and Their Sunday Resorts: Coney Island," *Sporting Times,* Aug. 8, 1867; "Owl Cars Run Again," *Chicago Times-Herald,* Nov. 20, 1896; Edwin L. Sabin, "The Owl-Car (3 A.M.)," *Puck,* Feb. 25, 1903.

33. Untitled editorial, *SRJ* 6, no. 8 (Aug. 1890): 389; "Steel Works Crippled," *Pittsburgh Post,* June 28, 1909.

34. "West Side in a Rage," *Chicago Times-Herald,* Nov. 16, 1896; "Curtailing the Night-Car Service," *Chicago Record,* Nov. 17, 1896; "Night Cars Must Run," *Chicago Times-Herald,* Nov. 17, 1896; "Yerkes Ignores the Law," *Chicago Times-Herald,* Nov. 18, 1896.

35. *The Boston Guide for Railroad, Steamboat and Horse Car Travel,* no. 1 (Boston: Railway Steam Printing Works, 1871); Rufus Blanchard, *Map of Chicago* (Chicago: Rufus Blanchard, 1857); "Hack Drivers," *NYT,* July 22, 1865; "Hacks and Their Drivers," *Omaha Herald,* Feb. 24, 1889.

36. "Toilers After Dark," *Milwaukee Sentinel,* Mar. 14, 1897; "Midnight Rides," *Sentinel,* Dec. 9, 1883; "The Night Hawk's Complaint," *Pittsburgh Commercial Gazette,* Jan. 5, 1886; "Street Characters," *St. Louis Globe-Democrat,* June 5, 1887; "A Growing Evil," *Wheeling Register,* Aug. 12, 1892; "Shall They Remain?" *Hartford Courant,* Nov. 13, 1901; Graham Russell Gao Hodges, *Taxi! A Social History of the New York City Cabdriver* (Baltimore: Johns Hopkins University Press, 2007), 12.

37. "Letters from the Railway Switchman, Number Four: The Night Side of Broadway," *NYT,* Nov. 8, 1852; "A Hackman's Confession," *NYT,* Apr. 2, 1882; "Life among the Night Hawks," *PI,* Nov. 18, 1892; "Trouble among City Hackmen," *NYT,* May 21, 1895; "Complains of Filthy Cabs," *NYT,* Feb. 17, 1901; "Decision Favors Cabby," *NYT,* Sept. 12, 1902; William Fearing Gill, "Cheap Cabs in New York," *NYT,* Jan. 11, 1903.

38. "Assaulted by Her Cab Man," *New Haven Register,* Jan. 14, 1888; "Cab Driver Charged with Assault," *PI,* Dec. 9, 1901; "Hold Driver for Assault," *NYT,* July 16, 1922; White, *Polite Society,* 115.

39. Mary Stewart Cutting, "A Little Surprise," *McClure's Magazine,* May 1903, 14–23; quotations at 17, 20, 22, and 23.

CHAPTER NINE

1. David E. Nye, *Electrifying America: Social Meanings of a New Technology* (Cambridge: MIT Press, 1990), 48–57; John A. Jakle, *City Lights: Illuminating the American Night* (Baltimore: Johns Hopkins University Press, 2001), 43; David Nasaw, *Going Out: The Rise and*

Fall of Public Amusements (New York: Basic Books, 1993), 6-9; "Chicago after Dark," *CIO*, Nov. 20, 1892.

2. Nasaw, *Going Out*, 3-5; Roy Rosenzweig, *Eight Hours for What We Will: Workers and Leisure in an Industrial City, 1870-1920* (Cambridge: Cambridge University Press, 1985), 179-80; Kathy Peiss, *Cheap Amusements: Working Women and Leisure in Turn-of-the-Century New York* (Philadelphia: Temple University Press, 1986), 5-6, 41-43.

3. Rowland Haynes, "Recreation Survey, Milwaukee Wisconsin," *Playground*, May 1912, 47-48; Richard Henry Edwards, *Popular Amusements* (New York: Association Press, 1915), 32-33.

4. A surprisingly archaic language of condemnation persisted in some descriptions. See, for instance, Helen Campbell, *Darkness and Daylight, or Lights and Shadows of New York Life* (Hartford, CT: A. D. Worthington, 1892), 209.

5. "The Electric Light," *PI*, Apr. 30, 1879; *Cleveland Herald* article of Apr. 30, 1879, reprinted as "The Electric Lamp," *Cincinnati Daily Gazette*, May 1, 1879. At the time, Public Square was called Monumental Park.

6. Wolfgang Schivelbusch, *Disenchanted Night: The Industrialization of Light in the Nineteenth Century* (Berkeley: University of California Press, 1995), 114-20; Jakle, *City Lights*, 40-47; Mark Jansen Bouman, "City Lights and City Life: A Study of Technology and Urbanity" (PhD diss., University of Minnesota, 1984), 274-88; "Table No. 1: Data concerning Street Illumination—Private Plants" and "Table No. 2: Data concerning Street Illumination—Municipal Plants," *Municipal Journal* 33, no. 19 (Nov. 7, 1912): 685-94.

7. Walter Firth, "Spring in New York," *Eclectic Magazine of Foreign Literature*, June 1895, 847; Bouman, "City Lights and City Life," 285-88; John S. Billings, *Report on the Social Statistics of Cities in the United States at the Eleventh Census: 1890* (Washington, DC: Government Printing Office, 1895), 20-22, 63-67; US Bureau of the Census, *Statistics of Cities Having a Population of Over 25,000: 1902 and 1903* (Washington, DC: Government Printing Office, 1905), 106; Lighting Committee of the Civic League, *Street Lighting in St. Louis* (St. Louis: Civic League of St. Louis, 1908), 12-13; Harold L. Platt, *The Electric City: Energy and the Growth of the Chicago Area, 1880-1930* (Chicago: University of Chicago Press, 1991), 21; Mark H. Rose, *Cities of Light and Heat: Domesticating Gas and Electricity in Urban America* (University Park, PA: University of Pennsylvania Press, 1995), 67-72.

8. David Graham Phillips, "The Bowery at Night," *HW*, Sept. 19, 1891; Campbell, *Darkness and Daylight*, 211-12; Firth, "Spring in New York," 847; Nye, *Electrifying America*, 54; Jakle, *City Lights*, 74-76; *Statistics of Cities Having a Population of Over 25,000: 1902 and 1903*, 106.

9. Nye, *Electrifying America*, 54-57, 69; Kate Bolton, "The Great Awakening of the Night: Lighting America's Streets," *Landscape* 23, no. 3 (1979): 41-47; Jakle, *City Lights*, 226-32; Bouman, "City Lights and City Life," 289-90; "Blaze of Lights Packs Asylum Street," *Hartford Courant*, Dec. 19, 1911.

10. "Crime and Electric Lights," *Boston Evening Transcript*, Mar. 4, 1887; Rollin Lynde Hartt, "The City at Night," *Atlantic Monthly*, Sept. 1901, 357; Civic League, *Street Lighting in St. Louis*, 35; "Electric Lights," *PPL*, Mar. 7, 1890; Bouman, "City Lights and City Life," 293-94, quotation at 294; "'Light Up Dark Streets and Cut Crime,' Sullivan," *Chicago Evening Post*, Oct. 30, 1920; Jakle, *City Lights*, 116-17; Mark J. Bouman, "'The Best Lighted City in the

World': The Construction of a Nocturnal Landscape in Chicago," in *Chicago Architecture and Design, 1923–1993*, ed. John Zukowsky (Munich: Prestel, 1993), 47.

11. Samuel Hopkins Adams, "A Metropolitan Night: Glimpses of New York with a Newspaper Reporter," *Frank Leslie's Popular Monthly*, June 1900, 4–18; James B. Carrington, "New York at Night," *Scribner's Magazine*, Mar. 1900, 326–36; Bouman, "City Lights and City Life," 276–89.

12. Bouman, "'Best Lighted City in the World,'" 38, 45; Platt, *Electric City*, 269, 277–78; Nye, *Electrifying America*, 299, 303–27.

13. Nye, *Electrifying America*, 49–57, 66–75; Bouman, "'Best Lighted City in the World,'" 34, 37–39; Robert M. Coates, "New York's Twenty-Four-Hour Corner," *NYT*, June 23, 1929.

14. Civic League, *Street Lighting in St. Louis*, 35; Juvenile Protective Association of Chicago, *Juvenile Protective Association, 1910–1911* (Chicago: Juvenile Protective Association, 1911), 13, 54–55.

15. Howard P. Chudacoff, *The Age of the Bachelor: Creating an American Subculture* (Princeton, NJ: Princeton University Press, 1999), 47–51, 58–59, 77, 126–31; Joanne J. Meyerowitz, *Women Adrift: Independent Wage Earners in Chicago, 1880–1930* (Chicago: University of Chicago Press, 1988); Campbell Gibson, "Population of the 100 Largest Cities and Other Urban Spaces in the United States: 1790 to 1990," Population Division Working Paper 27, US Census Bureau, June 1998, http://www.census.gov/population/www/documentation/twps0027/twps0027.html#citypop.

16. Michael Marks Davis, *The Exploitation of Pleasure: A Study of Commercial Recreations of New York City* (New York: Russell Sage Foundation, 1911), quotations at 3 and 5; Flo Farmington, "Cousin John and Others: A Pittsburgh Story," *Pittsburgh Post*, July 21, 1907; Anne O'Hagan, "A Summer Evening in New York," *Munsey's Magazine*, Sept. 1899, 864; Randy D. McBee, *Dance Hall Days: Intimacy and Leisure among Working-Class Immigrants in the United States* (New York: New York University Press, 2000), 21–22; Jane Addams, *The Spirit of Youth and the City Streets* (1909; Urbana: University of Illinois Press, 1972), 5, 13–15, 34, 47; Perry R. Duis, *The Saloon: Public Drinking in Chicago and Boston, 1880–1920* (1983; Urbana: University of Illinois Press, 1999), 3–5, 108–10; Peter C. Baldwin, *Domesticating the Street: The Reform of Public Space in Hartford* (Columbus: Ohio State University Press, 1999), 162–65; Evert Jansen Wendell, "Boys' Clubs," *Scribner's Magazine* 9, no. 6 (June 1891): 738–39.

17. Duis, *Saloon*, 93–95; W. D. Howells, *A Hazard of New Fortunes* (1889; Bloomington: Indiana University Press, 1976), 76. A brilliant exploration of this situation appears in George Chauncey's "'Privacy Could Only Be Had in Public': Forging a Gay World in the Streets," chap. 7 in his *Gay New York: Gender, Urban Culture, and the Making of a Gay Male World, 1890–1940* (New York: Basic Books, 1994).

18. Daniel T. Rodgers, *The Work Ethic in Industrial America, 1850–1920* (Chicago: University of Chicago Press, 1974), 94–124; Rosenzweig, *Eight Hours for What We Will*, 212; John F. Kasson, *Amusing the Million: Coney Island at the Turn of the Century* (New York: Hill and Wang, 1978), 4–8.

19. Richard Butsch, *The Making of American Audiences: From Stage to Television, 1750–1990* (Cambridge: Cambridge University Press, 2000), 74–80, 121–38; Davis, *Exploitation of Pleasure*, 25, 29–30, 35; Rosenzweig, *Eight Hours for What We Will*, 209; Edwards, *Popular Amusements*, 35, 43; "Moving Pictures Sound Melodrama's Knell," *NYT*, Mar. 20, 1910.

20. Butsch, *Making of American Audiences*, 103-6, quotation at 106.

21. Butsch, *Making of American Audiences*, 108-20, quotation at 113; Nasaw, *Going Out*, 19-32; Davis, *Exploitation of Pleasure*, 30; Edwards, *Popular Amusements*, 46-47.

22. Davis, *Exploitation of Pleasure*, 25, 30-32; Robert C. Allen, *Horrible Prettiness: Burlesque and American Culture* (Chapel Hill: University of North Carolina Press, 1991), 221-32; Edwards, *Popular Amusements*, 46; *Chicago by Day and Night: The Pleasure Seeker's Guide to the Paris of America* (Palmyra, PA: Diamond, 1892), 22, 33-35, 40-43; Chad Heap, *Slumming: Sexual and Racial Encounters in American Nightlife, 1885-1940* (Chicago: University of Chicago Press, 2009), 29-44.

23. Davis, *Exploitation of Pleasure*, 31; Edwards, *Popular Amusements*, 19-20; see also John Collier, "Moving Pictures, Their Function and Proper Regulation," *Outlook*, Oct. 1910, 233.

24. Nasaw, *Going Out*, 130-34, 154-73; Butsch, *Making of American Audiences*, 140; Davis, *Exploitation of Pleasure*, 10.

25. Edwards, *Popular Amusements*, 46-47; Nasaw, *Going Out*, 130-34, 154-73; *Harper's Weekly* quotation from Nasaw, *Going Out*, 168; Butsch, *Making of American Audiences*, 141-46; Peiss, *Cheap Amusements*, 146; Lauren Rabinovitz, *For the Love of Pleasure: Women, Movies and Culture in Turn-of-the-Century Chicago* (New Brunswick, NJ: Rutgers University Press), 107-11; Haynes, "Recreation Survey," 46.

26. Rabinovitz, *For the Love of Pleasure*, 122-36; *Tribune* quotation in ibid., 122; Louise de Koven Bowen, *Five and Ten Cent Theatres: Two Investigations by the Juvenile Protection Association of Chicago, 1909 and 1911* (Chicago: Juvenile Protective Association of Chicago, [1911]); Peiss, *Cheap Amusements*, 160-61.

27. Bowen, *Five and Ten Cent Theatres*; Bridgeport Vice Commission, *The Report and Recommendations of the Bridgeport Vice Commission* (Bridgeport, CT, 1916), 38; Addams, *Spirit of Youth*, 86; Vice Commission of Chicago, *The Social Evil in Chicago: A Study of Existing Conditions* (Chicago: Vice Commission of Chicago, 1911), 247-48; Vice Commission of Philadelphia, *A Report on Existing Conditions with Recommendations to the Honorable Rudolph Blankenburg, Mayor of Philadelphia* (Philadelphia: Philadelphia Vice Commission, 1913), 21; "The Campaign to Curb the Moving Picture Evil in New York," *NYT*, July 2, 1911.

28. Nasaw, *Going Out*, 221-35; Rapp quotation in ibid., 230; Rosenzweig, *Eight Hours for What We Will*, 192-93, 209.

29. *CT*, June 8, Nov. 27, 29, 1912, Apr. 3, 1915, Jan. 7, June 24, 1916, Jan. 10, 12, 14, June 26, 30, Oct. 30, Nov. 1, 1920, Jan. 14, 19, June 21, 1924; quotation from June 22, 1913. On the all-night theaters, see "Chicago's All-Night Theatre," *NYT*, Apr. 3, 1912; "Homeless Crowd All-Night Movie," *CT*, Jan. 3, 1914; "Sleeps in Chair and Cures Ills, Avers Ape Ape," *CT*, Jan. 4, 1916; Juvenile Protective Association of Chicago, *Twenty-First Annual Report* (Chicago: Juvenile Protective Association, 1922).

30. Elizabeth Aldrich, *From the Ballroom to Hell: Grace and Folly in Nineteenth-Century Dance* (Evanston, IL: Northwestern University Press, 1991), 119; McBee, *Dance Hall Days*, 53-55; "Little Girls in Danger," *Pittsburg Times*, Dec. 13, 1886.

31. Peiss, *Cheap Amusements*, 93; Davis, *Exploitation of Pleasure*, 15-17; Mrs. Charles Henry Israels, "The Dance Problem," *Outlook*, Oct. 1910, 243, 244; Vice Commission of Minneapolis, Report of the Vice Commission of Minneapolis to His Honor, James C. Haynes, Mayor (Minneapolis: Marion D. Shutter, 1911), 78; Louise de Koven Bowen, *The Public Dance Halls of*

Chicago, rev. ed. (Chicago: Juvenile Protective Association, 1917); Lewis A. Erenberg, *Steppin' Out: New York Nightlife and the Transformation of American Culture, 1890–1930* (Chicago: University of Chicago Press, 1981), 76–83.

32. Davis, *Exploitation of Pleasure*, 14; Mrs. Charles Henry Israels, "The Dance Problem," *Outlook*, Oct. 1910, 246; Peiss, *Cheap Amusements*, 100–104; Haynes, "Recreation Survey," 48.

33. McBee, *Dance Hall Days*, 55–61; Nancy Banks, "The World's Most Beautiful Ballrooms," *Chicago History* 2, no. 4 (Fall–Winter 1973): 206–15; Lewis A. Erenberg, "Ain't We Got Fun?" *Chicago History* 14, no. 4 (Winter 1985–86): 18; Collis A. Stocking, *A Study of Dance Halls in Pittsburgh, Made under the Auspices of the Pittsburgh Girls' Conference, 1925* (Pittsburgh: Pittsburgh Girls' Conference, 1925), 10–11.

34. Vice Commission of Philadelphia, Report, 73–74; first quotation from R. O. Bartholomew, in Edwards, *Popular Amusements*, 77–78; second quotation from unnamed investigator, quoted in Bowen, *Public Dance Halls of Chicago*; "'Dance of Death' for Many Girls," *Chicago Daily Journal*, Nov. 27, 1920; Stocking, *Dance Halls in Pittsburgh*, 17–18.

35. Stocking, *Dance Halls in Pittsburgh*, 28; Davis, *Exploitation of Pleasure*, 11, 14, 17; Edwards, *Popular Amusements*, 18; Israels, "Dance Problem," 241, 243; Erenberg, *Steppin' Out*, 79–84.

36. Stocking, *Dance Halls in Pittsburgh*, 18, 30; Ella Gardner, *Public Dance Halls: Their Regulation and Place in the Recreation of Adolescents*, US Department of Labor Bulletin 189 (Washington, DC: Government Printing Office, 1929), 32, 34; McBee, *Dance Hall Days*, 109; Christopher Morley, *Travels in Philadelphia* (Philadelphia: David McKay, 1920), 262–64.

37. McBee, *Dance Hall Days*, 91–94, 109–10.

38. On gender ratios in the dance halls, McBee, *Dance Hall Days*, 92, 94. On gender ratios in the movie theaters, Davis, *Exploitation of Pleasure*, 29–30; Rabinovitz, *For the Love of Pleasure*, 117; Talmage quotation in "The Influence of Club Houses," *BE*, Apr. 27, 1885.

39. Mark C. Carnes, *Secret Ritual and Manhood in Victorian America* (New Haven, CT: Yale University Press, 1989), 1, 4, 7–9, 23–29, 85–89; *Gopsill's Street Guide of the City of Philadelphia* (Philadelphia: James Gopsill's Sons, 1890); George Watson Cole diary, American Antiquarian Society, entries for Apr. 28, May 11, 1875.

40. Thomas J. Schlereth, *Victorian America: Transformations in Everyday Life, 1876–1915* (New York: HarperCollins, 1991), 225; Chudacoff, *Age of the Bachelor*, 107–9; *Report of the Vice Commission of the Cleveland Baptist Brotherhood* (Cleveland, 1911), 7; Duis, *Saloon*, 175–76, 184, 188–89; editorial, *PPL*, Mar. 19, 1890; Rosenzweig, *Eight Hours for What We Will*, 191; *Chicago by Day and Night*, 11.

41. Editorial, *PPL*, Mar. 19, 1890; "City News," *Chicago Times-Herald*, Nov. 19, 1896; "What I Know about Saloons," *Independent*, Sept. 10, 1908; "The Law on Closing the Saloons," *Pittsburg Times*, Nov. 19, 1885; Duis, *Saloon*, 105.

42. Francis G. Peabody, "Substitutes for the Saloon," *Forum*, July 1896, 595–606; George Esdras Bevans, "How Workingmen Spend Their Spare Time" (PhD diss., Columbia University, 1913), 19; Duis, *Saloon*, 106; Chudacoff, *Age of the Bachelor*, 109, 113–14; McBee, *Dance Hall Days*, 45–50.

43. Neil Larry Shumsky, "Tacit Acceptance: Respectable Americans and Segregated Prostitution, 1870–1910," *Journal of Social History* 19, no. 4 (Summer 1986): 665–79; Joseph Mayer,

"The Passing of the Red Light District—Vice Investigations and Results," *Social Hygiene* 4, no. 2 (Apr. 1918): 197–209; Chicago Vice Commission, *Social Evil in Chicago*, 78.

44. Vice Commission of Minneapolis, *Report*, 76–77.

45. Sarah Deutsch observes that some historians have carelessly allowed these sources to lead them into a "hypersexualized" understanding of youth culture. Sarah Deutsch, *Women and the City: Gender, Space and Power in Boston, 1870–1940* (New York: Oxford University Press, 2000), 91; Peiss, *Cheap Amusements*, 6, 58–59, 104–14.

46. "Ladies Insulted by Rowdies," *BE*, July 6, 1884.

47. Annual arrest statistics in Fairmount Park Commission, bound volume titled "Statistics," Philadelphia City Archives, FPC-64; on homosexuality, see also captain's reports for 1890, 1892, 1894, and 1896 in ibid.

48. Captain's reports for 1882, 1884, 1891, 1892, 1894, 1895, and 1897 in Fairmount Park Commission, "Statistics"; Vice Commission of Philadelphia, Report, 21; "To Protect Women Visitors to Park," *Philadelphia Evening Bulletin*, May 29, 1911.

49. "South Omaha Police Asked to Watch Park," *Omaha World Herald*, July 16, 1903; Roy Rosenzweig and Elizabeth Blackmar, *The Park and the People: A History of Central Park* (Ithaca, NY: Cornell University Press, 1992), 405; O'Hagan, "Summer Evening in New York," 856; Donald Barr, "Three Summertime Moods of the Park," *NYT*, Jan. 21, 1923; Chauncey, *Gay New York*, 182.

50. "Unrestrained 'Mashers,'" *PPL*, Sept. 24, 1900; Pittsburgh Bureau of Police Records, No. 10 Station (Hazelwood), ser. 16, "Record of Arrests, Oct. 18, 1919—Oct. 16, 1920," AIS, entry dated Dec. 14, 1919.

51. Pittsburgh Bureau of Police Records, No. 10 Station (Hazelwood), ser. 18, vol. 1, Complaint Book, May 1 to Dec. 31, 1922, AIS; quotations from the entries for May 3 and Oct. 25.

52. Vice Commission of Philadelphia, *Report*, 26.

CHAPTER TEN

1. Philip Davis, *Street-Land: Its Little People and Big Problems* (Boston: Small, Maynard, 1915), 62–63.

2. Davis, *Street-Land*, 19, 82–84; David Nasaw, *Children of the City: At Work and at Play* (New York: Oxford University Press, 1985), 49, 63, 74, 82–86, 123.

3. David I. Macleod, *The Age of the Child: Children in America, 1890–1920* (New York: Twayne, 1998), 65, 125–26, 131; Nasaw, *Children of the City*, 17–18, 20; Evert Jansen Wendell, "Boys' Clubs," *Scribner's Magazine* 9, no. 6 (June 1891): 738–39; John H. Finley, "The Child Problem in Cities," *Review of Reviews* 4, no. 24 (Jan. 1892): 685; Allen Hoben, "The City Street," in *The Child in the City*, papers presented at the conferences held during the Chicago Child Welfare Exhibit (Chicago: Chicago School of Civics and Philanthropy, 1912), 458–59; Percy Stickney Grant, "Children's Street Games," *Survey* 23 (Nov. 13, 1909): 235; William I. Hull, "The Children of the Other Half," *Arena* 17 (June 1897): 1045.

4. David Graham Phillips, "The Bowery at Night," *HW* 35 (Sept. 19, 1891): 710; Mrs. Schuyler Van Rensselaer, "Midsummer in New York," *Century Magazine* 62, no. 4 (Aug. 1901): 483–501; Hutchins Hapgood, *Types from City Streets* (1910; New York: Garrett Press, 1970), 75, 132–33.

5. Jane Addams, *The Spirit of Youth and the City Streets* (New York: Macmillan, 1909), 27.

6. Hall quoted in Howard P. Chudacoff, *How Old Are You? Age Consciousness in American Culture* (Princeton, NJ: Princeton University Press, 1989), 67; Joseph F. Kett, *Rites of Passage: Adolescence in America, 1790 to the Present* (New York: Basic Books, 1977), 217–20.

7. Macleod, *Age of the Child*, 20–21, 25–26, 30–31, 51; Karin Calvert, "Children in the House, 1890 to 1930," in *American Home Life, 1880–1930: A Social History of Spaces and Services*, ed. Jessica H. Foy and Thomas J. Schlereth (Knoxville: University of Tennessee Press, 1992), 75–93; Davis, *Street-Land*, 228–36; Mrs. Theodore W. Birney, *Childhood* (New York: Frederick A. Stokes, 1905), 47; F. S. Churchill, "The Effect of Irregular Hours upon the Child's Health," in *Child in the City*, 311.

8. Macleod, *Age of the Child*, 30–31; Davis, *Street-Land*, 84, 227, quotation on 29–30.

9. Kate Douglas Wiggin, "Children's Rights," *Scribner's Magazine* 12, no. 2 (Aug. 1892): 242–48; Edward T. Devine, *The New View of the Child*, National Child Labor Committee Pamphlet 71 (New York: National Child Labor Committee, 1908); Macleod, *Age of the Child*, 30.

10. Robert Owen Decker, *Hartford Immigrants: A History of the Christian Activities Council (Congregational), 1850–1980* (New York: United Church Press, 1987), 10–12, 117–18; "Historical Sketch of the Boys' Clubs in Hartford, Conn.," *Good Will Star* 5, no. 3 (Apr. 1908), in Hall Scrapbook 11, Stowe-Day Library, Hartford; *The Annual Report of the City Missionary to the City Missionary Society of Hartford, 1877–1878* (Hartford, CT: Case, Lockwood, and Brainard, 1878) 12–13; Howard Tooley, undated history of boys' club movement, CBC, box 1, folder 7; Edward T. Hartman, "The Massachusetts Civic Conference," *American City* 2, no. 1 (Jan. 1910): 31.

11. Tooley, history of boys' club movement; *First Annual Report of the Union for Home Work of Hartford* (1872–73) (Hartford, CT: Case, Lockwood and Brainard, 1873); Wendell, "Boys' Clubs;" Edith Parker Thomson, "A Remarkable Boys Club," *New England Magazine*, n.s. 19, no. 4 (Dec. 1898): 488–97; William Byron Forbush, *The Boy Problem*, 6th ed. (Boston, 1901), 68; *The Boys' Club Federation International, 1923 Year Book* (New York: Boys' Club Federation International, 1923), in CBC, box 43, folder 3.

12. Forbush, *Boy Problem*, 69–70; "Urge a Club for Boys of Street," *Chicago American*, Nov. 12, 1901; untitled typescript, n.d., CBC, box 1, folder 1; *Darkest Chicago and Her Waifs: Ninth Annual Report of the Chicago Boys' Club* (Chicago: Chicago Boys' Club, 1911); *Darkest Chicago and Her Waifs: Tenth Annual Report of the Chicago Boys' Club* (Chicago: Chicago Boys' Club, 1912), 17.

13. Forbush, *Boy Problem*, 73–74; untitled typescript, n.d., CBC, box 1, folder 1.

14. William I. Engle, "Supervised Amusement Cuts Juvenile Crime by 96 Per Cent," *American City* 21, no. 6 (Dec. 1919): 516; *Darkest Chicago and Her Waifs* (1911), 12; Luther Laflin Mills and Solon C. Bronson to Fred K. W. Moeller, Feb. 16, 1905, CBC, box 43, folder 2; *Building the Boy Right: Fourteenth Annual Report of the Chicago Boys' Club* (Chicago: Chicago Boys' Club, [1916?]); *Write Your Biography in the Life of a Boy* (Chicago: Chicago Boys' Club, 1928), CBC, box 26, folder 11; John H. Witter, "Memorandum to the Board of Directors at the Annual Meeting, January 1919," CBC, box 3, folder 5; *The Danger Hour* (Chicago: Chicago Boys' Club, 1918), CBC, box 26, folder 11.

15. Stephen Hardy, *How Boston Played: Sport, Recreation, and Community, 1865–1915* (Boston: Northeastern University Press, 1982), 105; Davis, *Street-Land*, 19.

16. "The Nation and Child Labor," *New York Times*, Apr. 24, 1904; Nasaw, *Children of the City*, 138–39; Felix Adler, "Child Labor in the United States and Its Great Attendant Evils," *Annals of the American Academy of Political and Social Science* 25, no. 3 (May 1905): 427–28.

17. E. N. Clopper, "Children on the Streets of Cincinnati," in *Child Labor and Social Progress: Proceedings of the Fourth Annual Meeting of the National Child Labor Committee*, suppl. to *Annals of the American Academy of Political and Social Science* 32 (July 1908): 113; Nasaw, *Children of the City*, 69–70, 103; Vincent Richard DiGirolamo, "Crying the News: Children, Street Work and the American Press, 1830s–1920s" (PhD diss., Princeton University, 1997), 308; "What of the Newsboy of the Second Cities?" *Charities* 10, no. 15 (Apr. 11, 1903): 368; Esther Lee Rider, "Newsboys in Birmingham," *American Child* 3, no. 4 (Feb. 1922): 315. H. M. Diamond, "Connecticut Study of Street Trades," *American Child* 4, no. 2 (Aug. 1922): 97; Bruce Watson, "Street Trades in Pennsylvania," *American Child* 4, no. 2 (Aug. 1922): 125; Sara A. Brown, "Juvenile Street Work in Iowa," *American Child* 4, no. 2 (Aug. 1922): 140; "Study of Newsboys in Springfield, Mass.," *Monthly Labor Review* 18, no. 1 (Jan. 1924): 97–98; Ernest Poole, *Child Labor—the Street* (New York: National Child Labor Committee, 1903), 1.

18. Myron E. Adams, "Children in American Street Trades," *Annals of the American Academy of Political and Social Science* 25, no. 3 (May 1905): 27; Poole, *Child Labor*, 2–3; "Child Labor Reform in New York," *Charities* 10, no. 2 (Jan. 10, 1903): 54–55; Scott Nearing, "The Newsboy at Night in Philadelphia," *Charities and the Commons* 17, no. 18 (Feb. 2, 1907): 778–79; Elsa Wertheim, *Chicago Children in the Street Trades* (Chicago: Juvenile Protective Association, 1917), 8–9.

19. Ernest Poole, "Waifs of the Street," *McClure's* 21 (May 1903): 40, 43; Poole, *Child Labor*, 4, 6; Nearing, "Newsboy at Night," 779.

20. Poole, *Child Labor*; Lewis W. Hine and Edward F. Brown, "An Investigation of the Street Trades of Wilmington, Delaware," (May 1910) in National Child Labor Committee papers, box 17, Columbia University Rare Books and Manuscripts Library.

21. Poole, *Child Labor*, 2, 17–18; "What of the Newsboy of the Second Cities?" 371; "Child Labor Reform in New York," 56; Florence Kelley, "The Street Trader under Illinois Law," in *Child in the City*, 296; Wertheim, "Chicago Children in the Street Trades," 6; Adams, "Children in American Street Trades," 33; Chicago Board of Education et al., *A Plea to Take the Small Boy and the Girl from the City Streets* (Chicago, [ca. 1911?]), JPAP, folder 82.

22. Owen R. Lovejoy, *Child Labor and the Night Messenger Service* (New York, 1910); "Pennsylvania Three-Ply Child Labor Campaign," *Survey* 25 (Mar. 18, 1911): 994.

23. H. H. Jones, "Night Messengers in Louisville, Ky., December, 1913," NCLC/LC, box 4, folder marked "Kentucky—Night Messengers—1913"; Edward Barrows, typescript titled "Extracts from Report on the Night Messenger Service in Pittsburgh, November, 1910," NCLC/LC, box 4, folder marked "Penn., Philadelphia—Night Messengers—1910"; H. M. Bremer, "Street Trades Investigation," Oct. 9, 1912, NCLC/LC, box 4, folder marked "New Jersey—Street Trades—1912"; Edward N. Clopper, *Child Labor in City Streets* (New York: Macmillan, 1912), 101–18; Hine and Brown, "Investigation of the Street Trades of Wilmington."

24. Chicago Board of Education et al., *Plea to Take the Small Boy and the Girl from the City Streets*, 3, 8; Edward N. Clopper, *Child Labor in City Streets* (New York: Macmillan, 1912), 65; testimony of W. O. Burr, "Stenographer's Notes of Public Hearings before the Joint Standing

Committee on Education," General Assembly of the State of Connecticut, January Session 1909, 200, Connecticut State Library; Clopper, "Children on the Streets of Cincinnati," 114.

25. Edward F. Brown, Jan. 26, 1912, remarks in stenographic record, binder marked "Proceedings, Eighth Annual Conference on Child Labor under the Auspices of the National Child Labor Committee, Louisville, Kentucky, January 25, 26, 27, 28, 1912," box 11, NCLC/LC; "Child Labor Laws in All States," *Child Labor Bulletin* 1, no. 2 (Aug. 1912): 1-79; "Map No. 3—Night Work Prohibited to Children," *Child Labor Bulletin* 1, no. 4 (Feb. 1913); Nasaw, *Children of the City*, 103; Clopper, *Child Labor in City Streets*, 132-33, 196; Charles G. Fitzmorris, "Circular Order No. 211," Dec. 7, 1921, JPAP, folder 86; Chicago Board of Education et al., *Plea to Take the Small Boy and the Girl from the City Streets;* Adams, "Children in American Street Trades," 40-42; James K. Pelting, "Enforcing the Newsboy Law in New York and Newark," *Charities* 14, no. 11 (June 10, 1905): 836-37; National Consumers' League, "Seventh Annual Report, Year Ending March 1, 1906," National Consumers' League papers, microfilm reel 3, Library of Congress; Lillian A. Quinn, "Enforcement of Street Trades Regulations," *Child Labor Bulletin* 1, no. 2 (Aug. 1912): 122-24; *Annual Report of the Juvenile Protective Association of Chicago: 1924* (Chicago, 1924), 7-9, 27-29; Jessie F. Binford to Morgan Collins, Mar. 16, 1925, JPAP, folder 88; "Violations of the Chicago Street Trades Ordinance," n.d. (ca. 1928), JPAP, folder 86; Letters from Morgan Collins to parents of newsgirls, Apr. 1, Sept. 24, 25, 1926, JPAP, folder 88.

26. Michael O'Malley, *Keeping Watch: A History of American Time* (Washington, DC: Smithsonian Institution, 1990), ix, 99-144, 164-71, 219.

27. J. J. Kelso, *Revival of the Curfew Law* (Toronto: Warwick Brothers and Rutter, 1896), quotation at 9; Andrew Jones and Leonard Rutman, *In the Children's Aid: J. J. Kelso and Child Welfare in Ontario* (Toronto: University of Toronto Press, 1981), 51, 64.

28. Wolfgang Schivelbusch, *Disenchanted Night: The Industrialization of Light in the Nineteenth Century* (Berkeley: University of California Press, 1988), 82; William Ruefle and Kenneth Mike Reynolds, "Keep Them at Home: Juvenile Curfew Ordinances in 200 American Cities," *American Journal of Police* 15, no. 1 (1996): 64-65; Charles S. Rhyne, *Municipal Curfew for Minors—Model Ordinance Annotated* (Washington, DC: National Institute of Municipal Law Officers, 1943), 2-3; Carl Smith, *Urban Disorder and the Shape of Belief: The Great Chicago Fire, the Haymarket Bomb, and the Model Town of Pullman* (Chicago: University of Chicago Press, 1995), 77-79; "Trouble with the Curfew," *NYT*, Mar. 22, 1896; untitled editorials, *Omaha Evening Bee*, Mar. 17, 18, 19, 1896.

29. "Col. Hogeland's Life Work Done," *Louisville Courier-Journal*, June 18, 1907; *Caron's Annual Directory of the City of Louisville for 1872* (Louisville: C. K. Caron, 1872), 261; Alexander Hogeland, *Ten Years among the Newsboys* (Louisville, KY: John P. Morton, 1883), quotation at 94-95.

30. "Found Dead, with the Gas Turned On," *Louisville Times*, June 17, 1907; "Col. Hogeland's Life Work Done," *Louisville Courier-Journal*, June 18, 1907; *First National Convention of the National Youths' Home and Employment Association of the United States* (Minneapolis: Swinburne, 1886); "Alexander Hoagland [sic] Found Dead at Omaha," *Indianapolis News*, June 17, 1907.

31. "A Curfew Ordinance," *Lincoln Evening Call*, Jan. 14, 1896.

32. *An Illustrated History of Lincoln County, Nebraska, and Her People*, vol. 1, ed. Ira L.

Bare and Will H. McDonald (Chicago: American Historical Society, 1920), 237–49; *History of Hall County, Nebraska*, ed. A. F. Buechler and R. J. Barr (Lincoln, NE: Western Publishing and Engraving Company, 1920), 118–19, 431–32. Neale Copple, *Tower on the Plains: Lincoln's Centennial History, 1859–1959* (Lincoln, NE: Lincoln Centennial Commission, 1959), 82–87.

33. Untitled editorials, *Omaha Evening Bee*, Mar. 17, 18, 19, 1896; untitled editorial, *Lincoln Evening Call*, Feb. 4, 1896; Kelso, *Revival of the Curfew Law*, 15–16; Rhyne, *Municipal Curfew for Minors*, 12–16; "Curfew," *Indianapolis News*, Dec. 21, 1897; "Curfew Ordinance Invalid," *NYT*, July 9, 1898.

34. "No Curfew Law for Chicago," *CT*, Feb. 21, 1896; "Curfew Law for Chicago," *NYT*, Jan. 1, 1898; "Curfew for New-York," *NYT*, Feb. 20, 1896; Mrs. John D. Townsend, "Curfew for City Children," *North American Review* 163 (Dec. 1896): 725–30; untitled editorial, *NYT*, Oct. 8, 1897; quotation from untitled editorial, *NYT*, July 10, 1898.

35. Police Judge's Docket, Lincoln Police Court, vol. 37, Nebraska State Library, Lincoln; "A Great Success," *Lincoln Evening Call*, Feb. 6, 1896; *Fourth Annual Message of Thomas Taggart, Mayor of Indianapolis* (Indianapolis, 1899), 223; *Fifth Annual Message of Thomas Taggart, Mayor of Indianapolis* (Indianapolis, 1900), 189–90; *Sixth Annual Message of Thomas Taggart, Mayor of Indianapolis* (Indianapolis, 1901), 195; *Seventh [sic] Annual Message of Charles A. Bookwalter, Mayor of Indianapolis* (Indianapolis, 1903), 190–91; *Second Annual Message of Charles A. Bookwalter, Mayor of Indianapolis* (Indianapolis, 1904), 218; *First Annual Message of John W. Holtzman, Mayor of Indianapolis* (Indianapolis, 1904), 186; "Curfew Will Not Ring," *Omaha Evening Bee*, Dec. 12, 1896; *Omaha Municipal Reports for the Fiscal Year Ending Dec. 31, 1897* (Omaha, 1898), 144; "Curfew Law to Be Revived," *Omaha Evening Bee*, June 11, 1907; "Social Welfare Plans Outlined by Local Forces," *Wilkes-Barre Times-Leader*, Feb. 29, 1916; "Police to Arrest Youths Who Roam the Streets at Night," *Times-Leader*, June 20, 1917.

36. Rhyne, *Municipal Curfew for Minors*, 12–16; "Porter Plans Curfew for Those under 15," *PI*, Apr. 3, 1915; Paula S. Fass, *The Damned and the Beautiful: American Youth in the 1920's* (Oxford: Oxford University Press, 1977); "A Grand Jury Speaks Out," *CT*, Nov. 2, 1920; Richard C. Lindberg, *To Serve and Collect: Chicago Politics and Police Corruption from the Lager Beer Riot to the Summerdale Scandal, 1855–1960* (Carbondale: Southern Illinois University Press, 1998), 153–67; "Fine Parents to Save Boys," *CT*, Nov. 13, 1920; "Delinquency Cut in Third, Juvenile Court Is Told," *Chicago Daily Journal*, Nov. 15, 1920; "Curfew Bill Passes Council without a Fight," *Chicago Evening Post*, Apr. 8, 1921; "Council Rings Curfew on Night Blooming Kids," *CT*, Apr. 9, 1921; "Chief Opposes Curfew Law," *Chicago Daily Journal*, Apr. 9, 1921; "Can't Enforce Curfew: Chief," *Chicago Daily Journal*, May 5, 1921; "One Lone 'Pinch' All New Curfew Nets First Night," *CT*, May 6, 1921; Genevieve Forbes Herrick, "Ask Collins O.K. on Curfew Law to Guard Girls," *CT*, Aug. 5, 1926; "Collins Pleased with Curfew: To Keep It Ringing," *CT*, Sept. 2, 1926. The *Daily Journal* series on the girl problem emphasized the temptations girls faced in dance halls and other public spaces; for instance, see "Dance of Death for Many Girls," *Chicago Daily Journal*, Nov. 27, 1920.

37. Alice Kessler-Harris, *Out to Work: A History of Wage-Earning Women in the United States* (New York: Oxford University Press, 1982), 110–12, 121–23, 143–44, 155; US Bureau of the Census, *Thirteenth Census of the United States Taken in the Year 1910*, vol. 4, *Population, 1910, Occupation Statistics* (Washington, DC: Government Printing Office, 1914), 37, 41.

38. "Bakers' Grievances," *Buffalo Express*, Mar. 28, 1886; Hazel Kyrk and Joseph Stancliffe Davis, *The American Baking Industry, 1849-1923, as Shown in the Census Reports* (Palo Alto, CA: Food Research Institute, Stanford University Press, 1925), 104; Illinois Bureau of Labor Statistics, *Seventh Biennial Report, 1892* (Springfield, IL: H. W. Rokker, 1893), 3.

39. Mary Van Kleeck, *Women in the Bookbinding Trade* (New York: Russell Sage Foundation and Survey Associates, 1913), 1-2, 26-42, 142-43, 254-55.

40. Kessler-Harris, *Out to Work*, 156-57; Nancy Woloch, "'Entering Wedge': Muller v. Oregon and Its Legacy," in *"Muller v. Oregon": A Brief History with Documents* (New York: Bedford Books, 1996), 6-7, 12-15.

41. Kessler-Harris, *Out to Work*, 183-84; Supreme Court of the United States, Louis D. Brandeis and Josephine Goldmark, *Women in Industry* (1908: Reprint, Arno Press and the New York Times, 1969), 7; National Industrial Conference Board, *Legal Restrictions on Hours of Work in the United States: A Reference Manual*, Research Report 68 (New York: National Industrial Conference Board, 1924), 23-24.

42. Kessler-Harris, *Out to Work*, 184-86; Lise Vogel, *Mothers on the Job: Maternity Policy in the U.S. Workplace* (New Brunswick, NJ: Rutgers University Press, 1993), 15-16.

43. Brandeis and Goldmark, *Women in Industry*, 22; US Supreme Court, *Muller v. Oregon*, 208 US 412 (1908), http://supreme.justia.com/us/208/412/case.html.

44. Woloch, "'Entering Wedge,'" 41-42; National Consumers' League, *The Eight Hours Day and Rest at Night by Statute* (1922), in National Consumers' League files, no. 5235, box 1, folder 18, KCCU; Kessler-Harris, *Out to Work*, 191; Van Kleeck, *Women in the Bookbinding Trade*, 137; John Thomas McGuire, "Making the Case for Night Work Legislation in Progressive Era New York, 1911-15," *Journal of the Gilded Age and Progressive Era* 5, no. 1 (Jan. 2006): 49; Women's Bureau, US Department of Labor, "Night Work Laws for Women," in American Association for Labor Legislation (AALL) pamphlets, no. 5001-P, box 4, folder 3, KCCU; National Woman's Party, "Night Work for Women" ([ca. 1924-29]), AALL records, box 4, folder 2, KCCU; National Industrial Conference Board, *Legal Restrictions on Hours of Work in the United States*, 12.

45. Kessler-Harris, *Out to Work*, 187; Emery R. Hayhurst, M.D., "Medical Argument against Night Work Especially for Women Employes" (*sic*), in AALL pamphlets, no. 5001-P, box 11, folder 4, KCCU; US Supreme Court, *Radice v. New York*, 264 US 292 (1924), http://supreme.justia.com/us/264/292/case.html.

46. Illinois Bureau of Labor Statistics, *Seventh Biennial Report*, xxxi-xxxii, 22; Kathryn Kish Sklar, "'The Greater Part of the Petitioners Are Female': The Reduction of Women's Working Hours in the Paid Labor Force, 1840-1917," in *Worktime and Industrialization: An International History*, ed. Gary Cross (Philadelphia: Temple University Press, 1988); Mary D. Hopkins, *The Employment of Women at Night*, Bulletin of the Women's Bureau 64 (Washington, DC: Government Printing Office, 1928), 57; Nebraska quotation from Josephine Goldmark, *Fatigue and Efficiency: A Study in Industry* (New York: Charities Publication Committee and Russell Sage Foundation, 1912), 432; Van Kleeck, *Women in the Bookbinding Trade*, 142-43; Hayhurst, "Medical Argument against Night Work."

47. Van Kleeck, *Women in the Bookbinding Trade*, 143; "Pioneer Work in Connecticut for Women and Children in Industry," undated typescript, ca. 1924, CLC, box 2, folder 9; quotation from untitled, undated typescript, CLC, box 2, folder 20; *Young Working Girls: A Sum-*

mary of Evidence from Two Thousand Social Workers, ed. Robert A. Woods and Albert J. Kennedy (Boston: Houghton Mifflin for the National Federation of Settlements, 1913), 23; Charlotte Molyneux Holloway, *Report of the Bureau of Labor on the Condition of Wage-Earning Women and Girls* (Hartford: State of Connecticut, 1916), 133–34; Alice Kessler-Harris, "The Paradox of Motherhood: Night Work Restrictions in the United States," in *Protecting Women: Labor Legislation in Europe, the United States, and Australia, 1880–1920*, ed. Ulla Wikander, Alice Kessler-Harris, and Jane Lewis (Urbana: University of Illinois Press, 1995), 350; Consumers' League of New York City, *Behind the Scenes in a Restaurant* (1916), reprinted in *The Consumers' League of New York: Behind the Scenes of Women's Work*, ed. David J. Rothman and Sheila M. Rothman (New York: Garland, 1987), 18.

48. National Woman's Party, "Night Work for Women" ([ca. 1924–29]), in AALL pamphlets, box 4, folder 2, KCCU; Judith A. Baer, *The Chains of Protection: The Judicial Response to Women's Labor Legislation* (Westport, CT: Greenwood Press, 1978), 86.

49. Hopkins, *Employment of Women at Night*, 50–56; Linda Gordon, "Single Mothers and Child Neglect, 1880–1920, *American Quarterly* 37, no. 2 (Summer 1985): 184; Holloway, *Report of the Bureau of Labor on the Condition of Wage-Earning Women and Girls*, 135; National Consumers' League, *The Waste of Industry: Overworked Women and Girls* (New York: National Consumers' League, 1915), in AALL pamphlets, box 10, folder 20, KCCU; "Women on the Night Shift," *Life and Labor*, Dec. 1914, in *America's Working Women*, ed. Rosalyn Baxandall, Linda Gordon, and Susan Reverby (New York: Random House, 1976), 160; Agnes De Lima, *Night Working Mothers in Textile Mills, Passaic, New Jersey* (N.p.: National Consumer's League and the Consumer's League of New Jersey, 1920), 8–10.

50. National Woman's Party, "Night Work for Women"; National Woman's Party, "Shall There Be Special Restrictive Laws for Women?" AALL pamphlets, box 4, folder 4, KCCU.

51. Kenneth Lipartito, "When Women Were Switches: Technology, Work, and Gender in the Telephone Industry, 1890–1920," *American Historical Review* 99, no. 4 (Oct. 1994): 1089–90; Holloway, *Report of the Bureau of Labor on the Condition of Wage-Earning Women and Girls*, 130–31; Helen Baker, *Women in War Industries* (Princeton, NJ: Industrial Relations Section, Princeton University, 1942), 37–38.

52. Van Kleeck, *Women in the Bookbinding Trade*, 143; "Pioneer Work in Connecticut for Women and Children in Industry," undated typescript, ca. 1924, CLC, box 2, folder 9; quotation from untitled, undated typescript, CLC, box 2, folder 20; *Young Working Girls*, 23; Holloway, *Report of the Bureau of Labor on the Condition of Wage-Earning Women and Girls*, 133–34.

53. Edward Hungerford, "The Night Glow of the City," *HW*, Apr. 30, 1910, 13; cover of the *New Yorker*, Oct. 11, 1930, in David E. Nye, *When the Lights Went Out: A History of Blackouts in America* (Cambridge: MIT Press, 2010), 38.

54. Max Seham, "Rest and Sleep," *Hygeia: The Health Magazine*, Oct. 1926; "Thoughts from the Country," *Anglo American*, Mar. 23, 1844; quotation from George T. W. Patrick, "The Psychology of Daylight Saving," *Scientific Monthly*, Nov. 1919, 387.

55. James Hervey, *Contemplations on the Night* (New York: James Rivington, 1774), 24.

56. Lynn Dumenil, *The Modern Temper: American Culture and Society in the 1920s* (New York: Hill and Wang, 1995); "City Night Life Found Moving Out to Country," *CT*, May 10, 1928; Virginia Scharff, *Taking the Wheel: Women and the Coming of the Motor Age* (New York:

Free Press, 1991); Beth L. Bailey, *From Front Porch to Back Seat: Courtship in Twentieth-Century America* (Baltimore: Johns Hopkins University Press, 1988); Chad Heap, *Slumming: Sexual and Racial Encounters in American Nightlife, 1885–1940* (Chicago: University of Chicago Press, 2009).

57. Heap, *Slumming*, 164–65, 180, 197; "Women to Drive Taxis for Women," *NYT*, Apr. 2, 1923; "The Taxi Lady Takes the Road," *NYT*, Apr. 22, 1923.

58. Martin Moore-Ede, *The Twenty-Four-Hour Society: Understanding Human Limits in a World That Never Stops* (Reading, MA: Addison-Wesley, 1993), 35–40, 71–77.

59. Ignatius Donnelly, *Caesar's Column: A Story of the Twentieth Century* (1890; Middletown, CT: Wesleyan University Press, 2003), quotation at 10.

INDEX

abortion and contraception, 42, 189
accidents, 10, 131, 134, 197
actors and actresses, 62, 66–68, 104, 151
Addams, Jane, 181–82
adultery, 7, 33, 96
advertising and business signs, 62, 88, 90, 100–101, 159, 161
African Americans, 5, 8, 11, 14, 19, 47, 105, 144, 202, 211n4
Albee, Edward F., 165
Allegheny, PA. *See* Pittsburgh, PA
alternating shifts, 128–32, 136
American Revolution, 3–7, 9, 212n6, 212n7
apothecaries, 1. *See also* drugstores
arc lights, 121, 158–59, 200
arson, 11, 21, 30, 32, 55
art galleries, 55
assaults, 10, 22, 26, 97, 150
Astor-Place Opera House, 70–71, 78
Atkinson, John F., 184
Auburn, NY, 199
automobiles, 3, 151, 172, 175, 202

bachelors, 163
bakery workers, 9, 109, 117, 195–96
ballet, 69, 90
ballrooms, 55, 76–83, 188
balls, 76–83, 151, 169
Baltimore, MD, 44; crime in, 22–23; nightlife and vice, 63–64, 92; police, 22–23; sanitary conditions, 106–7; street lighting, 16; street railways and workers, 125
Bangor, ME, 68
Barnum, P. T., 62, 70–71, 164
bartenders, 104, 150
Beecher, Henry Ward, 47, 51–52, 101
Beekley, Nathan, 59–63, 68, 94
beggars, 73, 181
Bell, Thomas, 130
Bennett, James Gordon, 78, 87
billiards and pool halls, 50, 83, 85, 88, 152, 156, 185, 193
binges, 131. *See also* sprees
Binghamton, NY, 186
blackface minstrelsy, 40, 64, 71, 89, 165
blackouts (interruptions of gas or electric service), 27–33, 160
bloods, 67, 76, 144, 181
blood sports, 99, 245n55. *See also* prize fighting
boardinghouses, 19, 35, 38–42, 46, 48, 53, 59, 94, 132–33, 163

bootblacks, 116, 183, 187
Boston, MA, 44; children, 179, 184, 186; crime, 11, 21, 30–32, 94, 160; fire of 1872, 30–32; food supply and distribution, 111; nightlife and vice, 57, 62, 67–68, 70, 87, 91–94, 99, 165, 174, 179; police and night watchmen, 21, 23, 27, 30, 32; public transportation, 150; publishing industry, 195; railroads and railroad workers, 123; sanitary conditions, 105, 107; street lighting, 16–18, 160; waterfront workers, 121
Boston Tea Party, 9
Bowen, Louise de Koven, 167–68, 170
Bowery (in New York City), 46, 87–88, 159
Bowery b'hoys, 39, 76
boys' clubs, 13, 180, 183–86, 190, 194
Brace, Charles Loring, 86
Brandeis, Louis, 196–97
brawling, 9, 84–85, 95, 97, 144, 150, 184, 202
Brewer, David, 196–97
Bridgeport, CT, 57, 130, 168, 173
bright lights districts, 24, 155–56, 160–61, 174, 180, 184, 187
Broadway (in New York City), 14, 16, 19–20, 24–25, 45, 56, 88, 90, 92–93, 97, 138–39, 142, 149, 161
Brooklyn, NY, 44, 46–47, 118; early closing movement, 50–52; food supply and distribution, 109, 112; nightlife and vice, 91; parks, 176–77; public transportation, 123, 141, 149–50; sanitary conditions, 105, 107; street lighting, 16; waterfront workers, 121
Brooklyn Bridge, 149
Brooks, William, 32
brothels, 9, 12, 40, 49, 69, 93–97, 143, 157, 189; attacks on, 9, 93–94. *See also* prostitution
Brown, Susan, 41
Buffalo, NY, 107, 124, 198
Bunner, H. C., 136
Buntline, Ned, 75, 89, 97
burglary, 10–11, 21, 24–25, 29, 55

burlesque shows, 166–67
Burton, Warren, 91–92

cabarets, 161, 171, 193, 202
call boxes, 24, 26
calling. *See* social calls
Cambridge, MA, 105, 150
camphene, 18
candles, 8–11, 15–16, 18, 28–29, 33, 36, 101
car-hook murder, 138–142
Castellanos, Henry, 11
cellars, 85, 88–90, 94, 99
Charleston, SC, 11
Chasteau, Louis, 177
Cheever, Tracy Patch, 66
Chicago, IL: children, 181, 184–90, 193–94; crime, 24, 100, 151, 160, 186, 193; demography, 163; food supply and distribution, 112; Great Fire of 1871, 191; lighting, 17, 151, 155–56, 160–62; nightlife and vice, 81, 92, 100, 151, 155–56, 162, 167, 169–71, 173–75, 202; police, 24, 167, 193; public transportation, 124, 149, 151; publishing industry, 195; railroads and railroad employees, 123; sanitary conditions, 105–7
Chicago Boys' Club, 184–86
child development, 179–83, 185–86, 190
child labor, 13, 109, 114, 116–17, 126–27, 135, 190
child labor laws, 180, 189–90, 194
child molesting, 13, 168, 177, 188
child prostitution, 91–93, 116, 188
children: and mass entertainment, 161, 165, 167–68, 175, 180–82; moral dangers to, 168, 175, 178; at play, 121, 133, 181, 186. *See also* boys' clubs
churches and church activities, 17, 24, 40, 43, 47, 50, 53, 56–58, 59, 69, 72, 130, 138, 144–45, 173, 183
Churchill, F. S., 182
Cincinnati, OH, 57, 81, 99, 102, 107, 144, 189
circadian rhythms, 203
circuses, 71

Civil War, 30, 109, 116, 125, 183
Clarke, William P., 127
class conflict, 2, 30, 67, 139–40. *See also* middle-class culture; wealth and poverty
Clement, Fanny, 60
clerks, 12, 34–35, 40, 43–56, 59, 86, 93, 100, 114, 117, 136–37, 156, 198–99
Cleveland, OH, 107, 120, 124, 149, 157–58, 171
clocks, 36–37, 53, 65, 138, 142, 190
clothing styles, 54, 56–57, 76–78, 83, 90, 93, 95–96, 137, 146–48, 153
clubs, 12, 86–88, 152, 170, 172
cockfights, 83, 99
coke production, 119, 133
Cole, George Watson, 57, 173
Columbus, OH, 159
concert saloons, 89–91, 97, 165, 172, 180
concerts, 40, 53, 61–62, 72–73, 157
Connecticut, 125, 200
corner loungers, 25, 39–40, 144–45, 181
Coughlin, John ("Bathhouse John"), 192
countercyclical work, 104–5, 246n2
courtship, 59–60, 96, 102, 132, 164, 202. *See also* flirtation; sexuality
Crane, Stephen, 136
Cranston, RI, 26
Crapsey, Edward, 96
crime and criminals, 7, 10–11, 13–15, 20–27, 43, 55, 75, 91, 93–95, 97, 100, 131, 138–44, 148–51, 191–93
crime rates, 22–23
crossing sweepers, 91
Crowninshield, Benjamin, 87, 99
Culyer, John, 176
curfews, municipal, 11, 13, 180, 190–94
Cutting, Mary Stewart, 153–54
Cuyler, T. L., 50

Dallas, TX, 192
Daly, Charles, 116
Daly, Maria Lydig, 116
dance halls, 42, 94–95, 151, 156–57, 163, 169–72, 175, 177–78

dancing, 37, 41–42, 55, 74–83, 90–91, 95, 97; in brothels, 9; in dance halls, 41–42, 169–72; in homes, 79; in taverns, 5
dandies and fops, 76, 89, 100, 144
Daniel, H., 70
darkness, 1, 10, 19, 33, 74, 153, 155, 177
Davids, L., 60
Davis, Michael, 163, 165–67
Davis, Philip, 180, 182–83
daylight, hours of, 36–37, 62–63, 233–34n24
daylight saving time, 200
Denver, CO, 192
Detroit, MI: food supply and distribution, 110, 113; nightlife and vice, 94, 157; public transportation, 149; sanitary conditions, 106; street lighting, 158
diary-keeping, 47, 50, 92
disease and health, 7, 80–81, 95, 134, 188–89, 196–98
Des Moines, IA, 124, 192
doctors, 9, 55
domesticity, 46, 58, 83, 88–89, 141, 154, 163–64, 173, 178, 182–84, 193, 199
Donnelly, Ignatius, 2–3, 203
downtown commercial districts, 12, 19, 24, 39, 94, 160–61, 167, 169–70, 173, 180, 198
drinking, 4–9, 23–24, 33, 41, 43, 48, 55, 67–69, 74–75, 80–91, 95–102, 117, 130–31, 138, 148, 150, 152–53, 171, 173–74, 177, 188–89, 193, 198, 202; arrests for drunkenness, 24; on the job, 21, 23, 84–85, 125, 131, 134; in theaters, 67–69, 71; by women, 5, 84–85, 171, 174, 198, 202
drugs, 189
drugstores, 1, 149, 151, 175, 199
Duval, Jeanne: and daughters, 138–39, 146, 148

early closing movement, 49–53
Eddy, Daniel, 100
eight hour movement, 124
electricity, 2, 12, 126, 135, 155–63, 181, 192, 194. *See also* interior lighting; street lighting

278 *Index*

elevated trains, 132, 153, 159, 164, 176
etiquette, 56, 60–61, 74, 77–78, 83, 146–49
European cities, 1–2, 141
evening schools. *See* night schools

Fall River, MA, 150, 184
family life, 55, 117, 121, 124, 127, 130–33, 136, 161, 163–64, 197–99, 203
Farley, Harriet, 42
ferries, 44, 110, 147, 149, 153
firehouses, 87, 91
firemen, 77–78, 82, 86, 91
fires, 6–7, 30–32, 82, 87, 179; dangers of, 7, 11, 149; efforts at prevention and control, 21, 25
Fisher, Samuel, 99
Fisher, Sidney George, 69, 76–78
Five Points (New York neighborhood), 14, 19, 24, 75, 94
Flint, MI, 132
flirtation, 68, 82, 87, 90, 93, 95–96, 143, 145, 176, 198
Floy, Michael, Jr., 47, 57
Foote, Henry Leander, 97–98, 101
Fort Wayne, IN, 192
Foster, George, 15, 68, 98
Foster, William, 138–42, 146, 257n4
fraternal organizations, 77, 150–51, 172–73

gambling and gamblers, 24, 48–50, 55, 75, 78, 83, 88–89, 97–102, 117, 152, 156–57, 186–88
gangs, 26–27, 186, 193
garbage collection, 108
garroting. *See* mugging
gas (illuminating), 16–17, 28–29, 37, 155, 158–59, 216n7. *See also* interior lighting; street lighting
Gaspee, 9
gas workers, 28–29, 118, 126–27, 135
gasworks, 3, 16, 28–29, 125–27
gender, ideas concerning, 82–83, 139–42, 144, 153–54, 196–202
glass industry, 119–120, 126–27, 195

glass workers, 120, 126–27, 195
Goldmark, Josephine, 196
Graham, Sylvester, 100
Grand Island, NE, 191–92
graveyards, 93
Greeley, Horace, 52
Green, Jonathan, 101–2
Griscom, John, 106
Griswold, Stephen, 46–47
grocers and groceries, 45, 50, 85, 110–12

hacks and cabs, 54, 151–52, 202
Hale, Sarah J., 81
Hall, G. Stanley, 182
Hall, Mary, 189
harbor thieves, 26–27
Harrisburg, PA, 151
Hartford, CT, 52, 159, 183–84, 198
Hayhurst, Emery, 197–98
Healey, Carolyn, 57
health and disease, 7, 80–81, 95, 134, 188–89, 196–97
heating, 37, 73, 149
Hervey, James, 7, 201
Hogeland, Alexander, 191–92
Holyoke, MA, 135
homosexuality, 91, 93, 177
Hone, Philip, 87
horses, 108, 111, 125, 149
hotels, 1, 17, 29, 49, 86–87, 116–17, 149, 152, 156, 172, 174–75, 188
hours of labor, 2, 7, 23, 33–56, 58–59, 61, 63, 65–66, 69, 75–76, 81–82, 102–37, 156–57, 172, 194–201
houses of assignation, 91, 96, 152
Howells, William Dean, 163–64
Hungerford, Edward, 200
Hutchinson family singers, 72
hygiene, personal, 38, 183, 191

ice cream parlors, 91, 163, 176
immigrants, 105, 108, 133–34, 136, 167, 181
incandescent lighting, 159–60

Indianapolis, IN, 106, 192–93
industrialization, 2
industrial workers, 44, 118, 125–26, 150, 177, 200. *See also* glass workers; iron and steel workers; paper mill workers; railroad workers; refinery workers; street railway workers; textile workers
Ingersoll, Ernest, 25
interior lighting, 12, 15, 17–18, 29; in ballrooms, 80; in brothels and gambling dens, 95, 100; in factories, 36–37, 126, 135, 158; in homes, 12, 17–18, 38, 58, 116, 159, 161, 192; in hotels and inns, 15, 17, 29, 158; in saloons, 50, 88–89, 156, 158; in shops, 12, 17, 28–29, 49, 52, 156, 158; on streetcars, 149, 153; in theaters, 28, 55, 62–63, 156, 158
Irish Americans, 9, 14, 33, 36
iron and steel production, 119–20, 125–33, 151, 195
iron and steel workers, 120, 128–35, 151, 195
Israels, Belle, 170–71
Italian Americans, 108

Jackson, Andrew, 78
Jenckes, Nathaniel, 5
Jersey City, NJ, 44, 50, 112
Jones, W. B., 51
Jones and Laughlin Steel Corp., 120
juvenile delinquency, 117, 133, 185–86, 188, 191, 193
Juvenile Protective Association of Chicago, 161, 167, 169

Kansas City, MO, 106, 157, 166, 192
Keith, Benjamin F., 165
Kelley, Florence, 188, 196
kerosene, 18, 159, 192
Kimball, Moses, 70

labor activism and strikes, 29, 32–33, 37, 43, 49–53, 121–22, 124, 126–27, 130, 135, 140, 151, 194–95

labor legislation, 13, 124, 180, 189–90, 194–201
lamplighters, 17, 21, 108
lantern smashing, 10, 14, 17, 30, 87, 97
lanterns, of pedestrians, 10–11, 21
Laughlin, Mary, 148
lectures, 35, 39–41, 47, 50–51, 53, 58
leisure, moral implications of, 35, 43, 49–53, 55, 164. *See also* sin
letter-writing, 47
Levee (district in Chicago), 94
libraries, 47, 49–51, 58
lighting. *See* interior lighting; street lighting
light pollution, 18
Lima, Agnes de, 136
Lincoln, NE, 191–93
Lippard, George, 16
Lochner v. New York, 196
loitering, 25, 39–40, 43, 48, 133, 167–68, 175, 181, 187
London, England, 22, 141
Long, Mason, 101
Long Island, NY, 110–11
longshoremen, 112, 121–23, 131–32, 134
long turn, 125, 128, 130–31
looting, 9, 30–33
Louisville, KY, 16, 56, 188, 191
Lovejoy, Owen, 188
Lowell, MA, 35–43, 48
lyceums, 40–41, 47, 58

Macready, William Charles, 70
marketmen, 110–14, 117. *See also* peddlers
mashing, 144–49, 177, 198. *See also* sexual harassment
masked balls, 96
Mathews, Cornelius, 15
McAllister, Ward, 80
McCabe, James, 17, 19–20, 92
McDowall, John R., 145
McGavock, Randal W., 92
meals, 38, 46, 48, 61, 81, 88, 99–100, 131–32, 134

Melville, Herman, 15
merchants, 46–47, 49–50, 54, 76
messengers, 187–89
middle-class culture, 19, 24, 54–55, 67–68, 70–73, 75–76, 84–85, 89, 141, 178, 180, 182–84
midwives, 9
Miles, Henry, 40
military guards, 6, 11
milk distribution, 109–10
Milwaukee, WI, 106–8, 152, 157, 167, 173
Minneapolis, MN, 23, 158–60, 175
moonlight, 10, 20
Morgan, Henry, 94
Morrill, F., 42
movie theaters, 117, 157, 163, 166–69, 172, 180, 187
mugging, 7, 10–11, 20, 22, 26, 28, 32, 149, 160, 181
Muller v. Oregon, 196–97
murders, 15, 26, 32, 43, 97, 138–41, 160, 193
museums, 41, 70–71, 91, 156
music, 1, 5, 9, 20, 40–41, 57–58, 60–61, 63–65, 67, 69–74, 95, 165, 170–71, 174–75, 183–84

Nast, Thomas, 139
National Child Labor Committee (NCLC), 188–89
Native Americans, 11
Newark, NJ, 93, 106
New Bedford, MA, 15, 56
New Haven, CT, 23, 56, 93, 101, 183
New Jersey, 36, 110–11, 120, 153–54
New London, CT, 68
New Orleans, LA, 11, 92, 149
Newport, RI, 4, 7, 11, 62
newsboys and newsgirls, 114–17, 167, 181, 184, 187–191
newspaper employees, 114–117, 150, 188, 191, 199
newspaper production, 45, 114–17, 187
New York City, NY, 19, 44; blackouts in the gaslight era, 28–30, 32–33; Blackout of 1977, 27; children, 181, 183–84, 187, 192–93; clerks, 43–53; crime, 11, 22, 26–27, 138–44; Draft Riot, 29–30; food supply and distribution, 109–13; Fourth Ward, 85–86; geography, 45–46, 48–49; industrial workers, 118; nightlife and vice, 56, 62–64, 68–75, 77–78, 81, 85–93, 104, 163, 165–67, 170, 173–74, 202; parks, 93, 107, 177; police and night watchmen, 11, 21–23, 27, 32, 90; printing industry and workers, 45, 86, 114–117, 195, 198; public transportation, 123–25, 138–42, 147–49, 152–53; railroads and railroad workers, 123; sanitary conditions, 105–8; street lighting, 9–10, 14–17, 20, 28–30, 32–33, 159–60, 200; waterfront workers, 121–22. *See also* Brooklyn, NY
Niblo, William, 64, 73
Nichols, Thomas L., 77, 80, 88
nickelodeons, 167–69
night hawks (hacks), 152
night in rural areas, 34, 46, 52
night schools, 8, 40, 47, 57–58, 184, 191
night soil collectors, 7, 104–8
night sounds, 1, 55, 107, 119, 123, 149
night watchmen, 6, 10, 21–22, 119, 132
night work by women, 13, 53, 117–18, 135–36, 142–43, 148, 180, 194–200
nocturnal culture, 6, 13, 103, 117, 142, 157, 178, 180, 188, 201–3
Norristown, PA, 59–60
North Platte, NE, 191–92

oil. *See* interior lighting; street lighting
oil refining, 44, 118, 120, 126
Omaha, NE, 104, 106–7, 177, 192
omnibuses, 72, 147–48, 150–52
Ontario, Canada, 190
opera, 63–65, 70, 153
opera houses, 12, 70–71
Orange Riot, 140
Otter, William, 8–9

owl cars, 123–24, 148–153, 174
oysters, 8, 88
oyster saloons, 49, 85, 88–89, 100, 152

Palmo, Ferdinand, 64
paper mill employees, 126, 131–32
paper mills, 126, 131–32, 135
Paris Commune, 29, 140
parks, 93, 107, 142, 145, 164, 176–77
parlors, 60, 95–96
parties: balls, 61, 77–83; dinners, 61; soirees, 61–62
Passaic, NJ, 136
Pastor, Tony, 165
Patterson, Henry, 50
Pawtuxet, RI, 4, 9, 212n7
peddlers, 38, 112–14, 116, 181, 187, 189. *See also* marketmen; public markets
peep shows, 166–67, 187
Philadelphia, PA, 11–12, 44, 105: children, 193; crime in, 29, 145; early closing movement in, 50; mashers, 146; nightlife and vice, 56, 59–60, 62, 90, 92–93, 98, 173–74, 177–78; parks, 145, 177; police, 21, 23, 26–27, 145, 177, 193; public transportation, 59, 123, 148, 152; publishing industry, 195; railroads and railroad workers, 122–23, 133; street lighting, 1, 10, 12, 17–18, 25, 29
pickpockets, 24, 39, 150, 153, 167, 181
pickups, 168, 171–72, 198
Pintard, John, 22
Pittsburgh, PA: crime, 178; Fifteenth Ward, 132–33; food supply and distribution, 110; glass industry, 126; iron and steel industry, 119–20, 126–29, 133, 151; nightlife and vice, 92, 169–171, 174; police and night watchmen, 119, 133; public transportation, 147, 151; railroads and railroad workers, 120, 123, 130; sanitary conditions, 106, 133; street lighting, 16, 159; view at night, 119
plumbing, 105–6

police, 15, 21–27, 73, 97, 114, 143, 175, 177, 189. *See also* night watchmen
Poole, Ernest, 187
Portland, ME, 105, 122
Portsmouth, NH, 191
post offices, 39, 114, 117
Potter, Eliza, 144
printing industry and employees, 28, 46, 86, 195, 198–99
privies, 105–6, 133
prize fighting, 86, 89, 99
produce markets. *See* marketmen; markets
Prohibition, 160, 174, 193, 202
prostitution, 5, 7, 9, 13, 24, 42–43, 49, 67–71, 76, 79, 88–98, 111, 141–43, 145, 151, 156, 162, 170, 174–75, 177, 186, 188–89; at dance halls, 94–95, 170–71; at theaters, 67–71, 168, 235n31
Providence, RI, 3–7, 26, 62, 93, 157, 159, 165, 183
public markets, 21, 110–14
public transportation. *See* omnibuses; street railways
purse snatching, 148
Putnam, Avery D., 138–40, 141, 148

Quincy, Anna Cabot Lowell, 67
Quincy, Josiah, 9

race relations, 6, 11, 14, 67, 94, 105, 202, 211n4
Radice v. New York, 197–99
railroad depots, 44, 94, 110, 114, 122, 152–53, 174–75, 200
railroad employees, 120, 122–23, 130–35, 151
railroad passengers, 137
railroads, 2, 44, 55, 72, 110, 112, 122–23, 133–34, 190
rape, 10, 13, 56, 143–46, 149, 152–53, 177–78, 181, 198
Rapp, George, 168
reading, 38–39, 47, 57–59, 86–88, 116, 124, 200

red light districts, 93–95, 162, 166, 175, 189, 201. *See also* brothels; prostitution
refinery workers, 118
restaurants, 12, 86–87, 97, 111, 114, 117, 152–53, 163, 174–75, 181, 198–200, 202
rioting, 2, 9, 27, 29–30, 67, 70–71, 94, 140; at theaters, 67, 70
Robinson, Richard P., 43, 48–49, 74
Ross, Joel, 20
Rothschild, Salomon de, 22–23
Royall, Anne, 12
Rural Electrification Administration, 161
rural night, 34, 46, 52, 110–11, 161, 200
Russell, William, 5

sailors, 4–9, 94–95
Salem, MA, 183
saloons and taverns, 4–5, 9, 24, 40, 47–50, 69, 83–91, 104, 111, 117, 152, 154, 156, 163–64, 172–75, 188, 193, 200, 202
Schmidt, Anna, 198–99
seamstresses, 117, 142–43, 148
serenading, 102
servants, 10, 110, 144–45, 148, 177
settlement houses, 184
sewer systems, 2, 105–6
sexual assault, 10, 13, 56, 143–46, 149, 152–53, 177–78, 181, 198
sexual harassment, 13, 24–25, 31, 39, 56, 67, 74, 93, 138, 142–49, 152–53, 177–78
sexuality, 7–8, 13, 33, 38, 42, 74–76, 80, 83, 88–98, 100, 102, 145, 151–53, 166, 168, 170–72, 175–78, 181–82, 188–89, 193, 198, 200
shipping, 5, 26, 112, 121–22
shopping, 35, 39, 45, 49, 51–52, 56, 92, 99, 146–47, 165
shops and shop windows, 1, 10, 12, 17, 35, 39, 44–45, 49–53, 92, 155, 159, 198
sidewalks, 10, 24, 39, 163, 172, 181, 200
sin, 7, 46–48, 66, 70, 74–75, 101, 201
skylines, 2, 35
slaves, 11
sleep, 7, 52, 55–56, 58, 61, 66, 69, 75–76, 107, 111, 117, 121, 124, 130–31, 136, 150, 163, 169, 179, 182, 187, 197, 199–200, 203
slumming, 166, 202
Smith, Jane Briggs, 56
Smith, Mary Bainerd, 56
Smith, Sol, 63
Smith, Virginia, 183
snow, 10, 213n11
Snowden, Thomas: and family, 59–60
social calls, 56, 59–61, 96, 148
soda fountains, 163, 172, 199
Sons of Liberty, 9
sporting men and fancy men, 49, 89, 99, 162, 241n33
sprees, 4–6, 12, 87, 96–98, 152, 181
Stem, Nathan, 59–60
Stewart, A. T., 45, 49
Storyville (district in New Orleans), 94
St. Louis, MO: nightlife and vice, 63–64, 94, 96; public transportation, 150; sanitary conditions, 105–6; street lighting, 159–60
St. Paul, MN, 94, 106, 110, 192
"Street Arabs," 116–17, 184, 187
street corners, 25–26, 98, 144, 149, 193
street lighting, 2–3, 9–10, 14–20, 35, 121, 145, 156–63: brightness of, 10, 16–17, 155, 157–60, 200; electric, 121, 156–63 ; gas, 14–20, 27, 49, 56, 155, 158–60; oil, 9–10, 16, 20; thought to deter crime, 2, 14–15, 19–20, 26–33, 160
street railway employees, 117, 138, 141–42
street railways, 2, 55, 72–73, 111, 123–25, 136, 192; cable cars, 123–24, 149; effects on urban geography, 19, 175; electric, 124, 149; employees, 124–25, 199; experience of riding, 138–42, 147–53 ; horse-drawn, 72, 123, 125, 147, 149; schedules of service, 72, 123–25, 148–49, 157. *See also* owl cars
street signs, 25–26
street sweeping, 108
streetwalkers, 11, 13, 24, 69, 91–94, 145, 181
Strong, George Templeton, 20, 24, 80, 82, 90, 116

sugar refining, 44, 118, 126
suicide, 55, 101
swells, 76, 78, 86, 100

Talmage, T. DeWitt, 55–56, 82, 173
Tammany Hall, 139
taverns. *See* saloons and taverns
taxis, 202
Taylor, Bayard, 95
technological change, 2–3, 12–13, 105–8, 126–28, 135, 194–95, 203, 212n5
telephone operators, 199–200
temperance movement, 9, 41, 43, 57, 84–85
ten-hour movement, 43
Terre Haute, IN, 56
textile mills, 36–40, 42–43, 125, 135
textile workers, 34–43, 135
Thaxter, Anna Quincy, 72–73
theaters, 8, 17, 24, 28, 40–41, 48–49, 55, 57, 62–74, 88, 97, 116, 152, 157, 161, 164–65, 167, 174–75, 178, 187, 198, 202, 236n43; audience behavior in, 62, 66–72, 165; criticism of, 40–41, 43, 62, 66, 69–70, 171; lighting of, 16, 28, 55, 62–63, 71, 156, 158, 161; prostitution in, 67–71; schedules of, 62–66, 69–72, 148, 169, 236n41
theft, 15, 26–27, 30–31, 100, 131, 184
Thompson, George, 91, 142–43
Thoreau, Henry David, 34
time consciousness, 65–66, 190, 194
Times Square (in New York City), 139, 161
Todd, John, 46–47, 69
Toledo, OH, 104–5
torches, 51, 114
tourists and travelers, 2, 12, 19, 95, 162
trash collection, 108
Trollope, Fanny, 57
Twain, Mark (Samuel L. Clemens), 140
Tweed, William M., 139–40

urban geography, 2, 12, 19, 24–25, 68, 94, 136, 150, 175
urban growth, 2, 12, 22, 43–44, 150

vandalism, 10, 17, 21, 53, 87, 97, 133
Van Kleeck, Mary, 198
vaudeville theaters, 157, 163–67, 169, 180
vice districts, 93–95, 162, 166, 175, 189, 201. *See also* brothels; prostitution
voluntary associations, 20, 47, 57, 77

wages, 121–22, 127
waiters and waitresses, 89–91, 96, 104, 151, 153, 198–99
Walker, Charles Rumford, 130, 133
walking: dangers of, 10, 20, 46, 56, 73–74, 142–44, 149, 151, 198–202; as mode of transportation, 39, 54, 59, 63, 136, 149, 201–2; recreational, 20, 24, 39, 46, 56, 145, 181
Walling, George, 26–27
Warren, Kate, 60
Washington, D.C., 56; food supply and distribution, 110; street lighting, 17–19
Washingtonians, 43, 85
watchhouses, 21–22
watchmen, 6, 10, 21–22, 119
waterfronts, 15–16, 44–45, 85–86, 95, 112, 121–22; crime, 26–27; lighting, 16, 121–22, 158; sanitary conditions, 106–7. *See also* shipping
waterfront workers, 112, 121–23, 131
Waterman, Zuriel, 4–5, 12
wealth and poverty, 6, 15–19, 30, 33, 52, 67, 75–80, 137, 143, 155
weapons, 6, 10–11, 27, 144
weather: effect on social calls, 59, 61; effect on street use, 3–5, 10; effect on theater attendance, 73
whale oil and spermaceti, 9, 15, 36
whites. *See* race relations
white ways, 159–61, 181. *See also* bright lights districts
Whitman, Walt, 49, 54, 91, 94–95, 136
window shopping, 146
Wood, Fernando, 24
Woolson, Abba Goold, 76

Worcester, MA, 93, 150, 168–69
working-class culture, 67–68, 85, 180, 182
World War I, 128, 199–200
World War II, 200

Yerkes, Charles, 152
YMCA (Young Men's Christian Association), 47, 53, 58
YWCA (Young Women's Christian Association), 53

SERIES TITLES, CONTINUED FROM FRONT MATTER

Colored Property: State Policy and White Racial Politics in Suburban America
by David M. P. Freund

Selling the Race: Culture, Community, and Black Chicago, 1940–1955
by Adam Green

The New Suburban History
edited by Kevin M. Kruse and Thomas J. Sugrue

Millennium Park: Creating a Chicago Landmark
by Timothy J. Gilfoyle

City of American Dreams: A History of Home Ownership and Housing Reform in Chicago, 1871–1919
by Margaret Garb

Chicagoland: City and Suburbs in the Railroad Age
by Ann Durkin Keating

The Elusive Ideal: Equal Educational Opportunity and the Federal Role in Boston's Public Schools, 1950–1985
by Adam R. Nelson

Block by Block: Neighborhoods and Public Policy on Chicago's West Side
by Amanda I. Seligman

Downtown America: A History of the Place and the People Who Made It
by Alison Isenberg

Places of Their Own: African American Suburbanization in the Twentieth Century
by Andrew Wiese

Building the South Side: Urban Space and Civic Culture in Chicago, 1890–1919
by Robin F. Bachin

In the Shadow of Slavery: African Americans in New York City, 1626–1863
by Leslie M. Harris

My Blue Heaven: Life and Politics in the Working-Class Suburbs of Los Angeles, 1920–1965
by Becky M. Nicolaides

Brownsville, Brooklyn: Blacks, Jews, and the Changing Face of the Ghetto
by Wendell Pritchett

The Creative Destruction of Manhattan, 1900–1940
by Max Page

Streets, Railroads, and the Great Strike of 1877
by David O. Stowell

Faces along the Bar: Lore and Order in the Workingman's Saloon, 1870–1920
by Madelon Powers

Making the Second Ghetto: Race and Housing in Chicago, 1940–1960
by Arnold R. Hirsch

Smoldering City: Chicagoans and the Great Fire, 1871–1874
by Karen Sawislak

Modern Housing for America: Policy Struggles in the New Deal Era
by Gail Radford

Parish Boundaries: The Catholic Encounter with Race in the Twentieth-Century Urban North
by John T. McGreevy